Gastrointestinal Motility

Tests and Problem-Oriented Approach

Gastrointestinal Motility

Tests and Problem-Oriented Approach

Edited by

Satish S. C. Rao
University of Iowa College of Medicine
Iowa City, Iowa

Associate Editors

Jeffrey L. Conklin
University of Iowa College of Medicine
Iowa City, Iowa

Frederick C. Johlin
University of Iowa College of Medicine
Iowa City, Iowa

Joseph A. Murray
Mayo Clinic
Rochester, Minnesota

Konrad S. Schulze-Delrieu
The University of Iowa College of Medicine
Iowa City, Iowa

and

Robert W. Summers
The University of Iowa College of Medicine
Iowa City, Iowa

Kluwer Academic / Plenum Publishers
New York, Boston, Dordrecht, London, Moscow

Including proceedings of the Sixth and Seventh Symposia on Gastrointestinal Motility, held March 13–15, 1998, in Dana Point, California, and March 19–21, 1999, in Cancún, Mexico, respectively

ISBN 0-306-46129-3

233 Spring Street, New York, N.Y. 10013

10 9 8 7 6 5 4 3 2 1

A C.I.P. record for this book is available from the Library of Congress.

Printed in the United States of America

PREFACE

Gastrointestinal motility has evolved from an esoteric laboratory tool into a sophisticated diagnostic technique that is now widely used clinically to guide in management of complex gastrointestinal problems. Today, it is the most rapidly growing subspecialty within gastroenterology. Previously, many of the gastrointestinal motility problems were either ignored or attributed to a disturbance of "psyche." But with the growing knowledge and understanding of how a dysfunction of the gastrointestinal muscle and nerves can cause disease, we are at the threshold of a revolution in our approach to the diagnosis and treatment of gastrointestinal motility disorders.

The purpose of this book is to serve as a useful, up-to-date reference manual and guide for the diagnostic and therapeutic approach towards common adult and pediatric gastrointestinal motility problems. In order to enhance the understanding of these disorders, a problem-oriented approach has been chosen, and wherever possible the authors have provided clinical case scenarios to illustrate their message. The emphasis has been on how to diagnose and treat motility disorders rather than to provide an encyclopedic information. The reference list at the end of each chapter should enable the enthusiast to seek further information. Some of the material presented in this book has been derived from the proceedings of the annual University of Iowa College of Medicine GI Motility Symposia.

Motility largely refers to the movement of muscles that line the gastrointestinal conduit. But the regulation of this movement involves many other components, principally the nerves, reflexes, visceral sensations, and neurohormones. Hence, this subspecialty is now more aptly described as neurogastroenterology and gastrointestinal motility. In the following chapters, we present some new information regarding the genesis, evaluation, and management of visceral hyperalgesia—the key problem which causes pain and discomfort related to several gastrointestinal organs. Because motility is a visual science, generous illustrations and tables have been provided throughout the text in order to enhance the understanding of common problems.

We hope that this book will serve as useful reading material for practitioners and trainees in gastroenterology, as well as for radiologists, internists, family practitioners, and general surgeons, each of whom encounter patients with motility problems on a daily basis. The "tests" section can be particularly useful for nurses and technical personnel that are involved in gastrointestinal motility. These diagnostic tests are evolving and there is no standardized approach. But the authors—who are experts in their field—have provided a succinct, practical, and balanced approach that can be used in any laboratory. We believe that users will gain information that will enable them to provide better care for their patients.

Given the increasing impositions of clinical work, and constraints on academic time, editing this book has not been easy. But this task was accomplished, because of the dedicated and uncompromising support of my wife Sheila, my children Priyanka, Anita and Nikilesh, and my secretary, Ms. Susie McConnell. I would like to especially thank the associate editors for helping me see the light when I was blindfolded, and each of the authors who have expended their valuable time in the service of humanity. Plenum Publishers, in particular Ms. MaryAnn McCarra and Jonathan Harmon, deserve special credit for their painstaking and meticulous efforts to produce this book in a timely fashion. Clearly, you the reader can only judge the quality of this work, and I hope you like it.

Satish S. C. Rao

CONTENTS

Section 6. Gastrointestinal Visceral Pain,
***Associate Editor*: Joseph A. Murray**

1

VIDEOFLUOROSCOPIC ASSESSMENT OF SWALLOWING

Bruce P. Brown*

Department of Radiology
University of Iowa Hospitals and Clinics
Iowa City, Iowa

INTRODUCTION

Because radiographic techniques of evaluating the upper, striated-muscle portion of the esophagus differ significantly from those of evaluating the lower, smooth-muscle portion of the esophagus, a radiographic assessment of swallowing should be directed by the patient's history. The history enables the examiner to focus on the anatomical segment that is most likely involved and to tailor the study accordingly. If the history suggests problems related to the upper esophagus, the patient will require rapid-sequence, cine-fluoroscopy of the oropharynx and hypopharynx with only a few spot films. In contrast, if the history suggests problems of the lower esophagus, then the patient will require cine radiographic evaluation of swallows in the prone position to assess motility, and double-contrast spot films to detect anatomic abnormalities.

NORMAL PHARYNGEAL ANATOMY (FIGURE 1)

The Cervical Esophagus

The pharynx is arbitrarily divided into the nasopharynx, oropharynx, and hypopharynx. The nasopharynx extends from the base of the skull to the soft palate, the oropharynx from the soft palate to the level of the vallecula, and the hypopharynx from the vallecula to the lower extent of the cricopharyngeus. The pharyngeal constrictors (superior, middle, and inferior) form the muscular posterior and lateral walls of the pharynx. On

* Phone: 319/356-4374, E-Mail: bruce-brown@uiowa.edu

Gastrointestinal Motility, edited by Rao *et al.*
Kluwer Academic / Plenum Publishers, New York, 1999.

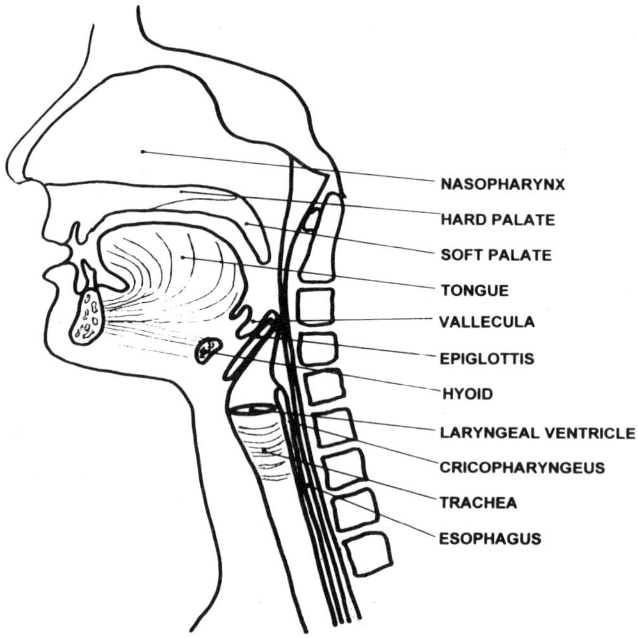

Figure 1. Normal anatomy of the pharynx.

either side and extending posteriorly around the laryngeal apparatus, there is space between the larynx and constrictors, the pyriform sinuses.

The Esophagus from the Cricopharyngeus Distally

The cricopharyngeus forms the major portion of the upper esophageal sphincter and is continuous with the caudal aspect of the inferior constrictor. It originates from the lateral aspect of the cricoid cartilages where it encircles the pharynx at approximately the level of C5–6. The cricopharyngeus is tonically contracted and relaxes as the constrictors above it force the food bolus into the upper esophagus.

Four to eight centimeters distal to the cricopharyngeus, the circular and longitudinal muscle layers of the esophagus change from striated to smooth muscle. The outer longitudinal layer crosses the gastroesophageal junction and merges with the outer layer of the gastric musculature. The inner circular layer thickens as it passes through the diaphragmatic hiatus and becomes part of the lower esophageal sphincter complex.

Normally, there is a thickening of the circular muscle at or slightly below the diaphragmatic hiatus, and this segment is referred to as the lower esophageal sphincter (LES). This sphincteric ring is tonically contracted but opens immediately after initiation of a swallow.

Slight distention of the esophagus with contrast or air reveals several radiologic and anatomic landmarks. The distal 2–4 centimeters of esophagus just above the gastroesophageal junction is often slightly wider than the more proximal body. This widened portion of the esophageal lumen has been termed the phrenic ampulla. Often, the esophagus just above the ampulla may be slightly and transiently narrowed after being distended. This narrowing of the proximal end of the phrenic ampulla is called the esophageal A ring or muscular ring.

With double contrast technique, a serrated transverse linear collection of contrast or Z line appears. This is the radiographic depiction of the squamo-columnar junction. At this same location, an area of mucosal thickening may also be seen in some patients on single contrast-distended views. This mucosal ring or B ring has also been termed a Schatzki's ring and most often produces no symptoms. However, if its diameter is less than 13 mm, it may produce dysphagia.

VIDEOFLUOROSCOPIC EXAMINATION OF SWALLOWING

Upper Esophagus

Normal swallowing involves a rapid transfer of the bolus proceeding from the naospharynx to the cricopharyngeus in less than one second.[1] To adequately visualize these events, high-speed video fluoroscopy and video recording equipment are needed to acquire images at frame rates of at least 30 per second. Such equipment should include on-screen time, date, and patient identification information, with slow motion and freeze-frame capabilities and a high-resolution playback monitor.

While some patients are able to cooperate and stand in the normal positions for fluoroscopy, many are weak or neurologically disabled. Under these circumstances, the examination can be performed by supporting the patient in the sitting position on a specially designed chair that allows anterior-posterior and lateral fluoroscopic views.

Evaluation of swallowing disorders may require the input of one or more specialists such as a gastroenterologist, radiologist, otolaryngologist, general surgeon, neurologist or speech pathologist each with varying degrees of involvement according to the patients needs. Generally, for complaints of cervical dysphagia, it is desirable to have a speech pathologist assist in the videofluoroscopic examination.

Because the examination must be tailored for each patient, there is no standard protocol. But there are several components that are common and can serve to illustrate the radiographic evaluation of dysphagia.

Evaluation of Swallowing Mechanics and Risk of Aspiration. During swallowing, if the oral contents are diverted into the airway passage up to the level of the vocal cords and no further, this event is termed "penetration." However, when oral contents transgress the vocal cord barrier, this event is termed "aspiration." When there is inadequate emptying of contrast from the esophagus, this event is termed "retention." If there is excessive retention in the oropharynx, initially, oral secretions and food may not be aspirated. However, further attempts at swallowing may produce aspiration into the airway from "overflow" of retained contrast.

In some series, 30–40% of patients with clinically unsuspected aspiration may show aspiration during videofluoroscopic evaluation.[2] This "silent-aspiration" may be clinically important because many otherwise healthy patients can present with unexplained recurrent pneumonia or chronic lung disease. Similarly, patients with generalized debility, strokes, or previous head and neck surgery may experience silent aspiration and present with unexplained pneumonia or chronic lung disease.

After determining the patient's ability to cooperate, it is preferable to start with the patient in the lateral position in order to obtain views of the oropharynx including the tongue, the hypopharynx including the cervical segments C5-C6, and to assess the cricopharyngeus. Suction equipment should always be available.

In those patients with suspected aspiration or problems of the upper esophagus, it is preferable to start with a teaspoon of dilute barium. *Caution*: For examinations to rule out aspiration, in our center we do not use water soluble agents such as Hypaque. These are iodinated contrast agents and have a high osmolarity. If aspirated, they can cause chemical pneumonitis and/or pulmonary edema. If a small amount of barium is aspirated, it is readily cleared and is unlikely to cause chemical pneumonitis.

In a given patient, if aspiration is less likely, then we begin with a moderate sized bolus of thin barium, e.g., 15 ml. On the other hand, if there is high index of suspicion of aspiration, a single teaspoon of contrast that is honey-like in consistency is recommended. Because this substance is more solid, it is less likely to cause massive aspiration. If the patient can satisfactorily handle small amounts of liquid barium, then it is appropriate to give either larger amounts of barium or allow the patient to drink at their own pace.

If the patient can handle liquid barium satisfactorily then the next step is to test the risk of aspiration from retention and overflow by using barium of a thicker consistency such as honey or paste-like. If these are handled well, the final step is to evaluate mastication and oral coordination by giving one half of a soda cracker with 5 ml of contrast (paste-like consistency).

If there is significant aspiration, the study is immediately discontinued and the airway is vigorously suctioned. However, if there is only penetration, a small amount of aspiration, or significant retention of contrast in the valleculae or pyriform sinuses, additional swallows may be performed after asking the patient to flex the neck into a chin-down position. In this position the airway is better protected from penetration by the food bolus, because with neck flexion the larynx moves superiorly and towards the base of the tongue. This maneuver may also lessen or eliminate aspiration in patients with poor elevation of the hyoid from muscle weakness or scarring.

Although the lateral position is the most useful for evaluating swallowing mechanics, problems with unilateral muscular weakness or vocal cord paralysis may be missed if only lateral views are performed. Thus, patients should also be studied in the anterior-posterior (AP) position.

To test for vocal cord paralysis, under fluoroscopic observation in the AP position, the patient is asked to raise the chin slightly and appose the vocal cords by saying "ee". In addition, this maneuver allows one to determine whether positioning the head to one side or the other may prevent retention or aspiration.

Thus, information gathered during fluoroscopic evaluation can not only help the physician to better understand the pathophysiology but also to prescribe food of appropriate consistency and devise maneuvers that may substantially reduce the risk of aspiration while maintaining oral nutrition.

Evaluation for Possible Organic Lesions of the Upper Esophagus. High-speed video fluoroscopy of the upper esophagus is essential for the evaluation of motility disorders and transient phenomena such as cricopharyngeal bars. However, characterization of anatomic abnormalities such as tumors and mucosal abnormalities requires double-contrast exami-

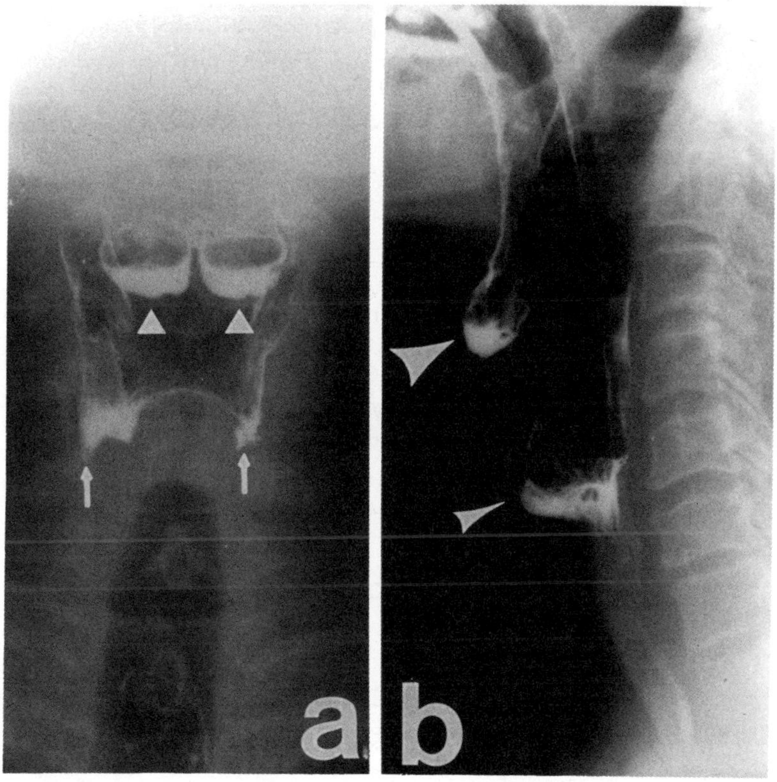

Figure 2. AP (a) and lateral (b) views of the hypopharynx after swallowing contrast in a normal subject. Anterior-posterior view demonstrates a small amount of contrast that pools in the valleculae (arrow heads) just above the epiglottis and in the pyriform sinuses (arrows) on either side of the vocal apparatus. Lateral view demonstrates the valleculae (large arrow head) and the pyriform sinuses (small arrow head).

nations. Videofluoroscopy, by itself is not adequate for this purpose, because small details can be obscured when taking pictures at high frame rates. In order to provide optimal information, high-density barium that can coat the oropharyngeal and hypopharyngeal mucosa is ingested just before photographing. Static AP and lateral spot films are taken (Figure 2). In some cases, abnormalities such as pharyngeal diverticuli can be identified by asking the patient to blow against pursed lips.

Distal Esophagus

Evaluation of Possible Motility Disturbance. Motility abnormalities of the distal esophagus cannot be evaluated in the standing or sitting position, because gravity aids transfer of the bolus distally. Therefore, motility of the distal esophagus is best investigated with the patient lying prone where gravity is not a factor in bolus transit. Filming is done using the same cine fluoroscopic video taping equipment as described for the upper esophagus. In the prone position, patients are asked to ingest a mouthful of contrast with a straw and not to swallow until the examiner is ready with the fluoroscope centered on the pharyngo-esohageal junction. When ready, the patient is asked to swallow once and only once. This single swallow technique eliminates the possibility that the normal peristaltic

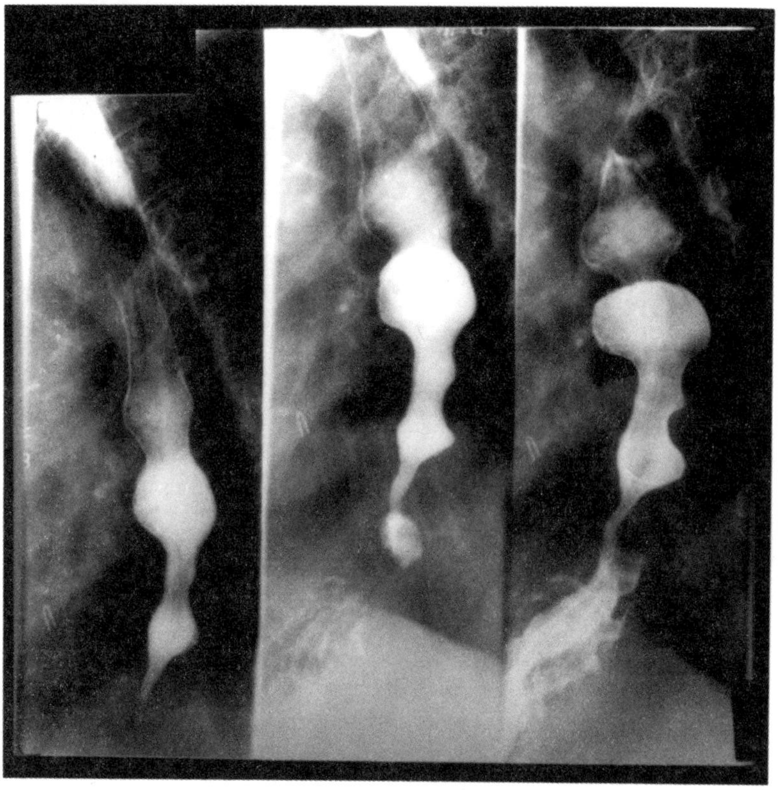

Figure 3. Three spot films of a prone swallow demonstrating simultaneous, deep, non-propulsive contractions of the distal esophagus. These contractions inhibit the flow of contrast distally. This is the radiographic picture of diffuse esophageal spasm.

wave propelling contrast distally will be interrupted by a second swallow before the effects of the first swallow have been fully observed.

The normal peristaltic wave appears as a coordinated stripping wave, an inverted "V" with its apex directed rostrally. This wave traverses the entire esophagus sweeping the contrast bolus through the phrenic ampulla into the stomach. Thus, study of the prone swallow allows diagnosis of esophageal motility disorders that might otherwise go undetected (Figure 3).

Evaluation for Possible Organic Lesion. Evaluation for possible structural lesions of the lower esophagus is best done with the patient in the standing position after having ingested effervescent crystals which distend the stomach with carbon dioxide. Subsequently, when the patient ingests liquid contrast, the lower esophageal sphincter relaxes with each swallow, and at the same time carbon dioxide escapes proximally distending the esophagus producing a double-contrast effect. This provides detailed images of the esophageal contour as well as the esophageal mucosa (Figure 4). Filling the stomach with gas also permits visualization of the distended gastric fundus in the upright view. This view should be performed in all esophagrams to ensure that the patient's dysphagia is not due to a lesion in the proximal stomach such as gastric cancer that is invading the lower esophagus.

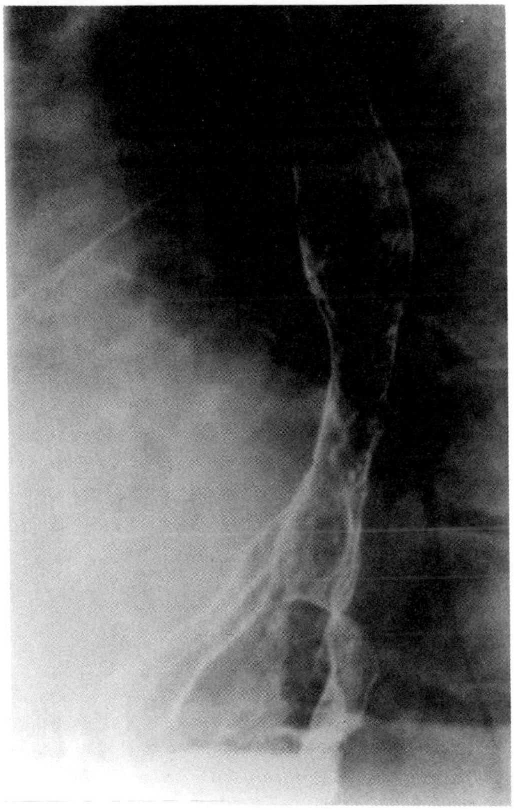

Figure 4. Double-contrast view of the distal esophagus demonstrating a smooth, hourglass configuration of the distal esophagus which represents an inflammatory stricture above a hiatal hernia in a patient with chronic gastroesophageal reflux and dysphagia. Notice that inflammation has produced an irregular mucosa.

In some cases, where it is difficult to evaluate the lower esophagus in the standing position, placing the patient prone may distend the gastroesophageal junction allowing identification of distal esophageal lesions.

Because there is a natural narrowing at the gastroesophageal junction it may be difficult to identify subtle strictures here. Hence, at the end of the examination, it is advisable to have the patient stand and swallow a barium tablet,13 mm in diameter, without chewing but with as much water as needed. Immediately, following these maneuvers, videofluoroscopy is performed and the passage of barium tablet into the stomach is monitored. If there is significant delay at the narrowed area, and the motility is adequate, then the narrowing is probably significant and may account for the patient's symptom of dysphagia.

REFERENCES

1. McConnel F, Cerenko D, Jackson RT, Guffin TN. Timing of major events of pharyngeal swallowing, Archives of Otolaryngology-Head and Neck Surgery1988;114,:1413–1418.
2. Linden P, Kuhlemeier K, Patterson C. The probability of correctly predicting subglottic penetration from clinical observations. Dysphagia1993;8(3): 170–179.

ESOPHAGEAL MANOMETRY

Jeffrey L. Conklin[1,*] and Joseph A. Murray[1,2]

[1]Department of Internal Medicine
College of Medicine
University of Iowa and Iowa City VA Medical Center
Iowa City, Iowa
[2]Mayo Clinic
Rochester, Minnesota

INTRODUCTION

The major functions of the esophagus and its sphincters are to transport from the pharynx to the stomach what is swallowed, to prevent the reflux of gastric contents into the esophagus and airways, and to clear the esophagus of gastroesophageal refluxate. Carrying out these tasks requires the precise integration of a bewildering array of neuromuscular processes in the central nervous system, the myenteric plexus, and the muscular components of the esophagus. Esophageal manometry gives the clinician insight into the function of these neuromuscular processes, and in some cases allows the diagnosis of specific abnormalities of esophageal motor function.

Functional Anatomy of the Esophagus (Figure 1)

The esophagus is a muscular tube composed of 3 distinct functional regions: the upper esophageal sphincter, the tubular esophagus and the lower esophageal sphincter. The physiological properties of each region can be evaluated by manometry.

The upper esophageal sphincter (UES) is a striated muscle structure that is constituted primarily by the cricopharyngeus, but receives contributions from the inferior pharyngeal constrictor and the adjacent esophagus. It is a slit-like structure that is closed at rest because the UES remains tonically contracted at rest and because the viscoelastic properties of adjacent structures tend to hold it closed. When it is approached by a peristaltic contraction of the pharynx the UES opens. The mechanisms contributing to its

* Phone: 319/356-2131, Fax: 319/353–6399, Email: jeffrey-conklin@uiowa.edu

Gastrointestinal Motility, edited by Rao *et al.*
Kluwer Academic / Plenum Publishers, New York, 1999.

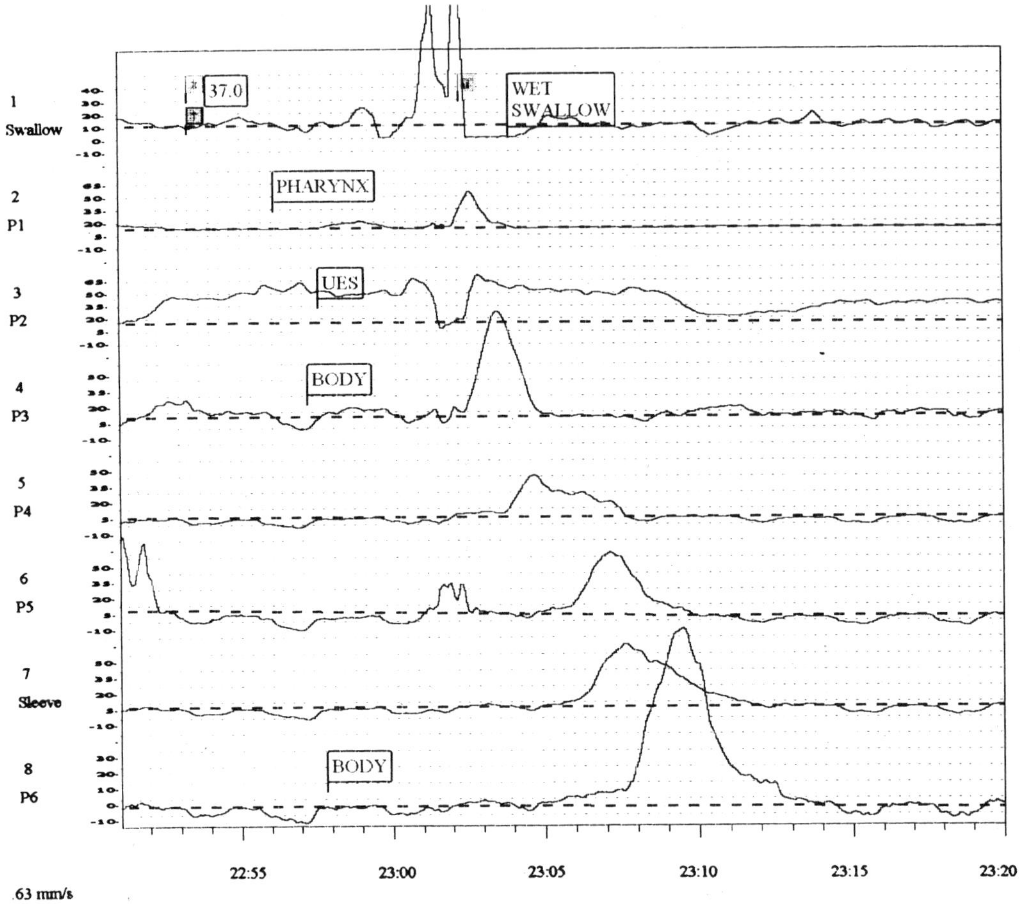

Figure 1. Manometric recording of normal esophageal motor function : On the Y-axis is pressure as measured by the manometry apparatus and on the X-axis is time. The catheter was positioned so that the most distal recording port was positioned in the stomach (gastric). The recording port was positioned in the LES, and the 5 recording sites above the LES were positioned at 3 cm intervals. The shaded areas indicate the intrabolus pressure, the pressure in the swallowed bolus that is being propelled in advance of the peristaltic contraction. The arrow indicates the arrival of the peristaltic contraction at the recording site. The upstroke in pressure indicated by the arrow correlates with closure of the esophageal lumen by the peristaltic contraction. Notice that the LES begins to relax well before the arrival of the peristaltic contraction, and stays relaxed until the contraction arrives.

opening include traction on the sphincter by contraction of the infrahyoid and suprahyoid muscles, relaxation of the cricopharyngeus and pressure in the swallowed bolus. Opening of the UES allows transfer of the bolus from the pharynx to the esophagus. Arrival of the pharyngeal contraction closes the UES.

The body of the esophagus is a 20–22 cm muscular tube that consists of an inner circular muscle layer and an outer longitudinal muscle layer. The muscle layers are named according to the axial orientation of their constituent smooth muscle cells. The top 5% of the esophagus is striated muscle, the middle 35–40% is a mixture of striated and smooth muscle with smooth muscle progressively replacing striated muscle until the bottom 50–60% is entirely smooth muscle. A neural network called the myenteric plexus is sandwiched between the muscular layers. The plexus receives neural input from the CNS and provides the terminal motor innervation to the smooth muscle of the esophagus. Esophageal peristalsis that is initiated by swallowing is called primary peristalsis which starts when the peristaltic contraction passes from the pharynx into the upper esophagus. Peristalsis in the esophagus consists of a ring of circular smooth muscle contraction that progresses uninterrupted along the length of the esophagus. Peristaltic contractions progress along the esophagus at a rate of 2–4 cm/sec. They strip the swallowed bolus from the esophagus into the stomach. As the peristaltic contraction enters the upper esophagus, the longitudinal muscle contracts to shorten the esophagus by 2.0 to 2.5 cm.

The lower esophageal sphincter (LES) is identified anatomically as a muscular thickening at the gastroesophageal junction. It is composed of 2 muscular components: one is a long bundle of muscle from a third, inner muscular layer that runs obliquely along the lesser curvature, loops around the greater curvature side of the esophago-gastric junction, and passes back along the lesser curve; the other is a short bundle of muscle fibers that is oriented transversely around the esophagus at the gastroesophageal junction. The muscles of the LES contract tonically at rest to produce an intraluminal pressure that is from 10 to 45 mmHg greater than intragastric pressure. The LES relaxes as the swallowed bolus reaches the upper esophagus, and it stays relaxed until the peristaltic contraction arrives at the gastroesophageal junction. The LES then contracts and regains its resting tone.

The spatial and temporal arrangement of these events and their derangement by pathological processes can be discerned with esophageal manometry.

MANOMETRIC EQUIPMENT

All gastrointestinal manometry setups consist of two pieces of hardware: a pressure sensor/transducer that is able to detect changes in intraluminal pressure and convert what is detected into an electrical signal, and a recording device that can amplify the signal and store it. Two general types of sensor/transducer manometry assemblies are commonly used: the water-perfused catheter system and the solid state pressure transducer. Both are reliable within limits, reasonably quantitative and available commercially. The perfused catheter is a bundle of thin polyvinyl tubes each of which has a small lateral opening that faces out. Each opening is at a different level equidistant from its neighbor so that pressure can be measured simultaneously at regular intervals along the catheter. A low-compliance pneumo-hydraulic pump is used to slowly perfuse the tubes with bubble-free water. The pressure in each tube is sensed and converted to an electrical signal by a volume-displacement transducer. When the flow of water through the side hole of a tube is impeded by contraction of the wall of the esophagus, the recorded pressure rises. While the amplitude and timing of the pressure change that is recorded reflect the force and timing of the circular muscle contraction, factors other than the force of contraction may influence pressure measurements. The physical characteristics of the recording system; i.e., the rate of perfusion, the compliance of the recording system, and the diameter of the catheter side hole all influence the fidelity of the recording (3,4). Pressure artifacts may occur if debris or compression occludes the catheter, if there are air bubbles in the manometry tubing or if movements of extraesophageal organs such as the heart, aorta, and diaphragm occlude the side hole. It is crucial for people performing manometry to be aware of these limitations and to try to allow for them when reading manometric tracings.

The other commonly used sensing device is an array of solid state strain gauge transducers spaced at regular intervals along a flexible tube. Many of the confounding factors that may confuse the data obtained with the perfused catheter are not encountered when using solid state devices. Solid state catheters are better able to record motor activity in the pharynx and upper esophageal sphincter because they can more faithfully record the rapid muscular events that occur in these regions. These devices have the disadvantages of being more expensive and more susceptible to damage.

It is difficult to obtain faithful recordings the motor events of the pharynx, UES and LES with either the perfused catheter or solid state catheter for a number of reasons. Since these structures are not concentrically symmetrical, the direction in which the lateral orifice or sensor faces greatly influences the pressure measured. For instance, the pressure recorded by a sensor facing in the anterior-posterior axis of the UES is greater than that recorded laterally because the sphincter has a slit-like shape. Both the UES and LES are displaced cephalad with swallowing: the UES because the pharynx shortens and the larynx elevates (5), and the LES because longitudinal muscle contraction shortens the esophagus (6,7). The manometry catheter; however, remains at a fixed length. If the point pressure sensor is positioned where the sphincter generates a maximal intraluminal pressure, cephalad movement of the sphincter repositions the sensor in a region of lower resting pressure: the esophagus when recording UES pressure and the stomach when recording LES pressure. Such positional changes are recorded as decreases in pressure that may be mistakenly interpreted as a relaxation. Configuring the perfused catheter with a Dent sleeve (8,9) can obviate this problem. The Dent sleeve is a 6 cm long silicone rubber membrane that covers a distal segment of the catheter. A perfusion port is located beneath the membrane. The devise registers the highest pressure occurring anywhere along its 6 cm length. The problem can also be overcome by spacing point sensors at short intervals along the distal part of the catheter. This assures that sensors will remain in the zone of high pressure as it moves. There are significant disadvantages in using the Dent sleeve. It has very limited frequency-response characteristics, making it unusable when rapid transients are to be recorded, and it misrepresents the timing of the end of LES relaxation because it simultaneously senses LES relaxation and pressures generated by peristaltic contractions in the distal esophagus.

Strip chart recorders and computers are commonly used as recording devices for esophageal manometry. Both have the frequency-response characteristics to reliable record the motor activity of the esophagus.

COMPONENTS OF ESOPHAGEAL MANOMETRY

Esophageal manometry is becoming a standard part of many clinical practices of gastroenterology. Unfortunately, most practitioners have little training in esophageal manometry and the methods are not standardized. This section provides one approach to the performance of esophageal manometry. The indications for manometry are shown in Table 1.

Patient Preparation

The patient should be kept fasting for at least six hours prior to the procedure. Ideally, they should avoid for 24 hours any medications that may alter esophageal motility. These drugs include caffeine, calcium channel antagonists, nitrates, atropine-like agents, prokinetic agents, beta-adrenergic antagonists and other vasoactive substances. Of course, there must be some exceptions: if the medication cannot be stopped because doing so will exacerbate a disease other than that for which the study is being performed, if stopping the medication will make the esophageal disease intolerable, or if the study is being done to study the effect of the medication on esophageal motor function. Sedation should be

Table 1. Indication for Esophageal manometry

Indicated for	
	Diagnoses of achalasia or nonobstructive dysphagia
	Locating LES before pH probe placement
	Identifying motor abnormalities associated with other systemic diseases ex: Scleroderma
	Preop evaluation for anti-reflux surgery
Not indicated for	
	Diagnosis of GE reflux
	Initial evaluation of noncardiac chest pain

avoided, but if it must be used because the patient cannot otherwise tolerate the procedure, the agent and its dose must be recorded contemporaneously as part of the manometric report. Local anesthesia may or may not be used, but its use should be documented. A swallow-recording device is desirable because it provides reliable identification of the timing of swallowing. A respiration monitor is also helpful because it helps to identify respiratory artifact, but may be considered optional.

Equipment Preparation

The type of catheter and a description of its configuration should be documented. Important information regarding the catheter includes whether it is a perfused catheter or a solid state catheter, the spacing between pressure sensors, and whether the catheter has a Dent sleeve. All of the recording devices should be turned on, and all of the appropriate connections made and documented. The recording equipment should be calibrated and the calibration should be documented on the recording. Prior to introducing the catheter into the patient, the equipment should be checked to assure that the recording system and the sensors are functioning accurately. A timing marker should be present as part of the recording, and the time interval indicated by the marks should be recorded on the manometry tracing.

Performance of the Study

The manometry catheter may be placed via the nares or the mouth. The route of introduction is most often determined by the preference of the operator, but should be documented as part of the record. Patents tend to tolerate intubation via the nares somewhat better, and fewer artifacts tend to occur when the catheter is positioned in this way. Placing the catheter by this route adds about 5 cm in length to the measurement sights. The patient is positioned supine after the catheter is passed, and if a perfusion system is used, the patient is situated so the pressure transducers are at the same level as the mid-thorax of the patient. This assures a reliable measure of baseline pressure. The head may be elevated slightly to facilitate the act of swallowing small volumes of water. How the patient is positioned is not as crucial when using solid state recording systems. The patient is allowed to rest for 10–15 minutes following the insertion of the tube so that they accommodate to its presence. Accommodation to the catheter is usually signified by the absence of coughing, repetitive and involuntary swallowing, gagging or excessive salivation.

Prior to beginning the study, the water level in the pump reservoir is checked, the perfusion line is checked for bubbles and the perfusion pump is turned on. The catheter should be placed so that at least the most distal recording site, and preferably more, is in the stomach. The effects of deep breathing on the pressure recording verify its position. If the sensor is below the diaphragm the pressure will increase whilst the patient inspires, and if it is above the diaphragm the pressure will decrease. These pressure changes reflect the negative pressure generated in the thoracic cavity and the positive pressure generated in the abdominal cavity during inspiration. The occurrence of a deep inspiration is documented on the record. Modern computer-based recording systems allow this to be done by simply clicking an icon on a menu bar. Fluoroscopy may be used to aid the positioning of the catheter, if its position cannot be verified. This may be particularly important when studying patients with the sigmoid-shaped esophagus and tightly closed lower esophageal sphincter of advanced achalasia, since in this situation the catheter tends to coil in the distal esophagus.

The study is accomplished as the catheter is withdrawn by the station pull-through technique. The catheter is pulled back in 0.5 cm to 1cm steps to perform the station pull-through. At least 30 seconds should be allowed at each station so that any reflex activity initiated by moving the catheter subsides. The position of the distal recording site relative to the incisors or nares is documented on the tracing as the station pull-through is being accomplished. Again, modern recording systems make this as easy as clicking an icon. At each step, deep breaths can be used to determine the positions of the recording sites relative to the diaphragm and swallows can be used to identify the LES. The timing of deep breaths and swallows should be indicated on the record.

EXAMINING THE REGIONS OF THE ESOPHAGUS (FIGURE 1)

The lower esophageal sphincter is defined as a zone of increased resting intraluminal pressure at the gastroesophageal junction that decreases with swallowing. The LES can be identified and its intraluminal pressure measured either during the station pull-through described above, by placing a Dent sleeve into the zone of high pressure or by a technique called the rapid pull-through. At least one, and preferably several of the recording sensors, is placed into the stomach prior to performing the rapid pull-through. The catheter is then pulled back at a rate of 1 cm/sec while the patient is holding their breath in end expiration and refraining from swallowing. The catheter is withdrawn so that the pressure sensor passes from the stomach into the esophagus. A pressure peak is seen on the manometry tracing as the sensor traverses the LES. The difference between the intragastric pressure and the maximum pressure recorded as the sensor is pulled across the zone of high pressure is the LES pressure. Pulling more than one sensor across the sphincter allows averaging of peak pressures. This technique gives reproducible results but may generate artifacts in the pressure because movement of the catheter in the esophagus triggers a reflex change in sphincter muscle tone. The rapid pull-through has its greatest utility when the zone of high pressure is difficult to locate, usually because the LES pressure is low.

Most often, during the station pull-through the LES pressure is determined by positioning several sensors in the zone of high pressure. The highest resting pressure recorded when each sensor is in the zone of high pressure can be used to determine an average LES pressure. The pressure recorded in the LES often varies cyclically, like a sine wave, with respiration. The convention is to take the mid-inspiratory or mid-expiratory pressure (the middle point in the sine wave) as the sphincter pressure. The resting sphincter pressure is the difference between this pressure and intragastric pressure. A continuous recording of resting LES pressure without swallowing should be made as part of the study.

The pressure inversion point, the location at which the recorded pressure deflection caused by inspiration changes from positive to negative, is also ascertained by the station pull-through. The pressure inversion point is the position at which the recording sensor is crossing the level of the diaphragm. This maneuver is done with catheters utilizing point sensors, and is not done with the Dent sleeve. In some situations, the pressure inversion point may be an important point of reference. Finding the LES located several cm above the pressure inversion point indicates the presence of a hiatal hernia.

To assess relaxation of the LES, a point sensor or the Dent sleeve is placed within the zone of high pressure. Since normal relaxation of the sphincter is considered to be a drop in pressure that approximates intragastric pressure, it is best to have a distal recording sensor in the stomach so that sphincter and gastric pressures can be measured simultaneously. Relaxation of the LES in relationship to swallowing should be evaluated with a minimum of five "wet swallows" using 5 cc of water at room temperature. Esophageal manometry has been standardized using 5 cc water swallows because wet swallows tend to give more vigorous and reliable results than "dry," saliva swallows (10,11). The frequency of swallowing during the study should be no more often than every 30 seconds to allow a return to baseline pressure and to allow recovery from any refractoriness in the neuromuscular system. The operator should document whether swallows are "wet" or "dry." A measurement of the residual pressure, that is the minimal LES pressure during sphincter relaxation is recorded. The operator notes whether the LES relaxes to intragastric pressure, and recognizes the transient pressure overshoot as the sphincter recovers from its relaxed state.

The motor function of the esophagus is examined using a catheter with at least three pressure sensors spaced between 3–5 cm apart. The catheter is positioned so that the most distal of these sensors is just above the LES. The function of the distal, smooth muscle, esophagus is evaluated by having the patient take at least 10 wet swallows. Again, at least 30 seconds are allowed to elapse between consecutive swallows. The station pull-through is continued until the upper esophageal sphincter is identified at the proximal recording site. The function of the upper esophagus; i.e., the striated muscle portion and the transition zone is also assessed by having the patient take 10 wet swallows no more frequently than every 30 seconds. The timing of each swallow and whether it is wet or dry is documented. If no peristalsis is apparent, the sensitivity of the pressure sensors and their function are checked by having the patient cough. Coughing produces a brief simultaneous pressure transient of about the same amplitude at each recording site when the sensors are working properly. If the response to a cough is slow or the amplitude is less in some recording sites than others, there may be a technical problem like air in the catheter, a plugged side hole, or a loss of pressure in the perfusion system

The *upper esophageal sphincter* is identified as a region of increased resting intraluminal pressure in the upper esophagus that decreases promptly with swallowing. Its function is assessed during the station pull-through by first identifying a zone of high pressure at the upper end of the esophagus and then by having the patient take wet swallows to demonstrate relaxation of the sphincter (Figure 2). The absolute pressure within the sphincter is not reliably recorded because it is a radially asymmetric structure, and the timing of the relaxation is not faithfully recorded by the perfused catheter system because its frequency-response characteristics are too slow to follow the rapid contractile events generated by striated muscles.

Some recent literature suggests that measuring intrabolus pressure in the pharynx during the pharyngeal phase of swallowing predicts which patients with oropharyngeal dysphagia will benefit from cricopharyngeal disruption (12). Pharyngeal pressures are re-

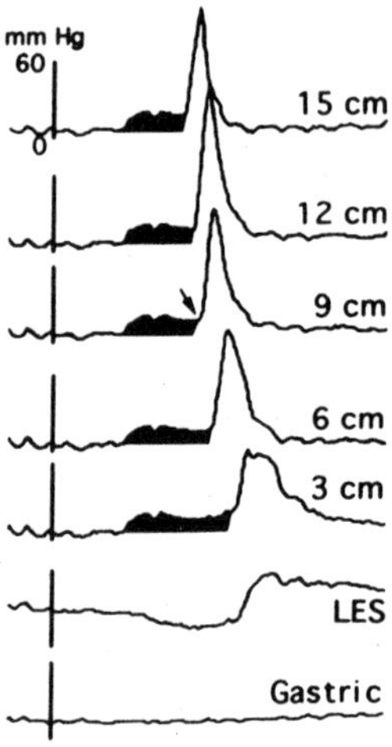

Figure 2. Manometric recording during a wet swallow showing normal pharyngeal contraction followed by upper esophageal sphincter relaxation and a normal primary peristalsis in the body of the esophagus.

corded by positioning at least 1 sensor above the UES during swallowing. The intrabolus pressure is the pressure in the swallowed bolus ahead of the peristaltic contraction. It is seen manometrically as a rise in intraluminal pressure prior to the sharp increase in pressure accompanying the peristaltic contraction (Fig. 1).

REPORTING THE ESOPHAGEAL MANOMETRY

The esophageal manometry report contains general information including the patient identifier, the date and time of the procedure, whether the catheter was placed via the mouth or the nares, the type and configuration of catheter, the type of recording device, documentation of calibration, and any medications used in the preceding 24 hours or during the procedure.

A description of the LES includes, both the distance of the high-pressure zone from the nares or incisors and the length of the high-pressure zone. The cephalad border of the high-pressure zone is documented since it is a landmark used when positioning a pH electrode for ambulatory 24 hour pH monitoring. The report includes the baseline LES pressure and the residual pressure during maximal swallow-induced relaxation. By convention, resting LES pressure is the highest pressure recorded in the zone of high-pressure zone minus the intragastric pressure (see above for a detailed description of LES pressure measurement). Normal resting sphincter pressure varies significantly depending on the technique and equipment used to measure it, but in general it ranges from 10mmHg to 40mmHg. Data adapted from one large study is shown in Table 2 (10).

The report of the motor activity of the esophageal body includes a description of the pressure waves generated by esophageal contractions. The peristaltic function of the esophagus is assessed according to the percentage of wet swallows that produce peristaltic pressure waves, by the velocity of propagation of the pressure waves and by the characteristics of the pressure waves. To be classified as a pressure wave produced by a contraction, the amplitude of a pressure increase should be least 20mm Hg. Normal peristaltic contractions propagate alone the esophagus at a rate of 2–4 cm/sec., a propagation rate of greater that 6 cm/sec is considered abnormal by most investigators because contractions moving faster than this do not propel the bolus effectively (13). Failed peristalsis occurs

Table 2. Normal values for esophageal manometry

	Recording site (cm above LES)	Wet swallows (95 subjects)
Amplitude of peristalsis (mm Hg, mean + 1 S.D)	18	62 ± 29
	3	109 ± 45
Duration of peristalsis (sec. ,mean ± 1 SD)	18	2.8 ± 0.8
	3	4.0 ± 0.9
Velocity of peristalsis (cm/sec., mean ± 1 SD)	Proximal	3.0 ± 0.6
	Distal	3.5 ± 0.9
Resting Lower esophageal sphincter pressure (mm Hg., Mean ± 1 SD)	RPT	29.0 ± 12.1
	SPT	
	End-inspiration	39.7 ± 13.2
	Mid-expiration	24.4 ± 10.1
	End-expiration	15.2 ± 10.7

Adapted and reproduced with permission from authors in reference No 10.
RPT = Rapid pull-through
SPT = Slow pull-through

when a wet swallow does not produce an esophageal contraction, when the peristaltic contraction dies out as it progresses down the esophagus, or when an initial peristaltic sequence travels part way down the esophagus and then ends in the distal esophagus as a simultaneous contraction (14). In each of these cases bolus transport is disrupted (15). The velocity of peristalsis is determined by identifying the beginning of the upstroke of the peristaltic pressure wave at adjacent recording sites, calculating the time from the upstroke of one pressure wave to the upstroke at the neighboring site, and then dividing this time by the distance between the recording sites. The measurement is done this way because the initial upstroke of the pressure wave correlates with occlusion of the esophageal lumen by the peristaltic contraction (15). Transient pressure increases that occur simultaneously at adjacent recording sites must not have the same morphology, i.e. the pressure increases and wave forms should not be identical, if they are to be considered the result of local esophageal contraction. Identical, simultaneous increases in pressure at multiple sites along the esophagus are isobaric pressure waves that are indicative of a rapidly propagated pressure wave in a common cavity. This is commonly seen in advanced achalasia. The velocity of peristalsis and these characteristics of the pressure waves are recorded for the upper and lower halves of the esophagus. The percentage of wet swallows that produce normal peristaltic contractions and the presence of isobaric pressure waves are documented.

The amplitude of esophageal contractions is calculated by determining the difference between the baseline resting pressure and the pressure at the peak of the pressure wave. The duration of the contraction is ascertained by measuring the time from the beginning of the upstroke of the pressure wave to the time when the pressure returns to baseline. Swallow-induced peristaltic pressure waves that are less than 35mm Hg in amplitude are considered by most to be feeble or hypotensive (10,14,16). Contractions with amplitudes of >180 mmHg are said to be hypertensive (10); however, some investigators argue that this number is too low and suggest that the cutoff should be around 200 mmHg. Repetitive contractions are those that have multiple (>2) peaks that occur at least 1 second after the preceding peak and have at least a 10 mmHg drop in pressure between peaks (17). The observer should document all of these characteristics of esophageal motor function.

The distance of the UES from nares or incisors and the presence of its relaxation with swallows are documented.

THE USE OF ESOPHAGEAL MANOMETRY IN CLINICAL PRACTICE

The clinical usefulness of esophageal manometry has been carefully and critically reviewed (18). Over the years, esophageal manometry has been used to study a number of esophageal problems like heartburn, dysphagia and noncardiac chest pain. While esophageal manometry can help in the diagnosis and management of these problems, its clinical usefulness is somewhat limited. Esophageal manometry lacks specificity for distinct pathophysiological entities, and there are relatively few patients with symptoms of esophageal disease who have manometrically identifiable and clinically significant motility disorders of the esophagus. None-the-less, there are a number of instances when esophageal manometry is useful and indicated.

Evaluation of Dysphagia

Esophageal manometry is indicated for patients who suffer from symptoms of esophageal dysphagia and who do not have a mechanical lesion obstructing the lumen of the esophagus or esophagitis. Accordingly, patients with dysphagia should undergo radiographic and/or endoscopic evaluation with esophageal mucosal biopsies before manometry is undertaken. Once inflammatory and mechanical lesions are ruled out the clinician can use manometry with the hope of identifying an abnormality of esophageal motor function. A few common manometric abnormalities and their characteristics are shown in Table 3. Up to half of patients complaining of dysphagia without an obvious underlying esophageal disease have abnormalities of esophageal motor function that do not meet the criteria for any specific disorder of esophageal motor function (19,20). The most common of these "nonspecific" abnormalities of esophageal motor function is peristaltic pressure waves of excessively high amplitude, the nutcracker esophagus. The nutcracker esophagus is found in 5% of patients complaining primarily of dysphagia (20). While these nonspecific abnormalities do very likely indicate motor dysfunction of the esophagus (16), their relevance to esophageal symptoms or the ability of the esophagus to transport a bolus is not clear. For example, bolus transport is not usually affected by the high amplitude peristaltic contractions of the nutcracker esophagus (21). There are two well characterized disorders of esophageal motor function that are associated with dysphagia and altered bolus transport in the smooth muscle esophagus, achalasia and diffuse esophageal spasm (13,22). Up to 19% of patients presenting with dysphagia as a prominent symptom have achalasia, and up to 7% have diffuse esophageal spasm (20). For practical diagnostic and therapeutic purposes the clinician is really using manometry to identify either achalasia or diffuse spasm in patients with unexplained dysphagia.

Evaluation of Noncardiac Chest Pain

Esophageal manometry also has a role to play in the evaluation of patients with noncardiac chest pain; however, it should never be considered as an initial test for this symptom. A number of clinical entities that may have nothing to do with esophageal motor function can present with chest pain and should be ruled out before looking for an esophageal motor problem. Patients with anxiety neurosis should first be treated for their psychiatric illness and patients with musculoskeletal pain should be treated appropriately. Patients with chest pain and either dysphagia or heartburn should have endoscopy looking for esophagitis or a lesion obstructing the esophageal lumen. Those with chest pain and no

Table 3. Manometric features of common esophageal motility disorders

Disease	LES	Esophageal body	UES and proximal esophagus
Achalasia	Impaired relaxation; Normal or hypertensive	Aperistaltic or spastic	Normal
Esophageal spasm	Normal; Hypertensive; Hypotensive (GERD induced); Normal relaxation	Some normal peristalsis; High amplitude propagated; High amplitude simultaneous. At least 20% Normal amplitude simultaneous	Normal
Scleroderma or other connective tissue diseases	Hypotensive or absent Normal relaxation	Aperistalsis or diminished amplitude	Normal
GERD	Normal or hypotensive Normal relaxation	Normal Increased proportion of peristaltic failure Diminished amplitude	Normal

esophageal symptoms may first be given a therapeutic trial with high doses of a proton pump inhibitor, since a large subgroup of patients with presumed esophageal chest pain respond to acid suppression (23,24). As with dysphagia, the chance of finding a specific motor disorder in a patient with noncardiac chest pain is relatively low: roughly 1% have achalasia and fewer than 10% have diffuse esophageal spasm (18). It is important to identify these disorders because there are several therapeutic options for their management. Nonspecific motor abnormalities will be seen in up to 50%, but once again the relationship between these abnormalities and chest pain is not at all clear.

Placement of Esophageal pH Probe

Ambulatory, 24-hour intraesophageal pH monitoring is commonly used in the evaluation of esophageal diseases. By convention, the pH electrode is placed in the esophageal lumen 5 cm above the upper border of the LES. Methods proposed as a means for accurately placing the pH electrode include endoscopy, observation of a pH step-up as the probe is pulled across the LES, placement based on the patient's height, fluoroscopic placement and manometry. Manometry is the most accurate and reproducible method for placing the pH probe (25–27). It is, therefore, the method of choice for determining where to position the pH electrode.

Pre-Surgical Evaluation

Esophageal manometry is widely accepted as an essential part of the pre-surgical evaluation of patients undergoing antireflux surgery (28). Intuitively, this approach makes sense because the surgeon forms a relative obstruction of the gastroesophageal junction with the fundoplication, and poor peristaltic function might then produce dysphagia. In fact, dysphagia is one of the most common untoward results of antireflux surgery (29,30). The literature, however, suggests that the motility abnormalities seen prior to fundoplication may be improved postoperatively, and that preoperative esophageal motor dysfunction does not predict a poor surgical result (30,31).

Perhaps the main reason that esophageal manometry continues to be recommended as an essential part of the pre-surgical evaluation prior to antireflux surgery is the concern that severe disorders of esophageal motor function may be missed. While the issue has not been formally addressed in the literature, it seems likely that performing a fundoplication on patients with achalasia, diffuse esophageal spasm or a connective tissue disease may result in disabling dysphagia or esophageal obstruction.

Evaluation of Systemic Diseases that Affect Esophageal Motor Function

Esophageal manometry is indicated if it may contribute to establishing the diagnosis of a systemic disease or if it may help in making treatment decision regarding that disease (18). This situation is most frequently encountered in patients with collagen vascular diseases (scleroderma, CREST, dermatomyositis, polymyositis etc.) and symptoms of esophageal dysfunction like dysphagia or heartburn (32). These patients often have aperistalsis of the esophagus and a hypotensive LES.

Evaluation of Gastroesophageal Reflux Disease

Esophageal manometry is not essential for making or confirming the diagnosis of gastroesophageal reflux disease. While manometric abnormalities (the hypotensive LES, nonspecific motor abnormalities of the esophagus, and increased episodes of transient LES relaxation) commonly accompany esophagitis, it is not entirely clear whether these abnormalities are the cause or consequence of esophagitis (14,33,34). It is also unclear whether identifying these abnormalities changes the clinical approach to the gastroesophageal reflux disease. The notable exceptions, as pointed out above, are for the placement of the pH probe and for the pre-surgical evaluation prior to fundoplication.

REFERENCES

1. Conklin JL, Christensen J. Motor Functions of the Pharynx and Esophagus. In: *Physiology of the Gastrointestinal Tract*, L.R. Johnson (Ed.), Raven Press, New York, NY, pp. 903–928, 1994.
2. Miller A, Bieger MD, Conklin JL. Functional Controls of Deglutition. In: *Deglutition and its Disorders: Anatomy, Physiology, Clinical Diagnosis and Management*, A. Perlman, K. Schulze-Delrieu (Eds.), Singular Publishing Group, Inc., San Diego, CA, pp. 43–97, 1996.
3. Dodds WJ, Steff JJ, Hogan WJ. Factors determining pressure measurement accuracy by intraluminal esophageal manometry. Gastroenterology 70: 117–123, 1976.
4. Arndorfer RC, Steff JJ, Dodds WJ, Linehan JH, Hogan WJ. Improved infusion system for intraluminal esophageal manometry. Gastroenterology 73: 23–27, 1977.
5. Kahrilas PJ, Dodds WJ, Dent J, Logeman JA, Shaker R. Upper esophageal sphincter function during deglutition. Gastroenterology 95: 52–62, 1988.
6. Dodds WJ, Stewart ET, Hodges D, Zboralske FF. Movement of the feline esophagus associated with respiration and peristalsis. J Clin Invest 52:1–13, 1973.
7. Edmundowicz S, Clouse RE. Shortening of the esophagus in response to swallowing. Am J Physiol 260: G512–516, 1991.
8. Dent J. A new technique fror continuous sphincter pressure measurement. Gastroenterology 71:263–267, 1976.
9. Linehan JH, Dent J, Dodds WJ, Hogan WJ. Sleeve devise functions as a Starling resistor to record sphincter pressure. Am J Physiol 248:G251-G256, 1985.
10. Richter JE, Wu WC, Johns DN, Blackwell JN, Nelson JL, Castell JA, Castell DO. Esophageal manometry in 95 health adult volunteers. Dig Dis Sci 32:583–592, 1987.

11. Dodds WJ, Hogan WJ, Reed DP, Stewart ET, Arndorfer RC. A comparison between primary peristalsis following wet and dry swallows. J Appl Physiol 35:851–857, 1973.
12. Ali GN, Wallace KL, Laundi TM, Hunt DR, DeCarle DJ, Cook IJ. Predictions of outcomes following cricopharyngeal disruption for pharyngeal dysphagia. Dysphagia 12:133–139, 1997.
13. Hewson EG, Ott DJ, Dalton CB, Chen YM, Wu WC, Richter JE. Manometry and radiology, complimentary studies in the assessment of esophageal motility disorders. Gastroenterology 98: 626–632, 1990.
14. Kahrilas PJ, Dodds WJ, Hogan WJ, Kern M, Arndorfer RC, Reece A. Peristaltic dysfunction in peptic esophagitis. Gastroenterology 91: 897–904, 1986.
15. Kahrilas PJ, Dodds WJ, Hogan WJ. The effect of peristaltic dysfunction on esophageal volume clearance. Gastroenterology 94:73–80, 1988.
16. Clouse RE, Staiani A. Contraction abnormalities of the esophageal body in patients referred for manometry; a new approach to classification. Dig Dis Sci 28: 784–791, 1983.
17. Clouse RE, Staiani A, Landau DW, Schlachter JL. Manometric findings during spontaneous chest pain in patients with presumed esophageal "spasms." Gastroenterology 85: 395–402, 1983.
18. Kahrilas PJ, Hogan WJ, Clouse RE. American Gastroenterological Association technical review on the clinical use of esophageal manometry. Gastroenterology 107: 1865–1884, 1994.
19. Benjamin SB, Richter JE, Cordova CM, Knuff TE, Castell DO. Prospective manometric evaluation with pharmacological provocation of patients with esophageal motility dysfunction. Gastroenterology 84: 893–901, 1983.
20. Katz PO, Dalton CB, Richter JE, Wu WC, Castell DO. Esophageal testing of patients with noncardiac chest pain or dysphagia. Ann Intern Med 106: 593–597, 1987.
21. Ott DJ, Richter JE, Chen YW, Wu WC, Gelfand DW, Castell DO. Esophageal radiology and manometry; correlation in 172 patients with dysphagia. Am J Roentgnol 149:307–311, 1987.
22. Massey BT, Dodds WJ, Hogan WJ, Brasseur JG, Helm, JF. Abnormal esophageal motility. An analysis of concurrent radiographic and manometric findings. Gastroenterology 101: 334–354, 1991.
23. Bancewicz J, Osugi H, Marples M. Clinical implications of abnormal esophageal motility. Brit J Surg 74: 416–419, 1987.
24. Achem SR, Kolts BE, Wears R, Burton L. Richter JE. Chest pain associated with nutcracker esophagus: a preliminary study of the role of gastroesophageal relfux. Am J Gastroenterol 88: 187–192,1993.
25. Singh S, Price JE, Richter JE. The LES locator: accurate placement of an electrode for 24 hour pH measurement with combined solid state pressure transducer. Am J Gastroenterol 87: 967–970, 1992.
26. Marples MJ, Mughal M, Bancewicz J. Canand esophageal pH electrode be accurately positioned without esophageal manometry? In: Siewert JR, Holscher AH, eds. *Diseases of the Esophagus.* Springer-Verlag, New York, 789–791, 1987.
27. Mattox HE, Richter JE, Sinclair JW, Price JE, Case ID. The gastroesophageal pH step up inaccurately locates the proximal lower esophageal sphincter. Dig Dis Sci 37: 1185–1191, 1992.
28. Joelsson BE, DeMeester TR, Skinner DB, LaFontaine E, Waters PF, O'Sullivan GC. The role of the esophageal body in the antireflux mechanism. Surgery 92: 417–424, 1982.
29. Spechler SJ. Department of Veterans Affairs Gastroesophageal Reflux Disease Study Group. Comparison of medical and surgical therapy for complicated gastroesophageal reflux disease. N Engl J Med 326: 786–792, 1992.
30. Mughal MM, Bancewicz J, Marples M. Esophageal manometry and pH recording does not predict the bad results of Nissen fundoplacation. Br J Surg 77: 43–43, 1990.
31. Escandell AO, Martinez de Haro LF, Paricio PP, Albasani JLA, Marcilla JAG, Cuenca GM. Surgery improves defective oesophageal peristalsis in patients with gastro-oesophageal reflux. Br J Surg 78:1095–1097, 1991.
32. Stevens, MB, Hookman P, Siegel CI, Esterly JR, Shulman LE, Hendrix TR. Aperistalsis of the esophagus in patient with connective-tissue disorders and Raynaud's phenomenon. N Engl J Med 270: 1218–1222, 1964.
33. Cohen S, Harris LD. Does hiatus hernia affect competence of the gastroesophageal sphincter? N Engl J Med 284: 1053–1056,1971.
34. Dodds WJ, Dent J, Hogan WJ, Helm, JF, Hauser R, Patel GK, Egide MS. Mechanisms of gastroesophageal reflux in patients with reflux esophagitis. N Engl J Med 301:1547–1552, 1982.

3

24 HOUR AMBULATORY pH TEST

Joseph A. Murray*

Department of Gastroenterology
Mayo Medical School
Rochester, Minnesota 55905

INTRODUCTION

The major function of the esophagus is to transport ingested food to the stomach. An equally important function is to prevent reflux of corrosive gastric contents into the upper aero-digestive tract. The lower esophageal sphincter prevents reflux of gastric acid into the esophagus. Failure of the lower esophageal sphincter to prevent excessive reflux may result in damage to the esophageal mucosa. The degree of damage depends on the frequency and duration of acid exposure, particularly when the pH is less than 4.

The advent of miniaturized probes has led to the development of methods to measure the intra-esophageal pH (1). In the early 70's, the advent of reliable recorders stimulated interest in the assessment of ambulatory pH (2). Today, prolonged esophageal pH monitoring is a widely used technique where an acid sensitive probe is placed in the lower esophagus, usually 5 cm above the manometrically defined lower esophageal sphincter. The probe is attached to a portable recording device. During the study, the patient can pursue activities of daily living. Thus, the study can be performed under physiological conditions. Subsequently, the recording is downloaded on to a computer and the analyzed data can either be viewed on a monitor or a hard copy can be printed (3).

EQUIPMENT

The necessary equipment for performing ambulatory pH study consists of a probe, a recording device and a computer for displaying and analyzing data.

Probe: Two types of pH sensitive electrodes are in common use; glass and antimony electrodes. Each of these have their own advantages and disadvantages (Table 1). How-

* 507-284-2467(Tel), 507-284-0538(fax), Email <murray.joseph@mayo.edu

Gastrointestinal Motility, edited by Rao *et al.*
Kluwer Academic / Plenum Publishers, New York, 1999.

Table 1. Comparison of glass and antimony electrodes

Glass electrodes	Antimony electrodes
Wide range of pH sensitivity	Cannot read values above pH 8
Fragile	Durable
Can be used 100's of times	Single or multiple(<8)
Precise	Less precise but adequate for clinical use
Hard to clean	Easy to clean/ disposable
Cost per use low, individually expensive	Disposable: high cost per use
Usually limited to a single sensor	One to three sensors

ever, the problems with cleaning and maintaining a delicate glass electrode probe has led to the antimony probe becoming more popular in the USA.

Antimony probes are relatively inexpensive, are less vulnerable to breakage but can only be used for a limited number of times. This leads to a high per use cost. Indeed many laboratories use single use antimony probes. All probes are subject to some drift and need to be carefully calibrated before and probably after completion of the study. The antimony probe has a limited range and cannot accurately record pH above 8. Clinically, this drawback is unlikely to be significant. As the pH sensitive probe in the esophagus is essentially completing an electrical circuit and this circuit can be completed either by using an internal reference electrode or less commonly by using an external cutaneous electrode placed on the thorax or upper abdomen. The external reference electrode should be securely fixed to the skin so that the electrical circuit is complete and there is no loss of conduction. The patient should be instructed to ensure that this electrode does not become dislodged. Recording systems may differ but currently they are all solid state with different methods of data compression and storage although earlier models used a tape recorder.

Although the method of storage makes little difference, apart from cosmetic advantages of size and weight, accuracy and reliability are absolutely essential for any recording device. Modern recording devices can save data even though its power supply has failed. One model was that of the radiometer capsule which was a self contained pH monitor contained in a capsule which was tethered by a string to the teeth or outside on the face and essentially suspended in the lower esophagus. It transmitted its data by telemetry to an extra-corporeal detection system and recorder. This system has not been widely used in North America.

INDICATIONS FOR 24 HOUR pH RECORDING

Twenty-four hour esophageal pH monitoring has been used for detecting, quantifying and characterizing acid gastroesophageal reflux in humans. The quantitative measure is limited to the duration and degree of pH change. It cannot measure the volume of reflux into the esophagus. This test has several clinical uses (Table 2). In addition to quantifying the frequency and duration of acid reflux it provides temporal correlation between gastroesophageal reflux and either typical symptoms such as heartburn or atypical symptoms such as chest pain, cough or wheezing (3, 4, 5). The utility of 24-hour pH monitoring in patients with obvious reflux symptoms is however quite limited. Even though most of the early studies have validated the use of ambulatory pH in patients with symptoms of gastroesophageal reflux such as heartburn and acid regurgitation, currently the potential uses for a 24 hour pH study include:

Table 2. Indications for ambulatory pH monitoring

Symptoms atypical for GERD	Asthma, hoarseness, cough
Atypical chest pain	Therapeutic failures
Non-ulcer dyspepsia	Pre- or post operative evaluation

1. to identify gastroesophageal reflux in patients with atypical chest pain;
2. to evaluate patients with non-ulcer dyspepsia;
3. to identify GERD-associated respiratory symptoms including asthma and laryngitis;
4. to investigate persistent symptoms after medical or surgical therapy for reflux;
5. possibly to predict the necessity and timing of long term anti-reflux therapy;
6. to possibly detect bile reflux into the esophagus(requires a special system) (4).

There has been a rather variable correlation between 24-hour pH studies and the presence of either endoscopic or microscopic esophagitis. This may be due to such factors as previous treatment of esophagitis, variability in the mucosal resistance factors and in the extent and volume of refluxate. A single sensor within the esophagus may not reliably distinguish the effects of these factors. The utility of the ambulatory pH test is not to identify esophagitis but rather to quantify the degree of acid reflux in the distal esophagus.

Several studies have described normative data in asymptomatic individuals and this has made it possible to compare data from patients with GERD(5). The discrepancies in the published normative data is possibly due to variations in normal subject selection, protocol, or equipment (3).

WHERE AND HOW TO PLACE THE pH PROBE

Conventionally, pH probes have been placed 5 cm above the upper border of the manometrically defined lower esophageal sphincter. This is to ensure that one can record the lower esophageal pH without dipping into the stomach, especially when the esophagus shortens during swallowing or during forced expiration. However, it is not always practical to perform manometry prior to pH study. Alternatives that have been used to assess the location of probe placement from the nares include the endoscopic measurement of the distance from the incisor teeth to the squamo-columnar junction and then adding an arbitrary distance, usually 5cm in the adult. Alternatively, an even less accurate method is the step-up pH method. The pH probe is placed into the stomach with the patient in the supine position. It should detect a low or acid pH of 1 and then it is slowly pulled back, stepwise, until there is a step up in the pH. From this level, the probe is pulled back by another 5 cm in to the lower esophagus. This method is fraught with difficulty as the probe may move out of the acid containing fluid prior to withdrawal through the lower esophageal sphincter. Also, if the patient has an incompetent lower esophageal sphincter, the lower esophagus may be bathed in acid. Some newer probes have incorporated into them a so-called lower esophageal sphincter locator device which is a single manometric probe device which can be utilized in a pull through fashion to identify the location of the lower esophageal sphincter. Some others have used fluoroscopy. The use of fluoroscopy is also fraught with problems particularly when there is coexistent hiatal hernia, which may not be readily appreciable with fluoroscopy, and therefore its use is largely limited to children.

Inaccurate placement of the probe may result in significant error, both by reducing sensitivity if placed too high or by producing falsely higher values if placed too low (6).

STUDY PROTOCOL

Assuming that the location is correct, and the probe and recording system are functional, one can proceed with the test. The probe must be carefully calibrated for the high and low pH values and the connections must be checked carefully. If an internal reference electrode is used any plastic covering over the internal reference electrode should be first removed. After applying topical lidocaine, the probe is inserted into the stomach through the nares. It is then withdrawn up to the desired location as measured previously by manometry or by using the LES locator. The patient is instructed to keep a detailed diary of events, symptoms and meals by completing a written diary. It is important to synchronize the patient's watch with the timer on the machine to ensure accurate description of the event. The recorder also has buttons that the patient may use to indicate events such as eating a meal, heartburn or lying down. Because, many patients cannot reliably identify the correct buttons especially at night, a symptom diary is certainly more simple and reliable. The patient is encouraged to undertake their usual daily activities and meals, except bathing. Towel washing may be done so long as care is taken not to wet the surface reference electrode or dislodge it or the tube. Patients may experience some throat discomfort during the test, but most can tolerate this adequately. The patient returns the next day. It is useful to check the recording for completeness before removing the probe, in case the recorder has malfunctioned and the test may have to be repeated. The accuracy of the diary should be verified with the patient. Some labs will re-calibrate the probe at the end of the test to check for drift that may have occurred.

INTERPRETATION

Assuming that a complete record has been acquired, then one can proceed with interpretation and analysis of data.

Visual Inspection

Initially the entire record should be displayed, either on a computer screen or on paper and the character of the pH tracing should be visually inspected. This will typically identify acid reflux episodes as abrupt drops in pH to a level below a pH of 4 (see Figure 1).

These abrupt drops in pH indicate reflux of acidic material. The recording system arbitrarily decides the sampling rate at which the pH data is recorded. The pH may remain low for a variable time period and then gradually, often in a step-wise fashion, increase back to baseline. This neutralization occurs both due to clearance of the refluxed acid back into the stomach either by gravity or esophageal clearance or due to repeated swallowing of alkaline saliva or ingested food. It is important to differentiate reflux episodes starting with an acute drop in pH from other drops in pH. Gradual acidification can be seen in patients who have esophageal retention for example achalasia (see Figure 2). If one relies exclusively on the numerical data, it is possible to erroneously diagnose acid reflux, when in fact what has occurred is intra-esophageal acidification. Occasionally, technical problems due to the dislodgment of the reference electrode may lead to an apparent drop in pH usually to 0 and

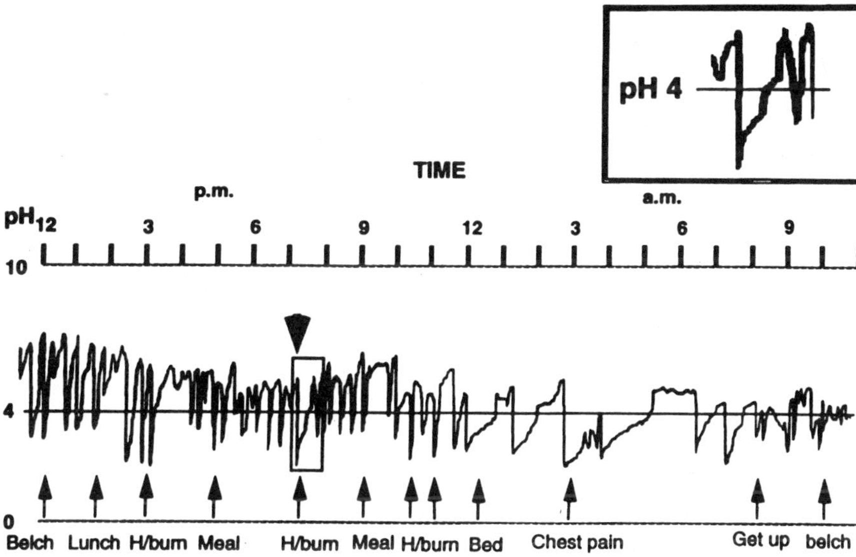

Figure 1. This is an ambulatory pH recording from a patient that shows typical reflux pattern, with reflux occurring both in the upright and supine periods. The inset demonstrates the cardinal feature of a reflux episode, the abrupt drop in pH. (reproduced with permission, copyright Singular Press).

these episodes need to be excluded from any calculation. One needs to analyze tracings visually and correlate changes in pH with diary events recorded by the patient, such as symptoms of heartburn, chest pain, respiratory symptoms, burping or belching, as well as changes in position including the time period in which the patient sleeps. Certain technical problems can produce apparent drops in pH. These could be a poorly connected reference electrode, faulty connection or probe failure. These may be analyzed by the computer as reflux events. These can be excluded from the analysis prior to computer calculation.

Computer Calculation

The analyst can enter diary events into the computer program so that computer-based analysis of various parameters can be made. Several parameters are routinely calculated by the software including: the frequency and duration of reflux episodes, the total

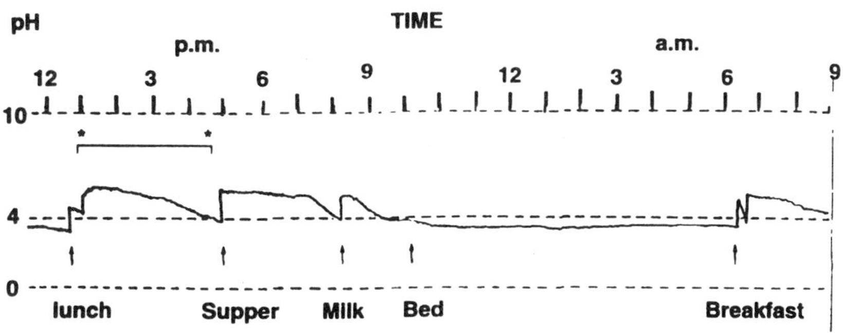

Figure 2. This tracing demonstrates a gradual drop in pH seen in a patient with untreated achalasia. This is due to in situ acidification of retained food contents(reproduced with permission, copyright Singular press).

Table 3. Normal values of 24 hour esophageal pH

	Mean	Stand.dev	Median	Minimum	Maximum	95th C.I.
% Total time pH<4.0	1.5	1.4	1.2	0	6.0	4.5
% Upright time pH<4.0	2.2	2.3	1.6	0	9.3	8.4
% Supine time pH <4.0	0.6	1.0	0.1	0	4.0	3.5
No of episodes	19.0	12.8	16.0	2.0	56.0	46.9
No of episodes ≥ 5 Min.	0.8	1.2	0	0	5.0	3.5
Longest episode (min)	6.7	7.9	4.0	0	46.0	19.8
Composite score	6.0	4.4	5.0	0.4	18.0	14.7

Reproduced with permission from Reference 3.

number of reflux episodes and those that last longer than 5 minutes, duration of the longest episode. These parameters are detailed in both the upright, supine and total time period of the recording. The single parameter that has been shown to best correlate with endoscopic damage has been the total time the pH is less than 4. The physiologic significance of a pH <4 lies in the fact that pH must be < 4 for damage to occur. Non acidic substances such as pepsin are not active unless the pH is <4. Known values have been established, both for hospitalized patients and subsequently for ambulatory patients. Ambulatory normal values tend to show a higher degree of variability during the daytime. This probably relates to normal variations in eating and physical activity, although in one study only 14.3% of subjects altered their routine activity (7). Normal values derived from 50 healthy subjects (2, 3) are shown in Table 3.

Symptom Correlation

While the aforementioned parameters provide some quantitative assessment of gastroesophageal reflux and correlate adequately with the presence and degree of esophagitis in untreated patients, they do not necessarily indicate that atypical symptoms are caused by acid reflux. Conversely it is not necessary to have esophagitis to have reflux induced pain (9).

Several parameters have been introduced to measure the likelihood of a causal association. The symptom index (SI) is a numerical scale that quantifies the percent association between symptoms and a pH <4. SI = symptom associated with reflux x 100/#symptom episodes expressed as a percentage (8). This index does not take into account all of the reflux episodes but only a small number of symptom-related events. For example, an individual with 200 reflux episodes but only a single episode of chest pain which happens to occur within 2 minutes of one of these reflux events may have a symptom index of 100 %.

To address this issue the symptom sensitivity index (SSI) was devised. SSI= #of symptomatic reflux episodes x 100/#of reflux episodes again expressed as a percentage. In the same example given above the SSI= 0.5% which is a very low value. In an attempt to address the shortcomings of both of these measures, several more complicated mathematical formulae have been developed. The easiest of these is the estimation of the symptom association probability using Fisher's exact test to study the probability that the reflux episode and the symptom occurred by chance. If there is a low possibility then one could infer that reflux is a cause of the symptom (9). If the SI is also low then it is unlikely to be the cause of the symptom, if it is high then it may be the only cause of the symptom.

SPECIAL METHODS

Short Duration Tests

There have been several efforts to shorten the duration of pH study, examining a different location of the probe within the esophagus or even pharynx and developing a probe that is sensitive to non-acidic refluxate. While there is good correlation between the amount of reflux which occurs during a three hour period after a standardized meal and gastroesophageal reflux in association with esophagitis, it is unlikely to be diagnostically important particularly in patients with infrequent or atypical symptoms (11, 12).

Dual Channel Recording

The placement of a pH probe in the upper esophagus or pharynx, in addition to the distal esophagus has been used to detect abnormal gastroesophageal or esophago-pharyngeal reflux (13). Through micro-aspiration, reflux of gastric material may cause chronic laryngitis. It must be recognized that normal values of proximal acid exposure may be quite different to those of distal esophageal acid exposure. Some have questioned the clinical value of measuring proximal esophageal pH (13). It is important to remember that the absence of proximal acid reflux does not necessarily rule out the occurrence of intermittent and severe acid reflux.

Studies in Children

Esophageal pH measurements have been widely used in infants and children. The indications are essentially similar to those in adults. The important difference is that the young infant may not report symptoms or maintain a diary. Apnea, bradycardia and Sandifer's syndrome (abnormal posturing) in infants may be associated with gastroesophageal reflux. Ambulatory pH testing is done both in the hospitalized and home settings and is usually well tolerated. It may be particularly useful in this population, as the child may be too young to relate typical symptoms of reflux. In children either fluoroscopy or the Strobel formula (5 + (0.0252X height in cms) is used to calculate the distance to a point 87% of the distance from the nares to the LES (14, 15, 16). Cooperation may be a problem and if the child is sedated then the data may not be physiological.

Treatment Failure

Ambulatory pH testing is particularly helpful in patients whose symptoms have not responded to standard therapy with proton pump inhibitors or surgery. For, post-surgical patients the protocol is similar to standard recording. For patients with symptoms in spite of treatment, a dual sensor pH study with one sensor located in the esophagus and another sensor in the stomach should be done. The patient must be instructed to stay on their usual medication during the study.

DETECTION OF BILE REFLUX

While acid is the major corrosive substance that causes esophageal injury, reflux of bile may play a role in either causing mucosal injury or in the severity of injury. Standard antimony probes are not sensitive enough to detect alkaline reflux nor can they detect neutral bile salt reflux. A spectrophotometric detection system which can detect the presence of bile in the esophagus has been developed (bilitec 2000) (17). The usefulness of this technique -outside of the research setting- has not yet been determined. Interestingly, the degree of bile reflux is reduced when acid production is inhibited possibly due to a reduction in the amount of gastric secretion present in the stomach (18).

SUMMARY

Ambulatory pH monitoring is a useful technique for quantifying acid reflux. It is being used widely, particularly for the diagnoses of GERD and atypical problems associated with GERD. In order to provide an accurate and clinically useful information, it is essential to perform and interpret the study meticulously.

REFERENCES

1. Miller FA. Monitoring of upper gastrointestinal tract pH: a new tool in the evaluation of acid-peptic disease. Pacific Med & Surg 1967;75(2):92–5.
2. Johnson LF. Demeester TR. Twenty-four-hour pH monitoring of the distal esophagus. A quantitative measure of gastroesophageal reflux. Am J Gastroenterol 1974;62(4):325–32.
3. Ambulatory pH monitoring 2nd edition ed:Richter Joel E. 1997 Williams and Wilkins Baltimore, Maryland.
4. Hewson EG, Sinclair JW, Balton CB, and Richter JE: 24 Hour Esophageal pH Monitoring: The Most Useful Test for Evaluating Non-Cardiac Chest Pain. Am J Med 1991;90(5):576–583.
5. Johnsson F. Joelsson B. Isberg PE. Ambulatory 24 hour intraesophageal pH-monitoring in the diagnosis of gastroesophageal reflux disease. Gut 1987;28(9):1145–50.
6. Mattox HE, The placement of the pH probe in: Ambulatory pH monitoring, ed Richter, Williams and Wilkins , 2nd edition, Baltimore, USA.
7. Mattioli S, Pilotti V, Spangaro M, Grigioni WF, Zannoli R, Felice V, Conci A, Gozzetti G: Reliability of 24 hour Home Esophageal pH Monitoring in Diagnosis of Gastroesophageal Reflux. Dig Dis Sci 1989;34:71–78.
8. Wiener GJ, Richter JE, Copper JB, Wu WC, Castell DO: The Symptom Index: A clinically Important Parameter of Ambulatory 24 Hour Esophageal pH Monitoring. Am J Gastroenterol 1988;83(4):358–61.
9. Shi G. Bruley des Varannes S. Scarpignato C. Le Rhun M. Galmiche JP. Reflux related symptoms in patients with normal oesophageal exposure to acid. Gut 1995;37(4):457–64.
10. Weusten BL. Roelofs JM. Akkermans LM. Van Berge-Henegouwen GP. Smout AJ. The symptom-association probability: an improved method for symptom analysis of 24-hour esophageal pH data. Gastroenterology 1994;107(6):1741–5.
11. McMahon M, Murray J, Hogan B Doyle JS. 3 hour vs 24 hour intra-esophageal pH monitoring, Gut 1989;A1492.
12. Rokkas T. Anggiansah A. Uzoechina E. Owen WJ. Sladen GE. The role of shorter than 24-h pH monitoring periods in the diagnosis of gastro-oesophageal reflux. Scandinavian J Gastroenterol 1986;21(5):614–20.
13. Wo JM. Hunter JG. Waring JP. Dual-channel ambulatory esophageal pH monitoring. A useful diagnostic tool? Dig Dis Sci 1997;42(11):2222–6.
14. Strobel CT. Byrne WJ. Ament ME. Euler AR. Correlation of esophageal lengths in children with height: application to the Tuttle test without prior esophageal manometry. J Ped 1979;94(1):81–4.
15. Vandenplas Y. Loeb H. The interpretation of oesophageal pH monitoring data. Eur J Ped 1990;149(9):598–602.
16. Sondheimer JM. Haase GM. Simultaneous pH recordings from multiple esophageal sites in children with and without distal gastroesophageal reflux. J Ped Gastroenterol & Nutri 1988;7(1):46–51.
17. Bechi P. Pucciani F. Baldini F. Cosi F. Falciai R. Mazzanti R. Castagnoli A. Passeri A. Boscherini S. Long-term ambulatory enterogastric reflux monitoring. Validation of a new fiberoptic technique. Dig Dis Sci 1993;38(7):1297–306.
18. Champion G. Richter JE. Vaezi MF. Singh S. Alexander R. Duodenogastroesophageal reflux: relationship to pH and importance in Barrett's esophagus. Gastroenterology 1994;107(3):747–54.

4

SCINTIGRAPHIC AND ULTRASOUND EVALUATION OF GASTRIC MOTILITY

Bruce P. Brown*

Department of Radiology
University of Iowa Hospital
Iowa City, Iowa

INTRODUCTION

Nausea, vomiting and dyspepsia are common symptoms of gastric motor dysfunction. However, they are not specific for gastric disorders. In order to elucidate an underlying gastric cause for such symptoms, it is important to objectively test gastric motor function.

The ideal test of gastric motor activity is one that preserves normal gastric physiology, is non-invasive, simple to perform, quantitative, and inexpensive. Attempts to satisfy these criteria have produced an array of tests from simple tube aspiration of gastric contents to MRI evaluation of gastric secretory rates. However, radionuclide scintigraphy remains the mainstay of evaluating gastric motor activity.

RADIONUCLIDE SCINTIGRAPHIC EVALUATION OF GASTRIC EMPTYING

Griffith and Owen (Ref. 1) in 1966 first described a radionuclide method of gastric emptying using a Cr-51 labeled meal. This provided a non-invasive, simple, quantitative evaluation of gastric emptying with only a small amount of radiation exposure to the patient. However, this study could not separate solid from liquid emptying and had the potential for error because of early label decay and adherence of label to the stomach lining long after the food had passed into the duodenum.

* Phone: 319/356-4374, E-Mail: bruce-brown@uiowa.edu

Gastrointestinal Motility, edited by Rao *et al.*
Kluwer Academic / Plenum Publishers, New York, 1999.

Improved labels have been developed that are specific for either liquids or solids and are firmly bound to specific food constituents. The most elegant example of a specific solid label was developed by Meyer in 1976.[2] By this method, Tc-99m was injected into live chickens prior to slaughter. The label was thus incorporated into the liver. After slaughter, the liver with the incorporated tracer was cooked and ingested by the patient as a marker for solid phase gastric emptying.

To simplify solid labeling, Leb (Ref. 3) achieved results similar to Meyer's by incorporating the Tc-99m label into the colloid matrix of scrambled eggs which was ingested as a solid-phase marker.

Stable markers for the liquid phase of the meal have also been developed. The best of these appear to be DTPA or EDTA chelates of Indium-111 or Gallium 67. Technecium 99m mixed into the liquid has also been used. When required, the separate liquid and the solid markers can be given in one meal which allows evaluation of liquid and solid emptying at the same time. Other food constituents such as fats now have specific labels, but are not in general use.

Methods

Unfortunately, even now, gastric emptying studies are not standardized between laboratories. Thus, it is best for each laboratory to establish its own emptying curves rather than to rely on published results.

For sold-phase emptying, we use scrambled eggs labeled with 1.0 millicurie Technecium 99m sulfur colloid. For liquid phase, we use 1 millicurie of Technecium 99m sulfur colloid in 300 ml water. For solid and liquid studies performed together, we use Technecium scrambled eggs, as in the single phase study, ingested with Indium 125 microcuries of Indium 111 DTPA in 300 ml water.

Because, evaluation of solid-phase emptying is the most sensitive test for gastric motor dysfunction, in our institution we commonly perform solid–phase gastric emptying test. Patients are brought to the laboratory after an overnight fast and, if possible, off all medications. The patient then ingests a standard test meal that consists of 2 large scrambled eggs containing Technecium 99m sulfur colloid (mixed into the egg white), one piece of toast with one square of butter, and 120 ml of water. Thereafter, with the patient sitting upright, two-minute anterior scans, followed by two-minute posterior scans are taken every 10–30 minutes for the next 180 minutes. If required, scanning is continued for longer periods.

Time-activity curves are plotted for the geometric means of the counts from the anterior and posterior scans. Each of the geometric means are corrected for decay, and tissue attenuation. When radionuclides for solid and liquid phases are used together, correction for "crosstalk" between the energy peaks must also be performed. Plots of the radioactivity over time are displayed and compared to normal curves developed in house.

Interpretation

Regardless of their fasting state, each patient, will have variable amounts of fluid in the stomach at baseline. Also, the test meal will have variable effects on gastric secretion. Thus, emptying cannot be expressed in absolute terms of ml/min. Instead, it is expressed in percent of the maximum counts seen immediately after ingestion of the radiolabeled meal. These percentage values are plotted against time.

Typically, liquids empty faster than solids and the liquid emptying curve takes an exponential shape. On the other hand, solids require trituration and mechanical dispersal before they empty, and the curve shows an initial lag phase followed by almost linear emptying over time.

Visual interpretation of such curves requires consideration that in some cases, there may be redistribution of tracer in the stomach without emptying. On a single view this may be misinterpreted as disappearance of tracer. This is corrected by referring to both the anterior and posterior views and using the geometric means of the anterior and posterior scans in the final time-activity plot. The simplest way to express gastric emptying is to identify the time it takes 50% of the radiotracer to empty from the stomach. This value has been shown to correlate nicely with the overall emptying time. Because this value can be influenced by the type of tracer, the size of meal and other factors, there is considerable variation in the gastric emptying half-time between laboratories.

Figure 1A shows a normal time-activity curve with two standard deviations from the mean for solid-phase emptying together with a plot from a patient with normal emptying. Figure 1B shows the anterior scans. The scans demonstrate an initial shift in the distribution of the radionuclide into the distal stomach and then gradual emptying into the duodenum. The half-time clearance (50% emptying) for this patient was normal at 97 minutes.

Figure 2A-B shows a similar plot and the corresponding anterior scans for a diabetic patient with nausea and vomiting. The early images show an initial shift of radiotracer within the stomach, but even after 180 minutes, there is only 10% emptying.

ULTRASOUND IMAGING OF GASTRIC MOTILITY

With the development of high-resolution, real-time ultrasonographic imaging methods, several investigators have used ultrasound to evaluate gastric emptying. Holt and Cervantes found a close correlation between ultrasound and scintigraphic emptying of a liquid test meal.[4] However, using ultrasound it is difficult to visualize and quantify the emptying of solids from the stomach. Therefore there is no data validating the emptying of a solid meal using ultrasound or comparing it with a rigorous gold standard. Thus, although ultrasound provides a physiologic method of visualization with virtually no risk to the patient, the inability to view solids in the stomach, and the technical difficulty in performing and analyzing studies together with the variable configuration and location of the stomach has limited its usefulness in clinical practice.

In our laboratory, ultrasound has been used primarily as a research tool for evaluating gastric emptying of a liquid meal with small particulate solids.[5] Our methods are similar to others who have used ultrasound to study liquid emptying of the stomach except that our subjects ingest measurable, digestible solid particles to study the effect of the physical breakdown of solid particles in a milieu which can be seen with the ultrasound probe. The meal consists of 500 ml of clear chicken broth taken with five garbanzo beans cut in half and swallowed whole.

Subjects are scanned in the sitting position with an Acuson 128 ultrasound machine (Acuson, Mountain View, California), using either a 5 or 7 MHz linear array transducer. Scans are obtained in the fasting state, and in the immediate postprandial state, and every twenty minutes thereafter for 160 minutes. All of the scanning images are recorded on a video tape for analysis later.

During each scan period, five minutes of scanning is performed in the transverse plane through the gastric antrum using the longitudinal axis of the superior mesenteric vein and the aorta as an internal landmark (Figure 3). This view provides quantification of the number, depth, and duration of contractions. Scanning of the antrum in the transverse

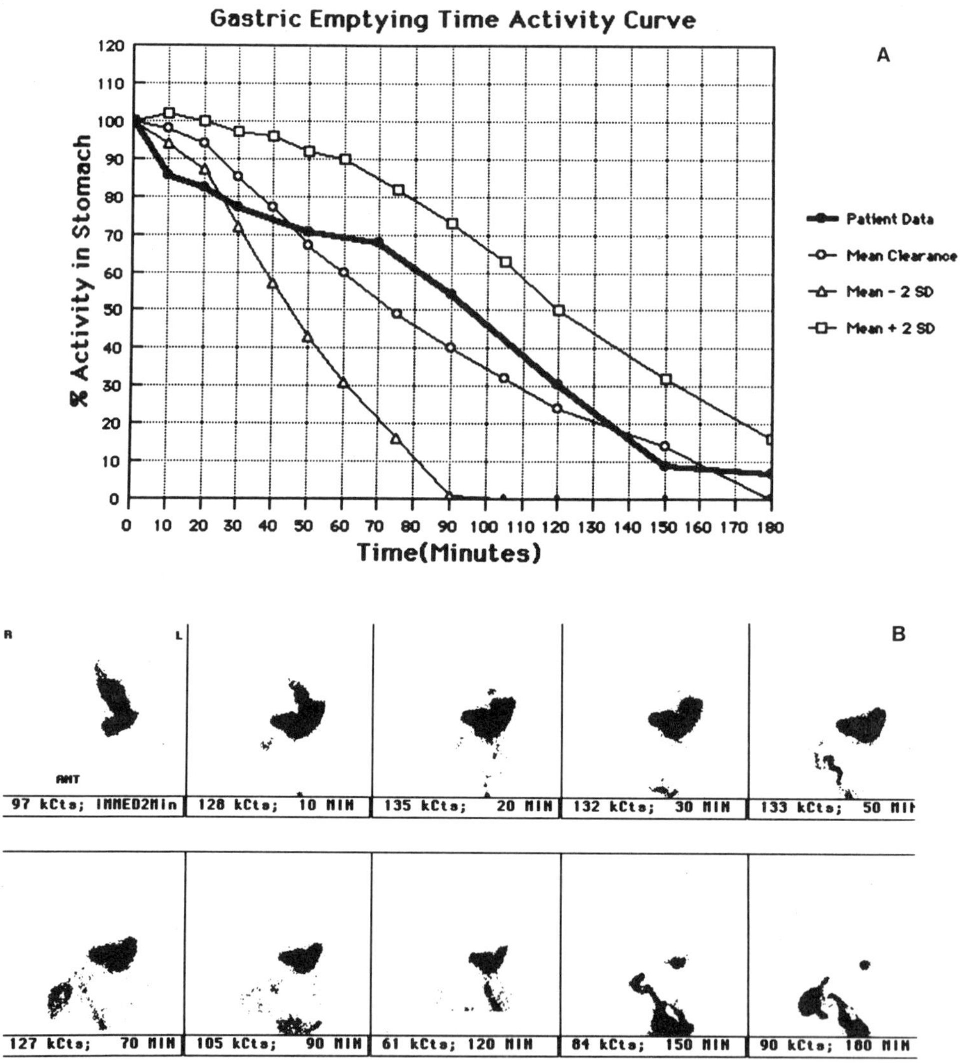

Figure 1. (A) Plot over time of the geometric means of anterior and posterior scans of a solid meal (Tc 99m-labeled scrambled eggs) in a normal patient in the sitting position (heavy black line) with accompanying plots of our normal in-house means ( ) of normal subjects +/- 2 SD (,). Note a small amount of early emptying followed by a normal emptying curve. (B) Scintigraphic camera views of the region of interest over the stomach and small bowel. Note that over time almost no label remains in the stomach.

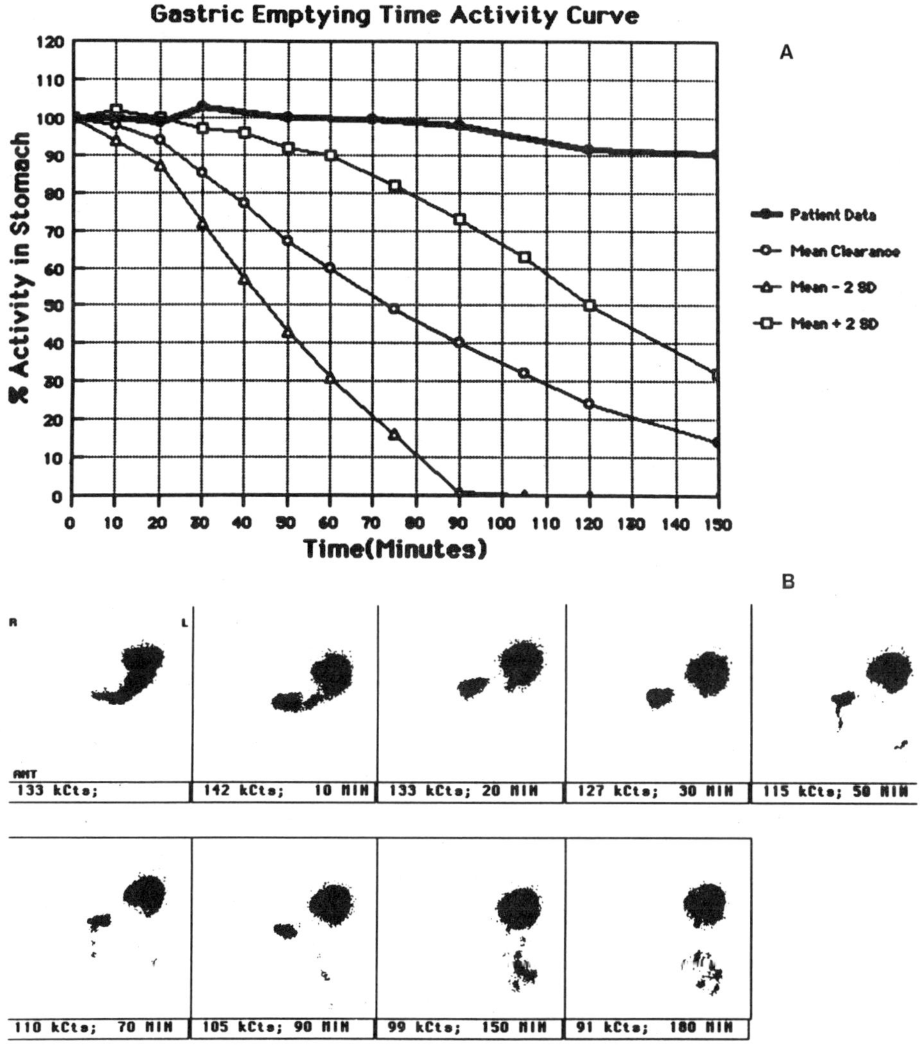

Figure 2. (A) Plot over time of the geometric means of anterior and posterior scans of solid meal emptying in a patient with diabetes with nausea and vomiting. Patient data (heavy black line) is superimposed on our normal in-house data. Note there is very little emptying over the 180 minute period of observation. (B) Scintigraphic camera views of the region of interest over the stomach and small bowel over time. Note that although there is shift of the configuration of the label in the stomach, there is little emptying into the small bowel.

plane is followed by two minutes of scanning in the longitudinal plane through the antro-duodenal junction (Fig. 3). This allows us to study the propagation of antral contractions as well as the flow through the pylorus. The last measurement in each period of study is the average maximum diameter of three beans in the antrum. This allows a rough estimate of the course of the physical breakdown of the gastric solids.

Figure 4 shows frames from a video tape recording of a normal subject during fasting and early post-prandial period. Corresponding line drawings are shown in Figure 5. In the transverse views of the antrum, one can identify the hypoechoic muscular wall, the

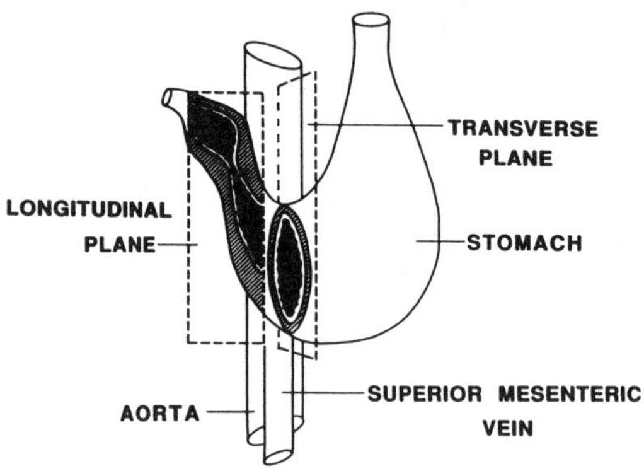

Figure 3. Orientation for ultrasound imaging of the gastroduodenal junction. Transverse scans of the antrum were performed through the plane defined by the long axes of the superior mesenteric vein and the aorta. Longitudinal scans of the gastroduodenal junction were performed in the long axis of the distal gastric antrum. (Used by permission of authors, Ref. 5.)

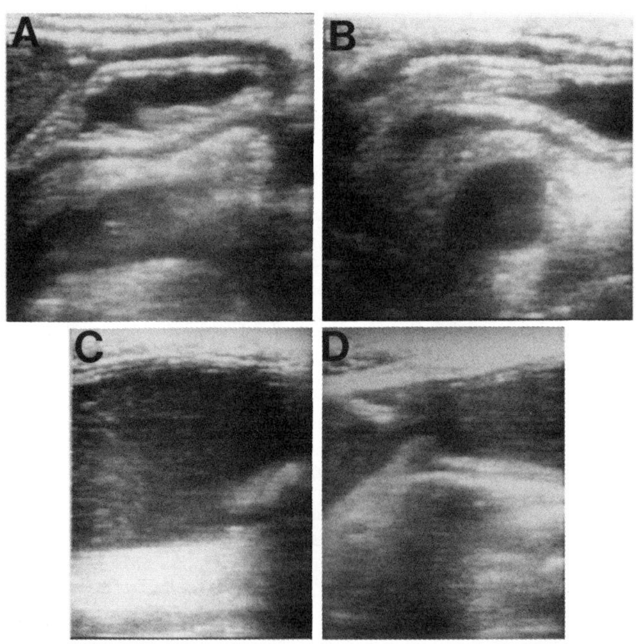

Figure 4. Ultrasonographic appearance of the normal gastroduodenal junction during fasting and early postprandial period (see the corresponding line drawings in Figure 5A-D). (A) A cross section taken through the gastric antrum in the plane of the superior mesenteric vein and aorta. The patient is fasting. The central lumen is reduced to a slit and contains a small amount of fluid (dark). The echogenic (bright) mucosa is folded into the lumen, and the outer hypoechoic (dark) muscular walls are thick and collapsed. (B) A longitudinal section in the plane of the long axis of the antro-pyloric junction. The pyloric sphincter is denoted by the focal thickening of the muscular wall that narrows the gastric outlet. Note that the mucosa occludes the lumen for some distance proximal to the distal pyloric sphincter. Both frames were obtained in the absence of superimposed contraction with no luminal or transpyloric flow. (C and D) Images taken during the transient, quiescent period immediately after ingestion. The antrum is distended with fluid; its mucosa is flattened, and its muscular wall has become thin (C). Note two beans settled along the dependent part of the gastric greater curve in the gastric sinus. Approximately 5 minutes later (D), liquids have begun flowing through the patent pyloric orifice and the prepyloric antrum is funnel-shaped. (Used by permission of authors, Ref. 5.)

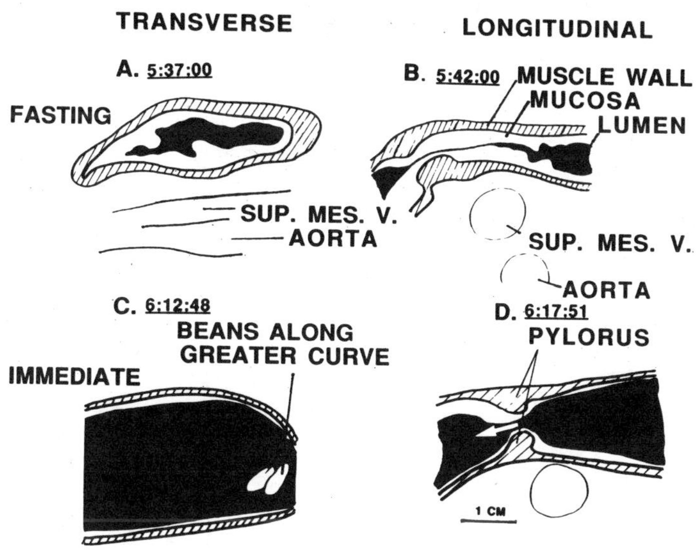

Figure 5. Line drawings of frames of a videotape from the subject shown in Figure 4. Real time in hours, minutes, and seconds accompanies each frame. The subject ingested the test meal between 6:05 and 6:10 a.m. The areas shaded with diagonal lines represent the hypoechoic muscular wall of the antrum. Areas in white represent the echogenic mucosa of the stomach. The areas outlined in black represent the fluid-filled lumen of the antrum. Fasting scans (A and B). The antrum is collapsed. The pylorus is closed. Contractions are infrequent. Immediately after ingestion (C and D) the antrum is maximally distended (C). The muscular wall and mucosa have become thin. Initially, there is a brief, transient, quiescent period with few contractions. Whole beans are retained along the greater curve of the stomach as liquids are emptied (decanted) over the beans (D). (Used with permission by authors, Ref. 5.)

echogenic mucosa, and fluid in the gastric lumen. In the longitudinal views, one can identify the muscular pyloric ring, the prepyloric antrum, and the base of the duodenal bulb.

In our study, on fasting scans, (Figure 4A-B; 5A-B) there were few contractions, and the antrum was collapsed with a closed muscular pylorus plugged by bright mucosa. Immediately postprandially, the beans lay along the dependent portion of the stomach in the gastric sinus, along the gastric greater curve, while liquids passed over the beans and through the open pylorus (Figure 4C-D; 5C-D). During this time, there was often antegrade flow across the pylorus not associated with contractions of the distal gastric antrum, suggesting that such flow was produced from contractions in the proximal stomach (5D).

At 20–60 minutes (Figure 6A-D) contractions reached a peak of 2–3 per minute. When beans would occasionally reach the distal antrum, pyloric closure blocked egress of whole beans (mean diameter 10–12 mm). In some cases, retrograde flow of liquid through the pylorus would force the bean back into the gastric body (Fig.6D). At 60–120 minutes, the terminal antrum had become tubular in shape with thickening of the muscular wall. The baseline diameter had decreased by approximately 50%. Now, with each gastric contraction the beans appeared in the antrum and the antral diameter further decreased. At the same time an echogenic paste was formed. The beans were able to pass through the pylorus only in this form, whereas larger particles were retained in the stomach, in part by pyloric closure. In this way, solid particles were retained in the stomach while smaller particles that were suspended in the antral liquid were decanted into the duodenum.

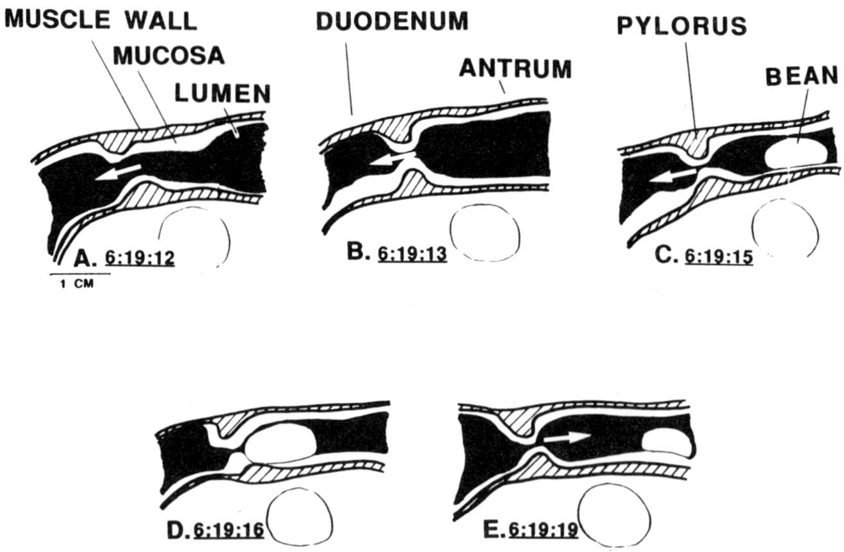

Figure 6. Early postprandial period. Example of retention of an intact bean by the pylorus. Shown is a five second sequence of longitudinal scans through the distal antrum, pylorus, and duodenal bulb. (A) The pylorus is relaxed, and there is free flow of liquids out of the stomach. (B) Pyloric closure begins as a contraction has reached the distal antrum. (C) The distal antrum is contracting, and a bean moves toward the pylorus. (D) The bean is retained at the pyloric orifice as its lumen has been reduced to barely a slit. Most beans remain intact and lay along the gastric greater curve. (E) Occasionally, the duodenal bulb becomes distended, and retrograde flow occurs through the partially relaxed pylorus. This sweeps the bean back from the pyloric orifice. Most beans remain intact and are only occasionally propelled from their position in the sinus into the distal antrum. (Used with permission of authors, Ref. 5.)

Thus, ultrasound provides a safe, non-invasive and useful method of scanning the stomach. It is not only possible to scan the stomach under physiological conditions and for prolonged periods but also to determine how the stomach separates liquids from solids in the upright position.

REFERENCES

1. Griffith GH, Owen GM, Kirkman S, Shields R, Measurements of rate of gastric emptying using chromium-51. Lancet 1966;1:244–1245.
2. Meyer JH, MacGregor IL, Gueller R, Martin P, Cavalieri R, Tc-99m-tagged chicken liver as a marker of solid food in the human stomach. Dig Dis 1976;21:296–304.
3. Leb G, Lipp RW. Criteria for labeled meals for gastric emptying studies in nuclear medicine. Eur J Nucl Med 1993;20:183–185.
4. Holt S, Cervantes C, Wilkinson AA, Wallace JHK. Measurement of gastric emptying rate in humans by real-time ultrasound. Gastroenterology 1986:90;918–923.
5. Brown BP, Schulze-Delrieu K, Schrier JE, Abu-Yousef MM. The configuration of the human gastroduodenal junction in the separate emptying of liquids and solids. Gastroenterology 1993;105:433–440.

5

ELECTROGASTROGRAPHY

Mark Pimentel and Henry C. Lin*

GI Motility Program
Department of Medicine
Cedars-Sinai Medical Center, and
UCLA School of Medicine
Los Angeles, California 90048

HISTORICAL INTRODUCTION

As the name implies, electrogastrography (EGG) is the study of the electrical activity of the stomach. The first EGG's were performed in the early 1920's by Walter Alvarez (1). During the time of Alvarez, the study of gastric electrical activity was crude and cumbersome making the analysis as well as the utility of data difficult. It was therefore not until the 1960's that the first abnormalities seen on EGG were described. Telander, et al. (2) described one of the first groups to attribute gastric distension in humans to faster than normal gastric rhythm, or tachygastria.

Over the last 20 years, there has been a resurgence in the study of EGG. With this resurgence came the improvement of technology used for recording and analysis. Computerized spectral analysis techniques have been major breakthroughs that have made EGG practical and reproducible. As a result of these improvements, EGG has advanced our understanding of the relationship between gastric rhythm and conditions such as peptic ulcer disease, non-ulcer dyspepsia, nausea, hyperemesis gravidarum, motion sickness and diabetes.

Of the many diagnostic procedures used in gastrointestinal motility today, EGG has the advantage of being non-invasive. However, because of its limited availability and the unsuccessful efforts to use EGG as a substitute for other diagnostic tests, its use as a diagnostic tool is still not fully established.

TECHNICAL ASPECTS

The basic premise behind the EGG is the measurement of the rhythm and amplitude of the gastric slow wave which originates from the gastric pacemaker under normal cir-

* Phone: 310-855-6143, Fax: 310-967-0125

Gastrointestinal Motility, edited by Rao *et al.*
Kluwer Academic / Plenum Publishers, New York, 1999.

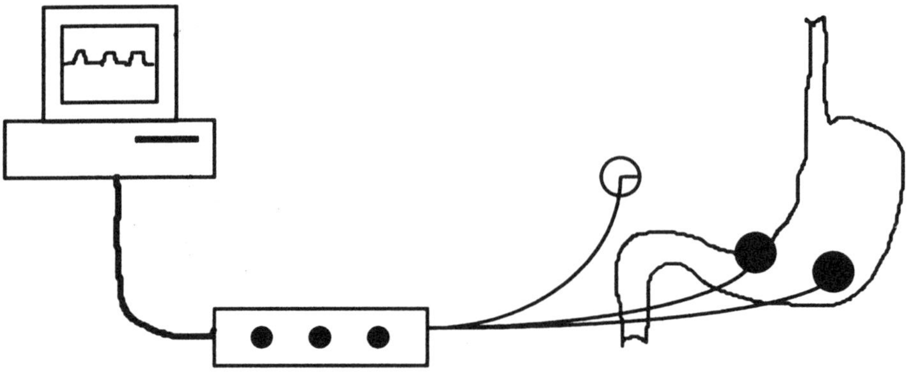

Figure 1. Technical diagram of electrogastrography.

cumstances. The gastric pacemaker is located along the greater curvature at the junction between the proximal 1/3 and distal 2/3 of the stomach. These electrical signals are normally transmitted in an antegrade fashion (similar in pattern to the electrical activity of the heart) toward the pylorus so that the distal 2/3 of the stomach becomes electrically coordinated or entrained. The gastric pacemaker fires at a rate of 3 cycles per minute (cpm).

From a technical point of view, the typical EGG requires three main components (Figure 1). These components consist of electrodes, a preamplifier and a computer with an analogue to digital converter board. The first of the components, the electrodes, are particularly important since many of the earlier electrodes were poor in quality leading to significant background noise. Despite the best technology, the electrical signal from the stomach is weak and noise must be minimized with careful skin preparation. Skin preparation reduces any impedance of low frequency signals. During EGG acquisition, the positioning of electrodes is also critical for signal quality. The leads should be placed over the antrum, located in the triangular area defined by the costal margin with the triangle apex represented by the xiphoid in the epigastric area. On occasion, the antrum must be localized by performing fluoroscopy or ultrasound for optimal placement.

The next component is the amplifier. The amplifier should have a built-in low-pass filter in order to exclude high frequency signals which include artifacts such as the heart signal as well as respiratory movement. This filter should be capable (if possible) of selectively sampling frequencies in the range of 0.05 Hz to 0.16 Hz. Such a filter would not exclude physiologically important tachygastric signals (3–4).

The third technical component for good EGG acquisition is a computer with appropriate software for data acquisition and analysis. After filtering, the data is passed into a computer and stored after conversion from analogue to digital signals. A sampling frequency of 1 Hz is sufficient for signal processing. The computer should be equipped with programs capable of analyzing EGG signals and performing Fast Fourier Transformation on the acquired data.

DATA ANALYSIS

The EGG signal in its raw form is sinusoidal in pattern (Figure 2). It is easy to appreciate the 3 cycles per minute (cpm) periodicity of the graph. Fast Fourier Transformation (FFT) is then applied to convert the signal to a power spectrum (5).

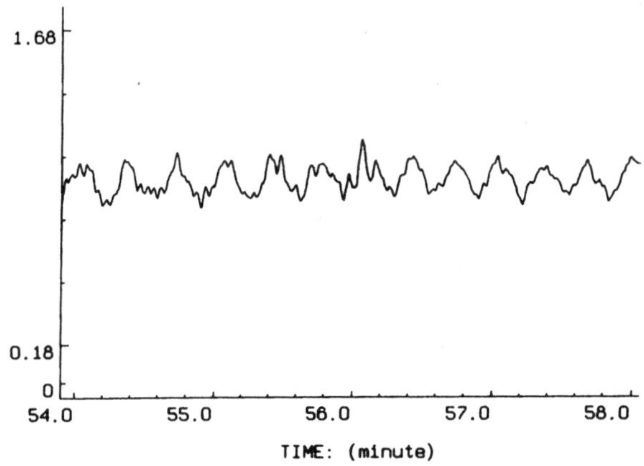

Figure 2. The raw EGG signal from a normal subject.

After FFT, two tools have been developed to simplify the interpretation of the EGG. These tools include the power and running spectral analyses (6). The power spectral analysis includes the compilation of all data during a period of time after the performance of FFT. An example of a power spectrum is provided in Figure 3a representing a single graph with a single large peak indicating the dominant frequency during the period of study. This dominant frequency is also given a power value. The power is reflective of the strength or amplitude of the EGG signal. Because power is so dependent on the quality of the signal it cannot be easily used to compare between individuals but can be used to compare responses within the same subject. Although there are alterations in power under physiologic conditions such as a meal, power is influenced by other factors such as abdominal wall thickness, distance from antrum, electrode type and even skin preparation. A change in power before and after a meal or a drug is often used for comparison.

Another limitation in representing the EGG as a power spectrum is that the format fails to provide information on the dynamics of change in gastric rhythm that occur on a moment-to-moment basis. For example, if the subject had a 1-hour EGG and during that time had a 10-minute run of tachygastria at 5 cpm, this would be lost in the power spectrum since it would be averaged into a single curve. The running spectral analysis was therefore developed to address this problem (6). An example of a running spectral analysis is seen in Figure 3b. The running spectral analysis is simply a series of power spectra each generated from data collected over short periods of time (as small as 2 minute intervals). In other words, instead of a single power spectrum representing a 30-min period of data, multiple, successive power spectra can be stacked one after another, each representing successive 2-min blocks of signal. The running spectral analysis can then be used to calculate the amount of gastric dysrhythmia over a period of time. Many studies using EGG now report the percent of time in an arrhythmic pattern. To do this, a running spectrum is used to tally the number of time intervals spent in tachygastria, bradygastria and normal rhythm and divide this by the total number of time intervals (e.g. for bradygastria: [total time lines with frequency <2 cpm / total time intervals]x 100 = Percent Bradygastria). Figures 4 and 5 demonstrate tachygastria and bradygastria. The criteria for defining tachygastria and bradygastria vary slightly. However, it is agreed that if the gastric electrical activity occurs at less than 2 cpm, this is considered to represent bradygastria. If the activity is greater than 4 cpm, tachygastria is described. In Figure 4, the subject demonstrates

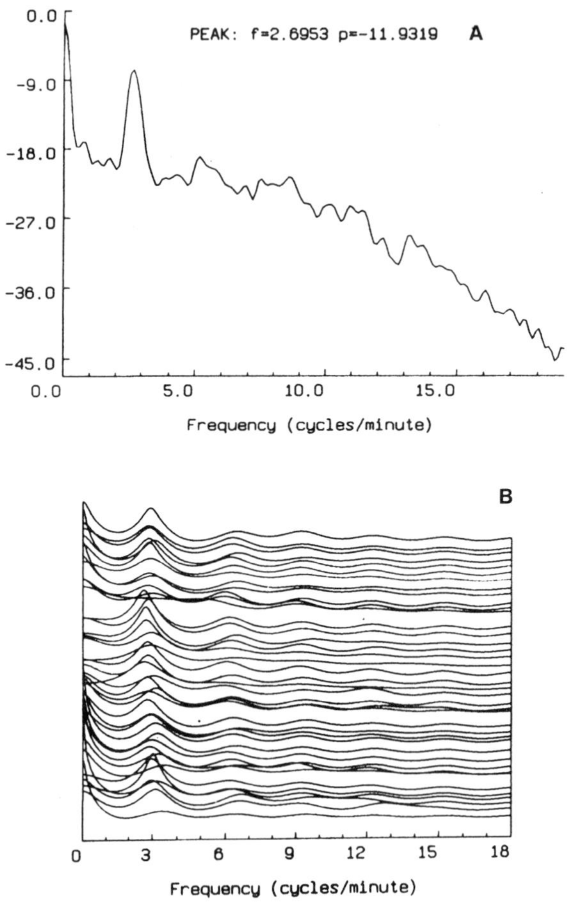

Figure 3. (A, Top) A power spectral analysis from a normal subject. (B, Bottom) Running spectral analysis from the same patient.

periods of tachygastria. Figure 5 demonstrates episodes of bradygastria. Both EGG's also demonstrate dysrhythmia since there is no consistent pattern as is seen in Figure 3b.

VALIDATION

Initial reports of EGG were met with skepticism as it was not believed that gastric electrical activity could be reliably measure percutaneously. The first question is whether the EGG signal is reproducible in the same patient. One study looked at healthy volunteers and performed EGG on three consecutive days in the same patient. There was a good correlation in the same patients indicating good reproducibility (7). When serosally obtained electrical activity was compared to that obtained cutaneously, there is synchronous signal acquisition validating the cutaneous technique (8–12). A more recent study compared intragastric electrical activity, manometric recordings and cutaneous EGG (13). In this study, the timing of gastric electrical activity was identical when intragastric and cutaneous pacesetters were compared. However, only 30–40% of increases in the amplitude of gastric electrical activity were associated with gastric contractions indicating that cutaneous measurements cannot predict motor events. Another study also confirmed that gastric

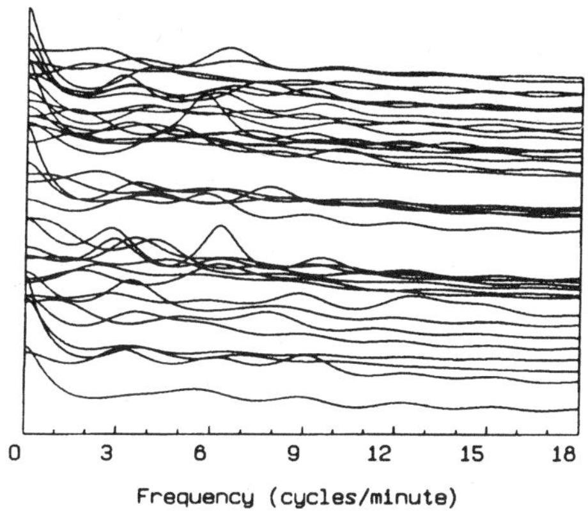

Figure 4. An example of a diabetic subject with tachygastria.

electrical activity does not correlate with the motor events and amplitude does not predict these motor events (14).

This data provides evidence that the cutaneous measurement of gastric electrical activity is a valid predictor of intragastric electrical events. However, the cutaneous measurement is unable to reliably predict which of the electrical events will result in a motor activity.

NORMAL ELECTROGASTROGRAM

Important to the interpretation of electrogastrography is the determination of the variation of slow wave activity during normal physiology. Standardization of EGG was recently performed by Pfaffenbach, et al in 1995 (15). In this study, young women had a dominant frequency of 2.9 (range, 2.1–3.3) as opposed to 2.8 (range, 2.1–3.3) in young men. The normal percent range for tachygastria and bradygastria were determined to be 7% and 5.3% respectively. This dominant frequency also varies with age. Table 1 demonstrates the changes in EGG from preterm infants to adulthood (16). At the other extreme, patients over 60 years are known to have a slightly, but statistically higher dominant frequency (3.3 versus 3.0 for subjects over 60 and less than 60 years, respectively) (17,18).

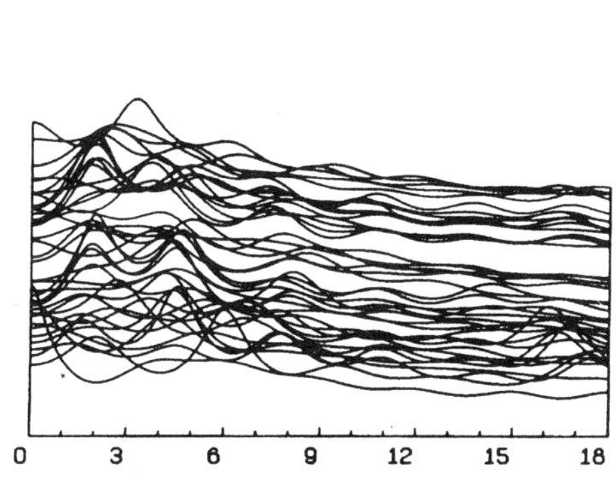

Figure 5. An example of a diabetic subject with bradygastria.

Table 1. Variation in time spent in expected normal range at varying age groups

Age group	Percent of time within the normal 2-4cpm range
Pre-term newborns	26.6 + 3.9
Full-term newborns	30.0 + 4.0
2-6 month infants	70 + 6.1
4-11 year old children	84.6 + 3.2
Adults	88.9 + 2.2

From Chen, et al Gastrointestin Liver Physiol, 1997.

Other normal physiologic conditions have been shown to influence EGG recordings. The menstrual cycle results in minor alterations in dominant frequency (15, 18–20). Gender is also deemed to be a factor in both fasting and post-prandial EGG signals (15). In another study, psychological stress was shown to significantly increase the power of the EGG (21).

A number of studies have been reported identifying pathologic conditions associated with changes in gastric slow waves. Some of these will be discussed in the section below.

Non-Ulcer Dyspepsia

Non-ulcer Dyspepsia (NUD) is a condition of unclear etiology. Therefore the diagnosis of this condition lacks a firm definition. Electrogastrographers have tried to identify subgroups of patients with NUD and characterize abnormalities in slow wave rhythm disturbances. In one study of adults with NUD, 68.75% had episodes of dysrhythmias as compared to 20% in controls (22). Others have confirmed the high rate of gastric arrhythmias in these patients (23–26).

Interestingly, in patients with true peptic ulcer disease, EGG also demonstrates tachygastria and a decrease in dominant frequency (27). These changes were restricted to those patients presenting with nausea and vomiting with their ulcer symptoms.

Motion Sickness and Nausea of Pregnancy

Where EGG has been particularly helpful is in determining the etiology of motion sickness and nausea of pregnancy. Tachygastric rhythms are the predominant defect

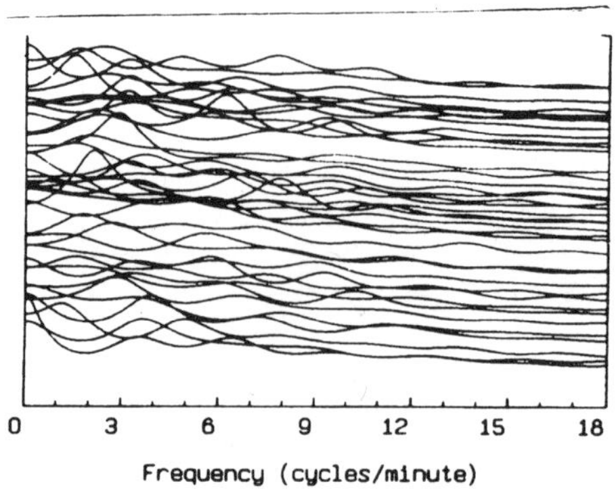

Figure 6. A running spectrum of a woman in the first trimester of pregnancy.

deemed to be the cause of motion sickness (28). During periods of circular vection using an optokinetic drum inducing nausea, frequencies of up to 9 cpm were observed. Interestingly, these same authors determined that vasopressin was able to reproduce this event and atropine was prophylactic indicating a vagal pathway for induction of the tachygastria.

In nausea of pregnancy, more than 50% of women are found to have tachygastria with most of the remainder having a variety of rhythm disturbances (29). However, a benign first trimester produces only mild aberrations in gastric electrical activity (30). In this case the mechanism appears to be related to gestational progesterone (31).

Post-Gastrectomy and/or Vagotomy

A great deal of work has been done in patients with the surgically altered stomach. Specifically, studies have been done in vagotomized stomachs. When highly selective vagotomy is performed, there appears to be no change in the dominant frequency. However, a dramatic drop in the power of the dominant frequency is seen (32).

The role of vagal control on the gastric pacemaker has also been confirmed (33). In a study using sham feeding, the amplitude of signals doubled during sham feeding. However, unlike ingestion of food during which the increased amplitude was seen for at least 30 minutes post-prandially, the duration of increase after sham feeding was only 10 minutes.

In patients with subtotal or total gastrectomy, a change from the 3 cpm dominant frequency to an 11 cpm frequency range is not uncommon (34). This frequency change was confirmed by manometric testing to represent the small intestine now in the location of the antrum post-operatively. It is speculated that this effect is due to antral removal or removal of the pacemaker, and resumption of pacemaker function by the remaining tissue. In dogs that have fundectomy only, there is no change observed in the pacesetter potential during the fasting state indicating preservation of pacemaker (35). However, post-prandially, these animals demonstrate a larger decrease in dominant frequency.

Diabetes

Of all EGG research, the most important work has been in the study of hyperglycemia and diabetes. Initial results did not seem positive. In 1995, Pfaffenbach (36), studied 18 patients with type I diabetes and 20 normal controls and found no obvious EGG differences. Approximately one year earlier, a group from the Netherlands had demonstrated that when patients with type I diabetes were glucose clamped to hyperglycemia more dysrhythmias were seen. Particularly, there was a prominence of tachygastric rhythms (37). This was the first indication that glucose itself may participate in the generation of gastric arrhythmias. In an elaborate study, Hasler, et al. performed glucose clamping in a group of normal patients (38). They determined that during hyperglycemia (glucose of 230 mg/dL), tachygastria was present for more than 20% of the time. Bradygastria was minimal. The threshold glucose level for the development of these gastric arrhythmias appeared to be 170 mg/dL. Interestingly, this study also demonstrated that this effect was blocked by the administration of indomethacin, suggesting a role for prostaglandins in this phenomenon. These findings later were correlated with gastric emptying indicating that in patients with gastroparesis, there was an increased incidence of electrical gastric slow-wave activity (39). Other groups have shown increased bradygastria in patient with diabetes (40).

Tachygastria versus bradygastria; both seem to be associated with patients with diabetes. In a recent study by our group, we confirmed that diabetes is associated with both tachygastria and bradygastria (41). The trend was that the tachygastria appeared to be associated with the acute level of hyperglycemia and the amount of bradygastria associated with the level of glycosylated hemoglobin indicating long term control (unpublished). Therefore for the first time the finding of both these rhythm disturbances can be accounted for in diabetes based upon the degree and duration of glycemic control. Interestingly, there are degrees to which diabetes exert effects on the stomach that can be measured by EGG (41). Specifically, some diabetic patients have relatively preserved pacesetter potential both fasting and post-prandially. Other diabetics have grossly abnormal fasting EGG with complete or partial correction with meal (we have called this meal-induced gastroversion). A third group has an abnormal EGG during fasting and post-prandially. It has also been determined that the degree of long term glycemic control may play a role in the type of presentation with higher glycosylated hemoglobin levels predicting the lack of ability to gastrovert.

Other

Many other medical problems have been studied using EGG technology. For example, patients with chronic renal insufficiency have a higher rate of dysrhythmias than controls (42). Also, in patients with intractable nausea and vomiting of unknown etiology, studies have demonstrated that many of these patients suffer from dysrrhythmias (43, 45). Specifically, these patients suffer from tachygastria.

There have been many conditions to which EGG correlations have been sought. Attempts have been made to determine conditions in which EGG may be useful as a diagnostic tool. Despite extensive study, EGG has not assumed this role and has to date remained an experimental procedure to all but the most experienced in the field. Nevertheless, these have been clear examples of how EGG has improved our understanding of many upper gastrointestinal tract problems.

CLINICAL APPLICATIONS

The following section will describe clinical examples of EGG findings seen in some of the conditions noted above.

Nausea

As mentioned, nausea is frequently associated with abnormalities in gastric electrical rhythm. Figure 7 demonstrates an EGG of a patient with recurrent transient episodes of nausea without vomiting. Notice the tachyarrhythmias with the increased number of curves demonstrating a greater than 3 cpm range. When the nausea corrects, the EGG usually is associated with a dramatic correction of the arrhythmia as was true in this patient.

Diabetes

Hasler, et al (38) has clearly shown that when normal patients are clamped to hyperglycemia there is a higher percentage of tachyarrhythmias. We have shown (41) this event in diabetic patients.

We have shown that there are degrees to the effects of diabetes on the gastric electrical activity. Patients with mild diabetes (low HgA1c), tend to have a preserved EGG before and after a meal. A second group of diabetics demonstrate an abnormal EGG in the fasting state that corrects post-prandially. A third group with poor glycemic control have an abnor-

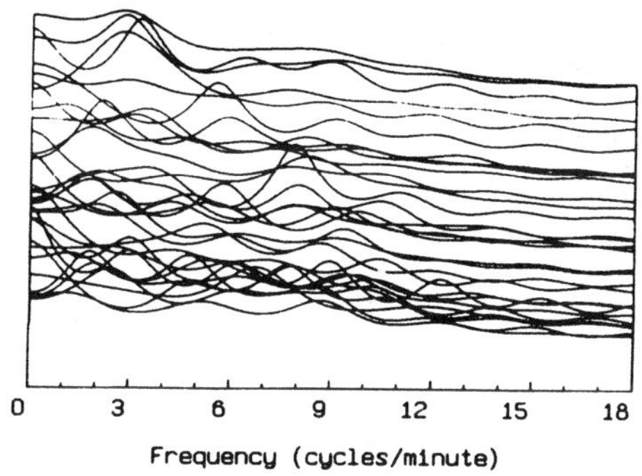

Figure 7. An example of a patient with episodic nausea.

mal EGG before and after meal. We term this correction with a meal, meal-induced gastroversion. An example of meal-induced gastroversion is seen in Figures 8a and 8b.

SUMMARY

As this chapter outlines, EGG provides a useful means of determining abnormalities of gastric electrical activities in the upper gastrointestinal tract. Its practical application is

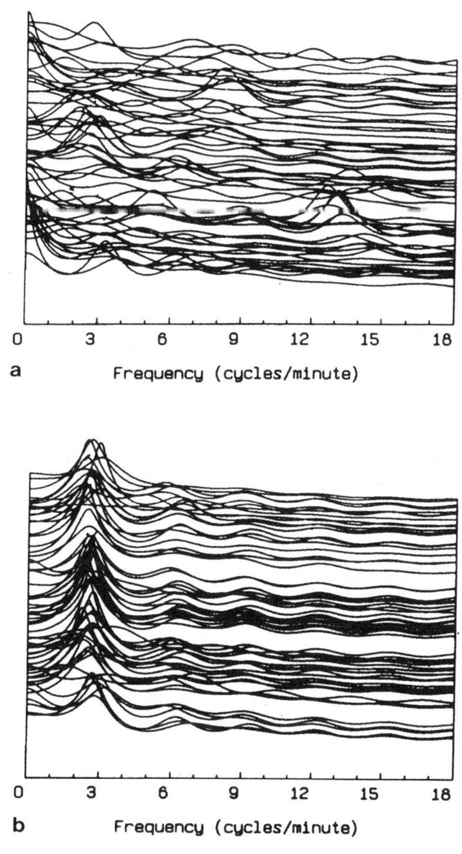

Figure 8. (a, Top) A diabetic subject with an abnormal fasting running spectrum demonstrating dysrhythmia. (b, Bottom) The same patient after a meal demonstrating normalization (meal-induced gastroversion).

primarily the diagnosis of gastric dysrhythmias. Since it is known that these dysrhythmias do correlate with decreased gastric emptying, EGG can be used to predict gastric emptying. EGG is particularly useful in cases of idiopathic nausea.

Diabetes is now the focus of interest in EGG research. Since hyperglycemia itself affects EGG, a dynamic tool that allows the measurements of a moment to moment nature is ideal. EGG can provide this information.

Lastly, EGG is a tool whose time has come. With better technology and understanding of the physiology of gastric electrical activity, EGG will play a major role as a non-invasive way of assessing gastric rhythm. Its use as a diagnostic tool has been lagging but with increased use, the future looks bright for the study of electrogastrography.

REFERENCES

1. Alvarez WC. The electrogastrogram and what it shows. JAMA 1922;78:1116–1118.
2. Telander RL, Morgan KG, Kreulen DL, et al. Human gastric atony with tachygastria and gastric retention. Gastro 1978;75:497–501.
3. You CH, Lee KY, Chey WY, et al. Electrogastrographic study of patients with unexplained nausea, bloating and vomiting. Gastro 1980;79:311–314.
4. Stoddard CJ, Smallwood RH, Duthie HI. Electrical arrhythmias in the human stomach. Gut 1981;22:705–712.
5. Cooley JW, Tukey JW. An algorithm for machine calculation of complex fourier series. Math Comput 1965;19:297–303.
6. van der Schee EJ, Grashius JL. Running spectrum analysis as an aid in the representation and interpretation of electrogastrographic signals. Med Biol Eng Comput 1987;25:57–62.
7. Riezzo G, Pezzolla F, Thouvenot J, Giorgio I. Reproducibility of cutaneous electrogastrography in the fasting state in man. Pathologie Biologie 1992;40:889–894.
8. Nelson TS, Kohatsu S. Clinical electrogastrography and its relationship to gastric surgery. Am J Surg 1968;116:215–222.
9. Smout AJPM, van der Schee EJ, Grashius JL. Postprandial and interdigestive gastric electrical activity in the dog recorded by means of cutaneous electrodes. In: Christiansen J, ed. Gastrointestinal Motility. New York Press: Raven Press, 1980;187.
10. Hamilton JW, Bellahsene BE, Reicherlderfer M, et al. Human electrogastrograms. Comparison of surgace and mucosal recordings. Dig Dis Sci 1986;31:33–39.
11. Stern RM, Koch KL, Stewart WR, Vasey MW. Electrogastrography: Current issues in validation and methodology. Psychophysiology 1987;24:55–64.
12. Familoni BO, Bowes KL, Kingma YJ, Cote Kr. Can transcutaneous recordings detect gastric electrical abnormalities. Gut 1991;32:141–146.
13. Mintchev MP, Kingma YJ, Bowes KL. Accuracy of cutaneous recordings of gastric electrical activity. Gastro 1993;104:1273–1280.
14. Sun WM, Smout A, Malbert C, et al. Relationship between surface electrogastrography and antropyloric pressures. Am J Physiol 1995;268:G424–G430.
15. Pfaffenbach B, Adamek RJ, Kuhn K, Wegener M. Electrogastrography in healthy subjects: Evaluation of normal values, influence of age and gender. Dig Dis Sciences 1995;40:1445–1450.
16. Chen JDZ, Co E, Clang J, et al. Patterns of gastric activity in human subjects of differed ages. Am J Physiol 1997; 272:G1022–1027.
17. Nishimura N, Hongo M, Yamada M, et al. Gastric myoelectrical activity in elderly human subjects-surface electrogastrographic observations. J Smooth Musc Res 1995;31:43–49.
18. Parkman HP, Harris AD, Miller MA, et al. Influence of age, gender, and menstrual cycle on the normal electrogastrogram. Am J Gastro 1996;91:127–133.
19. Parkman HP, Harris AD, Miller MA, et al. Influence of age, gender, and menstrual cycle on the normal electrogastrogram. Am J Gastro 1996;91:127–133.
20. Knight LC, Parkman HP, Brown KL, et al. Delayed gastric emptying and decreased antral contractility in normal premenopausal women compared with men. 1997;92:968–975.
21. Riezzo G, Porcelli P, Guerra V, Giorgio I. Effects of different psychophysiological stressors on the cutaneous electrogastrogram in health subjects. Arch Physiol Biochem 1996;104:282–286.

22. Lopez GA, Lopez DGA, Samiento A, Andrusch A. Abnormalities in gastric mechanic sensitivity, gastric emptying and electrogastrography in non organic dyspepsia (NOD). Acta Gastro Latinoamer 1997;27:7–10.
23. Cucchiara S, Riezzo G, Minella R, et al. Electrogastrography in non-ulcer dyspepsia. Arch Dis Child 1992;67:613.
24. Pfaffenbach B, Adamek RJ, Bartholomaus C, Wegener M. Dig Dis Sci 1997;42:2094.
25. Parkman HP, Miller MA, Trate D, et al. Electrogastrography and gastric emptying scintigraphy are complementary for assessment of dyspepsia. J Clin Gastro 1997;24:214.
26. Thor P, Lorens K, Tabor S, et al. Dysfunction in gastric myoelectrical and motor activity in Helicobacter pylori positive gastritis patients with non-ulcer dyspepsia. J Physiol Pharm 1996;47:469.
27. Geldof H, van der Schee EJ, Smout AJPM, et al. Myoelectrical activity of the stomach in gastric ulcer patients: An electrogastrographic study. J Gastro Motility 1989;1:122.
28. Kim MS. Chey WD, Owyang C, Hasler WL. Role of plasma vasopressin as a mediator of nausea and gastric slow wave dysrythmias in motion sickness. Am J Physiol 1997;35:G853.
29. Koch KL, Stern RM, Vasey JJ, et al. Gastric dysrhythmias and nausea of pregnancy. Dig Dis Sci 1990;35:961.
30. Riezzo G, Pezzolla F, Darconza G, Giorgio I. Gastric myoelectrical activity in the first trimester of pregnancy: A cutaneous electrogastrographic study. Am J Gastro 1992;87:702.
31. Walsh JW, Hasler WL, Nugent CE, Owyang C. Progesterone and estrogen are potential mediators of gastric slow-wave dysrrhythmias in nausea of pregnancy. Am J Physiol Gastrointest Liver Physiol 1996;33:G506.
32. Geldof H, van der Schee EJ, Van Blankstein M, et al. Effects of highly selective vagotomy on gastric myoelectrical activity: An electrogastrographic study. Dig Dis Sci 1990;35:969.
33. Stern RM, Crawford HE, Stewart WR, et al. Sham feeding: Cephalic-vagal influences on gastric myoelectrical activity. Dig Dis Sci 1989;34:521.
34. Homma S, Shimakage N, Yagi M, et al. Electrogastrography prior to and following total gastrectomy, subtotal gastrectomy, and gastric tube formation. Dig Dis Sci 1995;40:893.
35. Wilbur BG, Kelly KA, Code CF. Effect of gastric fundectomy on canine gastric electrical and motor activity. Am J Physiol 1974;226:1445.
36. Pfaffenbach B, Wegener M, Adamek RJ, et al. Antral myoelectrical activity, gastric emptying, and dyspeptic symptoms in diabetics. Scand J Gastroenterol 1995;30:1166.
37. Jebbink RJA, Samsom M, Bruus PM, et al. Hyperglycemia induces abnormalities of gastric myoelectrical activity in patients with Type I Diabetes Mellitus. Gastroenterology 1994;107:1390.
38. Hasler WL, Soudah HC, Dulai G, Owyang C. Mediation of hyperglycemia-evoked gastric slow-wave dysrhythmias by endogenous prostaglandins. Gastro 1995;108:727.
39. Chen JD, Lin Z, Pan J, McCallum RW. Abnormal gastric myoelectrical activity and delayed gastric emptying in patients with symptoms suggestive of gastroparesis. Dig Dis Sci 1996;41:1538.
40. Manitides A, Stafanides G, Kioulanis J, et al. Cutaneous electrogastrography for the assessment of gastric myoelectrical activity in type I diabetes mellitus. Am J Gastroenterology 1997;92:1190.
41. Mathur R, Pimentel M, Barnett P, Lin HC. Relationship of Acute and Chronic Glycemic Control on Gastric Myoelectrical Activity in Type 2 Diabetes. American Diabetes Association, Chicago, 1998.
42. Lin X, Mellow MH, Southmayd L, et al. Impaired gastric myoelectrical activity in patients with chronic renal failure. Dig Dis Sci 1997;42:898.
43. You CH, Chey WY, Lee KY, et al. Ann Intern Med 1981;95:449. Geldof H, van der Schee EJ, van Blankstein M, Grahius JL. Electrogastrographic study of gastric myoelectrical activity in patients with unexplained nausea and vomiting. Gut 1986;27:799.
44. You CH, Lee KY, Chey WL, Menguy R. Electrogastrographic study of patients with unexplained nausea, bloating, and vomiting. Gastroenterology 1980;79:311.

6

ANTRODUODENOJEJUNAL MANOMETRY

Robert W. Summers and Satish S. C. Rao

Department of Internal Medicine
4545 JCP
James A. Clifton Center for Digestive Diseases
University of Iowa College of Medicine
200 Hawkins Drive
Iowa City, Iowa 52242

INTRODUCTION

Small bowel manometry has been used as a diagnostic tool for at least two decades. Although there is no uniform methodology or technique, over the last few years there has been tremendous enthusiasm for automating the analysis and for developing appropriate and user-friendly techniques.

The physiology of small bowel motor activity is well defined and accepted and several abnormal motor patterns have been described (1). In addition to demonstrating normal physiology, antroduodenojejunal manometry has also been used to examine the effects of pharmacologic agents, hormones or surgery on small intestinal motility.

WHEN IS ANTRODUODENOJEJUNAL MANOMETRY USEFUL?

Table 1 lists the most common indication for this procedure.

a) In many situations, a normal study is just as useful as an abnormal study. A normal fasting pattern, fed pattern and sleep pattern definitively excludes a clinical suspicion of small bowel dysmotility [1,2]
b) In patients with gastric emptying disorders it is helpful to know whether the small bowel function is normal or if there is diffuse motor dysfunction of the gastrointestinal tract [1,2]. This information is particularly useful if therapeutic maneuvers such as enteral feeding or gastrectomy are being considered [3,5]. Similarly, manometry is helpful when contemplating subtotal colectomy for severe constipation. A normal small bowel motor function suggests that the colo-

Gastrointestinal Motility, edited by Rao *et al.*
Kluwer Academic / Plenum Publishers, New York, 1999.

Table 1. Indications for antroduodenojejunal manometry

• Unexplained nausea, vomiting or prolonged fullness after a meal
• Unexplained abdominal pain, particularly right upper quadrant pain
• To identify underlying mechanism(s) for upper gut stasis, i.e., myopathy vs. neuropathy
• Suspected pseudo-obstruction
• Patients with constipation prior to colectomy
• Patients with chronic intestinal pseudo obstruction prior to transplantation
• Identify the efficacy of response to a prokinetic drug

nic problem is not a diffuse gut disorder and that problems with enteral nutrition is unlikely in the post-operative period.

c) Certain small bowel motility patterns are highly suggestive of mechanical small bowel obstruction. So-called "clustered contractions" (short bursts of regular pressure waves preceded and followed by quiescent activity), are frequently associated with mechanical impediments to intestinal flow [1, 4, 6]. Appropriate contrast radiographs should be carefully reviewed for evidence of mechanical blockade and in some cases, it may even be appropriate to perform exploratory laparotomy to search for mechanical impediments to flow.

d) Small bowel manometry may be useful in the work up of chronic, severe, and unexplained upper abdominal symptoms such as bloating, distension, crampy abdominal pain, and dyspepsia [1, 2, 4, 6].

Manometry may be useful to investigate the effects of potential therapeutic agents in patients who have abnormal motility patterns. If one cannot demonstrate any change in motor activity after administration of erythromycin, octreotide or cisapride, then it is rather unlikely that the particular agent will produce clinical improvement.

TYPES OF MANOMETRY PROBES AND RECORDING SYSTEMS

Water Perfused Catheters

Low compliance, water-perfused catheters are most often used because of their low cost and ready availability [2]. They can be fashioned in countless different configurations to accomplish specific purposes, but often are constructed with three closely-spaced side holes (1–2 cm apart) to be placed with fluoroscopic guidance in the region of the antrum. Three or four additional side holes are then placed at 10–15 cm intervals distal to the antral side holes to record from the duodenum and proximal jejunum. The catheter is perfused with de-gassed distilled water using a pneumohydraulic perfusion pump. Small volume displacement external strain gauges are attached to the perfusion pump's capillary tubes and the manometry catheters. The signals from the strain gauges are often amplified and filtered using analog physiologic recorders and displayed on strip chart recorders. The system can be used for recording from other digestive hollow organs and is relatively inexpensive. Antroduodenal manometry is especially useful if disordered gastric motility is suspected, however, it must be monitored by a nurse or technician and it is not convenient for use for more than 4–6 hours.

Solid State Catheters

Solid state catheters have the pressure sensing strain gauges incorporated into the catheter itself and the electronic output is usually sampled digitally using a computer program [5,7]. The advantage of this system is that ambulatory recordings can be made so that the subject can go about their daily routine and fasting, fed and sleep patterns can easily be recorded for periods greater than 24 hours. A wide variety of variables such as exercise, sleep, mental and physical stress, or medications can be tested [5,7]. Prolonged recordings with perturbations can be extremely useful in observing physiological, pathological or pharmacological influences [5,7]. Once, the advantages of this technique are realized many laboratories are switching to this approach. The high cost of the catheters is partly outweighed by the lack of need for monitoring. One other disadvantage is that constant movement of the catheter with motion and time may make it difficult to record from the antral region.

PROBE PLACEMENT

Patients are usually asked to report in the morning after an overnight fast. Unless therapy is being evaluated by manometry, medications that might affect motility should be withheld for at least 48 hours before the procedure [2]. Perfused catheters can be introduced over a wire because they have a central lumen. The wire can be introduced by endoscopy or by fluoroscopic guidance and manipulated into the stomach and small bowel. However, current solid-state catheters do not have a central lumen. These catheters can also be placed by fluoroscopic guidance. However, reduced motor activity or anatomic abnormalities may make this approach unacceptably lengthy or impossible. In this case, the endoscope can be used to direct the catheter into the duodenum by using a biopsy forceps to grasp a silk thread at the tip and manipulate the catheter into the distal duodenum. A balloon at the tip can be inflated at this point to anchor the catheter in the duodenum while the scope is withdrawn. Peristaltic activity carries the catheter into the proximal jejunum. Recordings are begun as soon as the catheter is in place and the subject is asked to keep a diary of activity and the timing of meals, symptoms, medications, sleep, waking and other events significant to the study.

A STANDARD PROTOCOL FOR SMALL BOWEL MANOMETRY?

Although a standard protocol for performing recording is highly desirable, there is no consensus regarding the most appropriate method [1,2,5,6,7]. Most laboratories that perform a water-perfused antro-duodenal manometry, record fasting activity for 3–4 hours and then administer a meal and record post-prandial activity for 1–2 hours. In contrast, laboratories that use solid state manometry systems, record motility of the duodenum and jejunum continuously for at least 24 hours [7]. Fasting activity is often recorded for 6–7 hours before a meal is given. After the meal, the patient is encouraged to be up and about and to avoid going to sleep for an additional 5–6 hours. The size and caloric load of the meal is also important. The recent AGA consensus statement recommends that at least a 400 Kcal meal should be administered (2). The recordings during sleep seem to be especially useful in assessing whether or not the motor patterns are normal or abnormal. The influence of the central nervous system on motility is markedly reduced during sleep and the stages of the migrating motor complex can easily be recognized if they are normal. It is very convenient to record motor activity for additional hours after waking and before removing the catheter to evaluate the variables noted above, especially medications. Many laboratories will use this time to observe the effect of medications, especially those that might be useful therapeutically.

ANALYSIS OF ANTRODUODENJEJUNALMANOMETRY

Analysis is currently performed manually by observing the recordings, although several computer programs have been developed [6,8,9]. During the fasting period, the most important pattern to identify is the migrating motor complex (MMC). Phase III or the activity front is a pattern of regular pressure waves which usually originates in the antrum (Figure 1). It may begin in the duodenum or the proximal jejunum, last from 3–6 minutes and then migrates aborally. The pressure waves of the activity front occur at a maximal rate of 3 per minute in the antrum, but change to a frequency of about 11 per minute in the duodenum. The migration is noted as the regular pressure waves cease in the proximal sensor and then sequentially appear in the more distal sensors. The activity front is normally followed in each sensor by a variable period of 5–30 minutes without pressure waves and this is termed phase I of the MMC. During phase II, the pressure waves reappear and occur randomly for 20–45 minutes until phase III or the activity front reappears and the cyclical MMC repeats itself. During phase II, the pressure waves often occur in groups with 3–6 regular waves in a row and these are termed "clusters". This is a normal pattern during the fasting MMC.

After a meal and usually within minutes, the MMC pattern is replaced by a fed or post-prandial pattern (Figure 2). This pattern of pressure waves resembles phase II in that the pressure waves appear to occur at random intervals at about 5–7 per minute. It is important to realize that the pressure waves are regulated by electrical oscillations called slow waves that occur in the proximal bowel at a rate of about 11 per minute. About half of the slow waves are followed by pressure waves in the duodenum for several hours after a meal. This is usually termed the fed or digestive pattern and no MMC's occur during this time[2;10–12]. During the early period of the fed pattern, clustered pressure waves do not occur frequently. If they do, partial mechanical bowel obstruction should be suspected. From 3–6 hours after a meal, the fed pattern is suddenly replaced by the fasting MMC pattern which continues until the next meal occurs.

The pattern of contractions changes when the patient falls asleep [6,13]. If the pattern has reverted to the fasted pattern, there is a qualitative difference in the three phases of the MMC. Phase I becomes longer and is the longest phase during which essentially no contractions occur. Phase II either becomes much shorter or may essentially disappear. Phase III or the activity front changes very little except that it migrates more rapidly from sensor to sensor. The duration of the MMC becomes shorter which results in more MMC's occurring per unit of time. This qualitative change is thought to represent withdrawal of the influence of central nervous system and emergence of the basic patterned program of the enteric nervous system.

INTERPRETATION: NORMAL OR ABNORMAL?

A detailed discussion of normal and abnormal patterns of the small bowel is beyond the scope of this presentation. However, there are several critical questions which should be asked during the analysis of the motor activity of the small bowel.

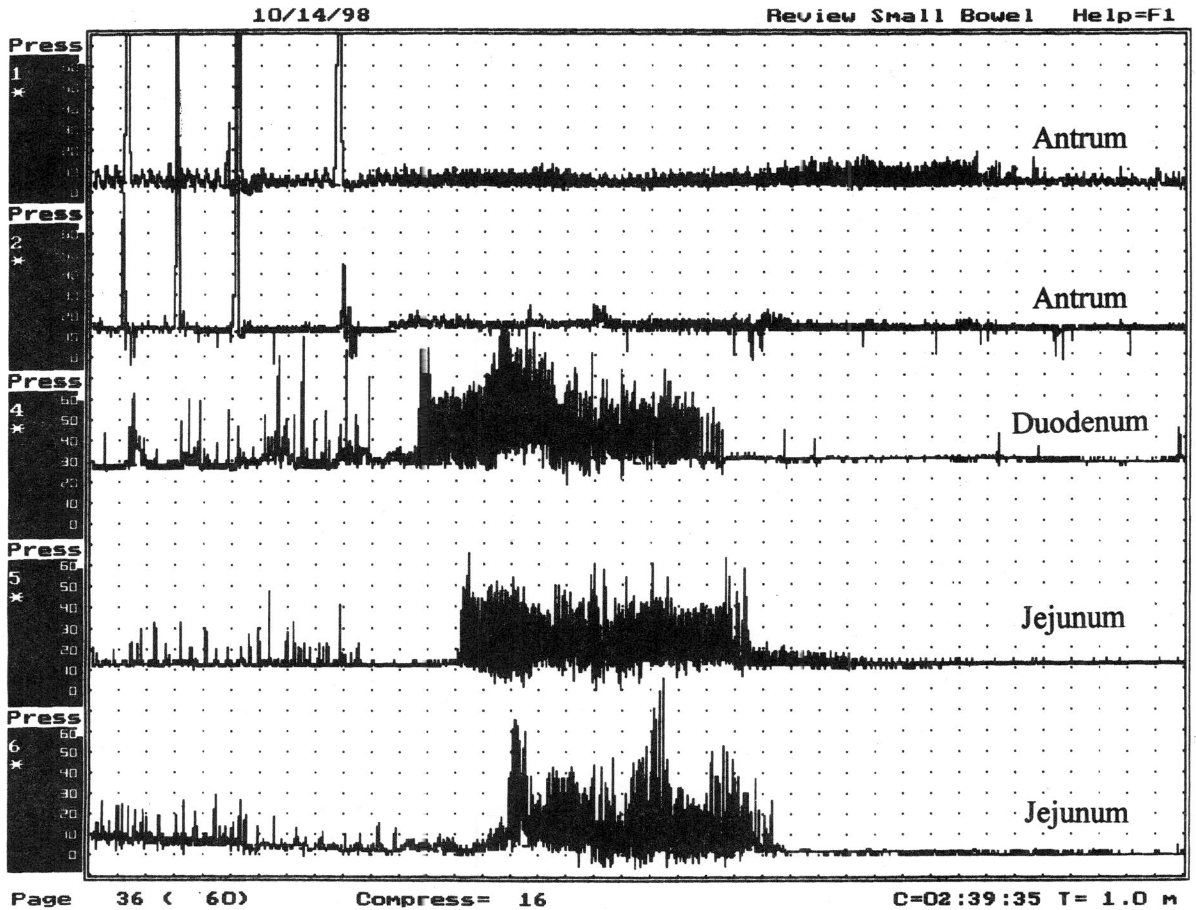

Figure 1. Antroduodenojejunal manometry recording showing a normal fasting pattern consisting of Phase I, a period of quiescence, Phase II, a period of intermittent random contractions and Phase III, a high frequency, migrating, tonic and phasic activity.

Table 2. Duration (in min) and percent duration (%) of the various phases of the MMC, and velocity of migration of phase III during fasting and sleep in men and women (mean ± SEM)

	MMC	Phase 1 (min)	Phase 2 (%)	Phase II (min)	Phase II (%)	Phase III (min)	Phase III (%)	Velocity (cm/min)
Fasting								
Men	95.3±7.7	23.6±2.4*	31.0±3.4*	64.9±7.4	56.1±4.4	6.8±0.5	12.8±2.4	5.33±1.07
Women	79.0±7.5	8.1±2.3	12.8±3.2	65.0±7.2	72.0±4.3	6.7±0.5	15.1±2.3	4.5±0.92
Sleep								
Men	72.4±7.5	23.4±2.3	37.3±3.2*	40.3±7.1	46.1±4.3*	8.6±0.5	15.7±2.3	3.72±0.64
Women	83.3±9.0	18.3±2.8	22.4±3.9	57.6±8.7	66.7±5.1	7.3±0.6	11.0±2.8	3.12±0.72

*$p \leq 0.05$ men *versus* women.

Reproduced with permission from Soffer, et al. (14).

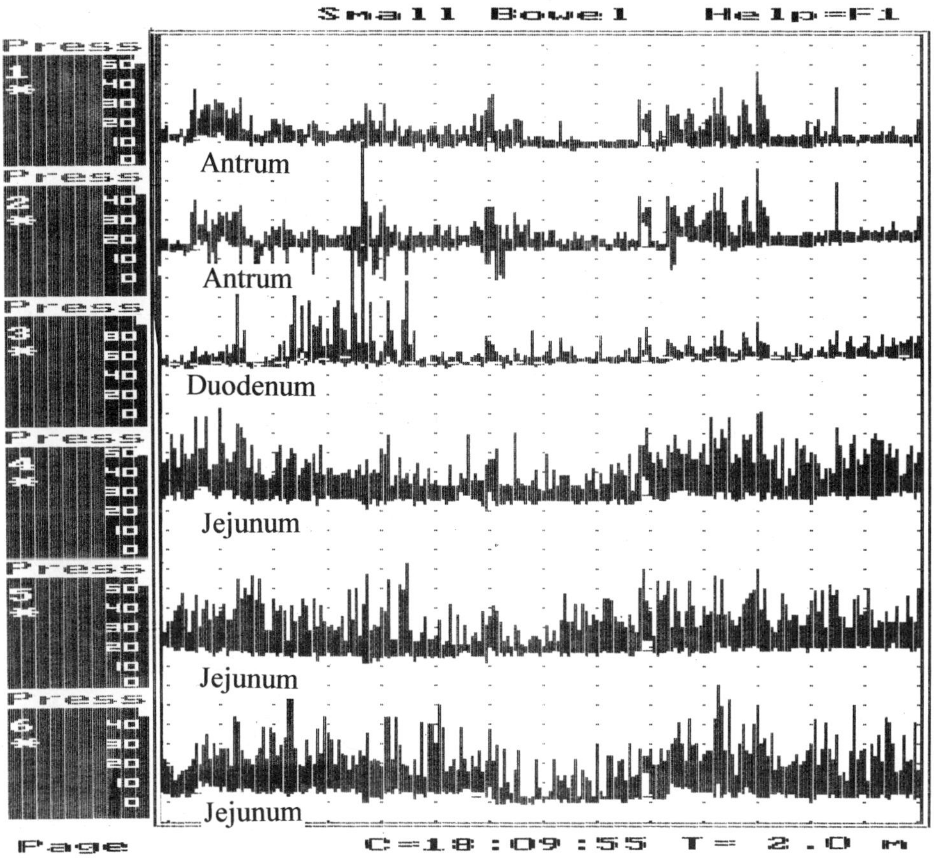

Figure 2. A normal fed response characterized by multiple phasic pressure waves, throughout the antral, duodenal and jejunal leads many of which propagate abroad.

1. During the fasting period are phase I, II and III present? Are they normal in terms of pattern (Figure 1).
2. Do the pressure waves have normal amplitudes in comparison with normal controls [5]?
3. Do the activity fronts (phase III's) migrate aborally and in sequence?
4. Is the fasting MMC pattern replaced by a fed pattern for at least two hours after a 400 Kcal meal (2), (Figure 2)?
5. Do clustered pressure waves occur frequently during the early period after a meal [1, 4, 6]?
6. Is there a sleep pattern with prolonged phase I and shortened phase II?
7. Are there any "abnormal" motility patterns [1, 2, 5, 6]? Example, Figure 3.

If all of the questions except #5 and #7 can be answered as yes, the recording is probably normal.

NORMAL VALUES AND EFFECT OF GENDER

There is limited and conflicting information regarding normal values, in part because of differences in methodology and probe design. A recent study of prolonged ambulatory

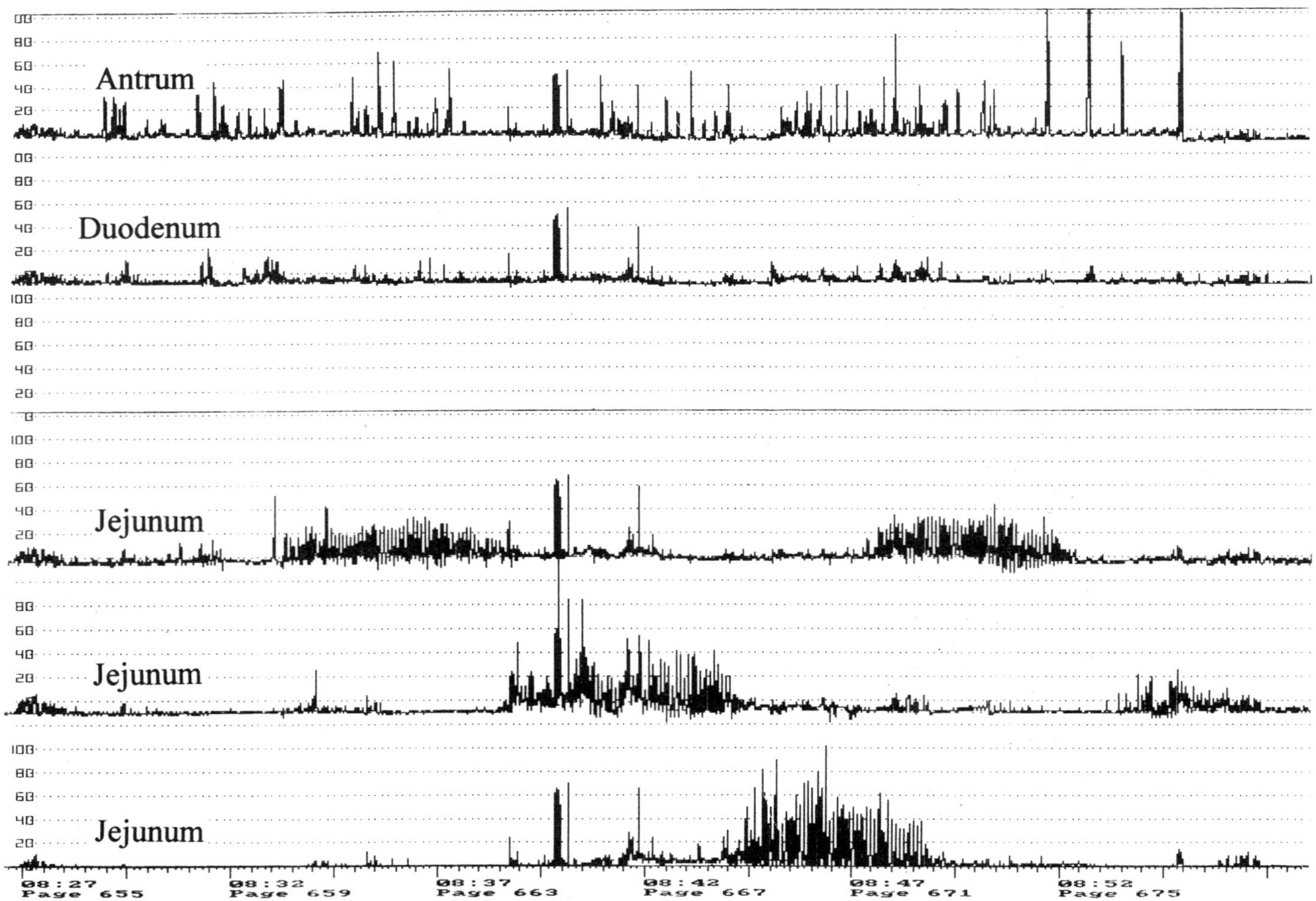

Figure 3. An abnormal pattern of antroduodenal manometry. The antrum, duodenum and jejunum exhibit discordinated and chaotic activity. An apparent Phase 3 like activity is seen in the jejunal leads, without similar antecedent activity in the antral or duodenal leads. Following a brief pause (Phase 1) there is again a short burst of intense phasic activity. This irregular pattern of activity is typical of small intestinal neuropathy.

duodeno-jejunal manometry, using solid state transducers has described normal values in healthy controls.(14). During fasting, there were (mean ± SEM) 3.5±0.3, Phase III, MMC activity per subject (range 0–6) and during sleep there were 3.8±0.3, Phase III, MMC activity per subject (range 0–5). There was no gender difference for the incidence of these patterns. During sleep, the cycles tended to be shorter, with a longer phase I and a shorter phase II component when compared to daytime (awake) recordings. Meal ingestion was associated with an increase in the frequency of phasic contractions (4.5±0.2 vs. 7.2±0.2) and motility index but no change in the amplitude or duration of contractions (14).

CONCLUSION

Although antroduodenojejunal manometry has evolved into a useful physiological test, the interpretation of small bowel motor patterns continues to be partially subjective and open to controversy. Differences in opinion persist and differences in instrumentation, catheter design, protocols and perturbations will remain until there is standardization and cooperation among centers to allow the scientific process to supersede individual and local approaches to this procedure and its interpretation. Nevertheless, it is invaluable when faced with a patient with unexplained, nausea, vomiting, or pain, in patients with pseudo obstruction, and to assess the therapeutic utility of prokinetic agents.

REFERENCES

1. Summers RW, Anuras S, Green J: Jejunal manometry patterns in health, partial intestinal obstruction, and pseudoobstruction. Gastroenterology 1983;85:1290–1300.
2. Camilleri M, Hasler WL, Parkman HP, Quigley EMM, Soffer E. Measurement of Gastrointestinal motility in the GI laboratory. Gastroenterology 1998;115:747–762.
3. vonSchonfeld J, Evans DF, Renzing K, Castillo FD, Wingate DL: Human small bowel motor activity in response to liquid meals of different caloric value and different chemical composition. Dig Dis Sci 1998;43:265–269.
4. Quigley EM: Gastric and small intestinal motility in health and disease (Review). Gastroenterol Clini North America 1996;25:113–145.
5. Soffer, E, Thongsawat S: Clinical value of duodenojejunal manometry. Its usefulness in diagnosis and management of patients with gastrointestinal symptoms. Dig Dis & Sci 1996;41:859–863.
6. Wingate DL: Small bowel manometry (Review). American Journal of Gastroenterology 1995;90:536–539.
7. Soffer EE, Thongsawat S: Small bowel manometry: short or long recording sessions? Dig Dis & Sci 1997;42:873–877.
8. Andrioli A, Wilmer A, Coremans G, Vandewalle J, Janssens J: Computer-supported analysis of continuous ambulatory manometric recordings in the human small bowel. Med & Biol Engin & Comput 1996;34:336–343.
9. Holland R, Gallagher MD, Quigley EMM: An evaluation of an ambulatory manometry system in assessment of antroduodenal motor activity. Dig Dis & Sci 1996: 41:1531–1537.
10. Read NW, Al-Janabi MN, Edwards CA, Barber DC: Relationships between postprandial motor activity in the human small intestine and the gastrointestinal transit of food. Gastroenterology 1984; 87:721–727.
11. vonSchonfeld J, Evans DF, Wingate DL: Daytime and night time motor activity of the small bowel after solid meals of different caloric value in humans. Gut 1997;40:614–618.
12. Wilmer A, Andrioli A, Coremans G, Tack J, Janssens J: Ambulatory small intestinal manometry. Detailed comparison of duodenal and jejunal motor activity in healthy man. Dig Dis Sci 1997;42:1618–1627.
13. vonSchonfeld J, Evans DF, Goebell H, Wingate DL: Comparison of the small bowel motor response to solid and liquid meals in man. Digestion 1997;58:402–406.
14. Soffer EE, Thongsawat S, Ellerbrock S. Prolonged ambulatory duodeno-jejunal manometry in humans: Normal values and gender effect. Am J Gastroenterol 1998;93:1318–1323.

7

SPHINCTER OF ODDI MANOMETRY

Glen A. Lehman*

Division of Gastroenterology/Hepatology
Department of Medicine
Indiana University Medical Center

HISTORICAL BACKGROUND

In 1887, Ruggero Oddi, described the sphincter of Oddi (SO) (1). But until recently, problems related to sphincter of Oddi dysfunction were either poorly diagnosed or were treated empirically with surgery. The introduction of endoscopic retrograde examination of the biliary and pancreatic ducts provided an opportunity for evaluating the sphincter function. This endoscopic technique has subsequently facilitated the development of biliary and pancreatic manometry (2).

ANATOMY, PHYSIOLOGY, AND PATHOPHYSIOLOGY

The sphincter of Oddi is a small complex expansion of smooth muscle that surrounds the terminal portion of the common bile duct, main (ventral) pancreatic duct (of Wirsung), and the common channel (ampulla of Vater), when present. It has both circular and figure of 8 components. The length of the high pressure zone generated by the sphincter is approximately 4–10 mm. long. Its role appears to be to regulate bile and pancreatic exocrine juice flow and to prevent duodenum-to-duct reflux (i.e., maintain sterile intraductal environment). The SO has both a variable basal pressure and variable phasic contractile activity. Sphincter regulation is under both neural and hormonal control. Phasic wave activity of the sphincter is closely tied to the migrating motor complex (MMC) of the duodenum. Innervation of the bile duct does not appear to be essential as sphincter function has been reported to be preserved following liver transplantation (3). Although regulatory processes vary among species, cholecystokinin and secretin appear to be most important in causing sphincter relaxation while nonadrenergic, noncholinergic neurons

* Telephone: 317-274-4821, Facsimile: 317-278-0164

Gastrointestinal Motility, edited by Rao *et al.*
Kluwer Academic / Plenum Publishers, New York, 1999.

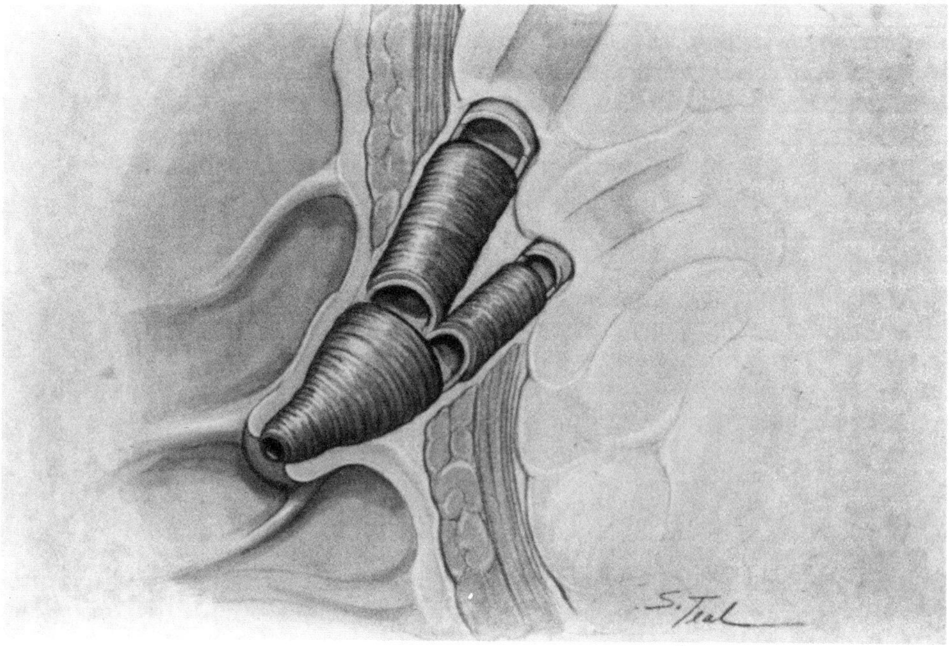

Figure 1. Sphincter of Oddi has 3 components: 1) Sphincter Ampullae (surrounds common channel at duodenal side of papilla). 2) Sphincter Choledochus (surrounds terminal bile duct). 3) Sphincter Choledochus (surrounds terminal pancreatic duct).

which at least in part transmit vasoactive intestinal peptide (VIP) and nitric oxide also relax the sphincter (4). The role of cholecystectomy in altering these neural pathways needs further definition.

Wedge specimens of the SO obtained at surgical sphincteroplasty from Sphincter of Oddi dysfunction (SOD) patients, show evidence of inflammation, muscular hypertrophy, fibrosis or adenomyosis within the papillary zone in approximately 60% of patients (5). In the remaining 40% with normal histology, a motor disorder is suggested. Less commonly, infections with cytomegalovirus or Cryptosporidium, as may occur in AIDS patients, or Strongyloides have caused SOD.

It is suspected that SOD causes pain by impeding the flow of bile and pancreatic juice resulting in ductal hypertension. Alternatively, ischemia arising from spastic contractions and "hypersensitivity" of the papilla have been proposed.

SPHINCTER OF ODDI MANOMETRY (SOM)

The most definitive development in our understanding of the pressure dynamics of the Sphincter of Oddi came with the advent of SOM. SOM is the only clinically available method to measure sphincter of Oddi motor activity. Although SOM can be performed intraoperatively or percutaneously, it is most commonly done along with an ERCP. SOM is considered by most authorities to be the gold standard for evaluating patients with sphincter dysfunction (2,6). The use of manometry to detect motility disorders of the sphincter of Oddi is similar to the use of manometry to identify motor dysfunction in other parts of the gastrointestinal tract. Unlike other areas of the gut, SOM is more technically demanding and hazardous. Questions remain as to whether these short-term observations (2–10 m-inute recordings per pull-through) reflect a "24-hour profile of dysfunction" of the sphincter. Despite such concerns, SOM has gained more widespread clinical application.

PRE-PROCEDURAL CONSIDERATIONS

Sphincter of Oddi manometry is usually performed at the time of ERCP. All drugs which relax (anticholinergics, nitrates, calcium channel blockers, and glucagon) or stimulate (narcotics or cholinergic agents) the sphincter should be avoided for at least 8–12 hours prior to manometry and during the manometry session. The current data indicate that benzodiazepines do not affect the sphincter pressure and therefore are acceptable sedation for SOM. Recent data suggest that meperidine, at a dose of ≤1 mg/kg, does not affect the basal sphincter pressure (although it does increase the phasic wave frequency (7). Since the basal sphincter pressure is generally the only manometric criterion used to diagnose sphincter of Oddi dysfunction and determine therapy, it was suggested that meperidine could be used to facilitate conscious sedation for manometry. If glucagon must be used to achieve cannulation, an 8–10 minute (at least) waiting period is required to restore the sphincter to its basal condition.

INDICATIONS

Sphincter of Oddi manometry (SOM) is recommended for patients with idiopathic pancreatitis or unexplained disabling pancreaticobiliary pain with or without hepatic enzyme abnormalities. An ERCP is usually performed (if an adequate study is not available) immediately before or after the SOM to exclude other potential causes for the patient's symptoms. Indications for the use of SOM have also been developed according to the Hogan-Geenen (8) SOD Classification System (Table 1). In Type I patients, SOM may be useful in documenting SOD, it is not an essential diagnostic study prior to endoscopic or surgical sphincter ablation. Such patients uniformly benefit from sphincter ablation regardless of the SOM results (see below). Type II patients demonstrate SO motor dysfunction in 50 to 65% of cases. In this group of patients, SOM is mandatory as the results of the study can predict outcome from sphincter ablation. Type III patients have pancreaticobiliary pain without other objective evidence of sphincter outflow obstruction. SOM is strongly preferred to confirm the presence of SOD. Although not well studied, it appears that the results of SOM may predict outcome from sphincter ablation in these patients.

EQUIPMENT

Routinely, a 5 French catheter should be used, because virtually all normal standards have been established with this size of catheter. Triple lumen catheters are state of the art and are available from several manufacturers. A variety of catheter types can also be utilized. Catheters with a long intraductal tip may help secure the catheter within the bile duct, but such a long nose is commonly a hindrance if a pancreatic manometry is required. Over the wire (monorail) catheters can be passed after first securing one's position within the duct with a guidewire. Some triple lumen catheters will accommodate a .018-inch di-

Table 1. Hogan-Geenen sphincter of Oddi classification system related to the frequency of abnormal sphincter of Oddi manometry and pain relief by biliary sphincterotomy

Patient group classifications	Approximate frequency of abnormal sphincter manometry	Probability of pain relief by sphincterotomy if manometry		Manometry before sphincter ablation
		Abnormal	Normal	
Biliary I patients with biliary-type pain, abnormal SGOT or alkaline phosphatase >2 x normal documented on two or more occasions, delayed drainage of ERCP contrast > 45 minutes, and dilated CBD > 12 mm diameter.	75-95%	90-95%	90-95%	Unnecessary
Biliary II patients with biliary-type pain but only one or two of the above criteria.	55-65%	85%	35%	Recommended
Biliary III patients with only biliary-type pain and no other abnormalities.	25-55%	55-65%	20%	Mandatory

ameter guidewire passed through the entire length of the catheter. This can be used either to facilitate cannulation or to maintain position within the duct. Guidewire tipped catheters are being evaluated. Aspiration catheters in which one recording port is sacrificed to permit both end- and side-hole aspiration of intraductal juice are particularly recommended for performing pancreatic manometry (9). Most centers perfuse the catheters with a rate of 0.25 ml/channel, using a low-compliance pneumohydraulic pump. Lower perfusion rates may give accurate information regarding basal sphincter pressures, but may not give accurate information regarding phasic wave information. The perfusate is generally bubble-free distilled water, although physiologic saline merits further evaluation.

TECHNIQUE

Sphincter of Oddi manometry requires selective cannulation of the bile duct and/or the pancreatic duct. After cannulation, if the aspirate within the catheter looks yellow (in the endoscopic view) then it indicates that the bile duct has been entered. In contrast a clear aspirate indicates that the pancreatic duct has been entered. Once deep cannulation is achieved and the patient is acceptably sedated, the catheter is withdrawn across the sphincter at 1–2 mm intervals by using standard station pull-through techniques. Current data indicate that an abnormal basal sphincter pressure may be confined to one side of the sphincter in 35–65% of patients with abnormal manometry(10–13). Thus, one sphincter may be dysfunctional whereas the other could be normal. Raddawi and colleagues(10) reported that an abnormal basal sphincter was more likely to be confined to the pancreatic duct segment in patients with pancreatitis and to the bile duct segment in patients with biliary-type pain and elevated liver function tests.

Abnormalities of the basal sphincter pressure should ideally be observed for at least one minute in each lead and should be seen on two or more separate pull-throughs. From a practical clinical standpoint, we settle for one pull through (from each duct) if the readings are clearly normal or abnormal. In some centers, three sets of rapid pull-through are performed in addition to the station pull through although its clinical use has not been defined. Once the baseline study is completed, agents to relax or stimulate the sphincter can be given (example – cholecystokinin) and manometric and/or pain response can be monitored. For every day use, the value of these provocative maneuvers needs further study before they can be widely recommended.

INTERPRETATION

Criteria for interpreting SOM tracing are relatively standard. However, they may vary somewhat from center to center. Some areas where there may be disagreement in interpretation include the required duration of basal SO pressure elevation, the number of leads in which basal pressure elevation is required, and the role of averaging pressures from the three (or two in an aspirating catheter) recording ports (2). At Indiana University, the standard method for reading the manometry tracings is to first define the "zero" duodenal baseline both before and after the pull-through maneuver. Alternatively, intraduodenal pressure can be continuously recorded from a separate intraduodenal catheter attached to the endoscope. Next, identify the highest basal pressure that is sustained for a least 30 seconds and take the four lowest amplitude points in that zone. The mean of these readings is taken as the basal sphincter pressure for that lead and for that pull-through maneuver. By averaging the basal sphincter pressure for all interpretable observations, it is possible to calibrate the final basal sphincter pressure. The amplitude of phasic wave contractions is measured from the beginning of the slope of the pressure increase to the peak of the contraction wave. The number of phasic waves per minute and the duration of phasic waves can also be determined. Most authorities read only the basal sphincter pressure as an indicator of pathology of the SO. However, data from Kalloo and colleagues (14) suggest that intraductal biliary pressure (which is easier to measure than SO pressure) correlates with SO basal pressure. In this study, intrabiliary pressure was found to be significantly higher in patients with SOD than those with normal SO pressure (20 vs. 10 mmHg; $p<.01$). Although, this finding needs to be confirmed, it supports the theory that increased intrabiliary pressure can be a cause of pain in patients with SOD.

NORMATIVE DATA

The best study which described normal values for sphincter of Oddi manometry was reported by Guelrud and associates (15). Fifty asymptomatic control patients were evaluated and repeated on two occasions in 10 subjects. This study established normal values for intraductal pressure, basal sphincter pressure, and phasic wave parameters (Table 2). Moreover, the reproducibility of SOM was confirmed. Various authorities interchangeably use 35 mmHg or 40 mmHg as the upper limit of normal for mean basal sphincter of Oddi pressure.

COMPLICATIONS

Several studies have indicated that pancreatitis is the most common complication after SOM (9). Using standard perfused catheters, pancreatitis rates as high as 31% have been reported. Such high complication rates have initially limited more widespread use of SOM. These data also emphasize that manometric evaluation of the pancreatic duct, particularly in patients with chronic pancreatitis, is associated with a high complication rate. Rolny and associates reported an 11% incidence of pancreatitis following pancreatic duct manometry (16). Twenty-six percent of chronic pancreatitis patients undergoing SOM developed pancreatitis.

Table 2. Suggested standard for abnormal values for endoscopic sphincter of Oddi manometry obtained from 50 volunteers without abdominal symptoms

Basal sphincter pressure*	> 35 mm Hg
Basal ductal pressure	> 13 mm Hg
Phasic contractions	
Amplitude	> 220 mm Hg
Duration	> 8 sec
Frequency	> 10/min

Notes: (1)Values were obtained by adding 3 standard deviations to the mean (mean obtained by averaging the results on 2-3 station pull-throughs).
(2)Data combine pancreatic and biliary studies.
*Basal pressures determined by: (1) reading the peak basal pressure (i.e., highest single lead as obtained using a 3 Lumen catheter); (2) obtaining the mean of these peak pressures from multiple station pull-throughs.
See reference 15.

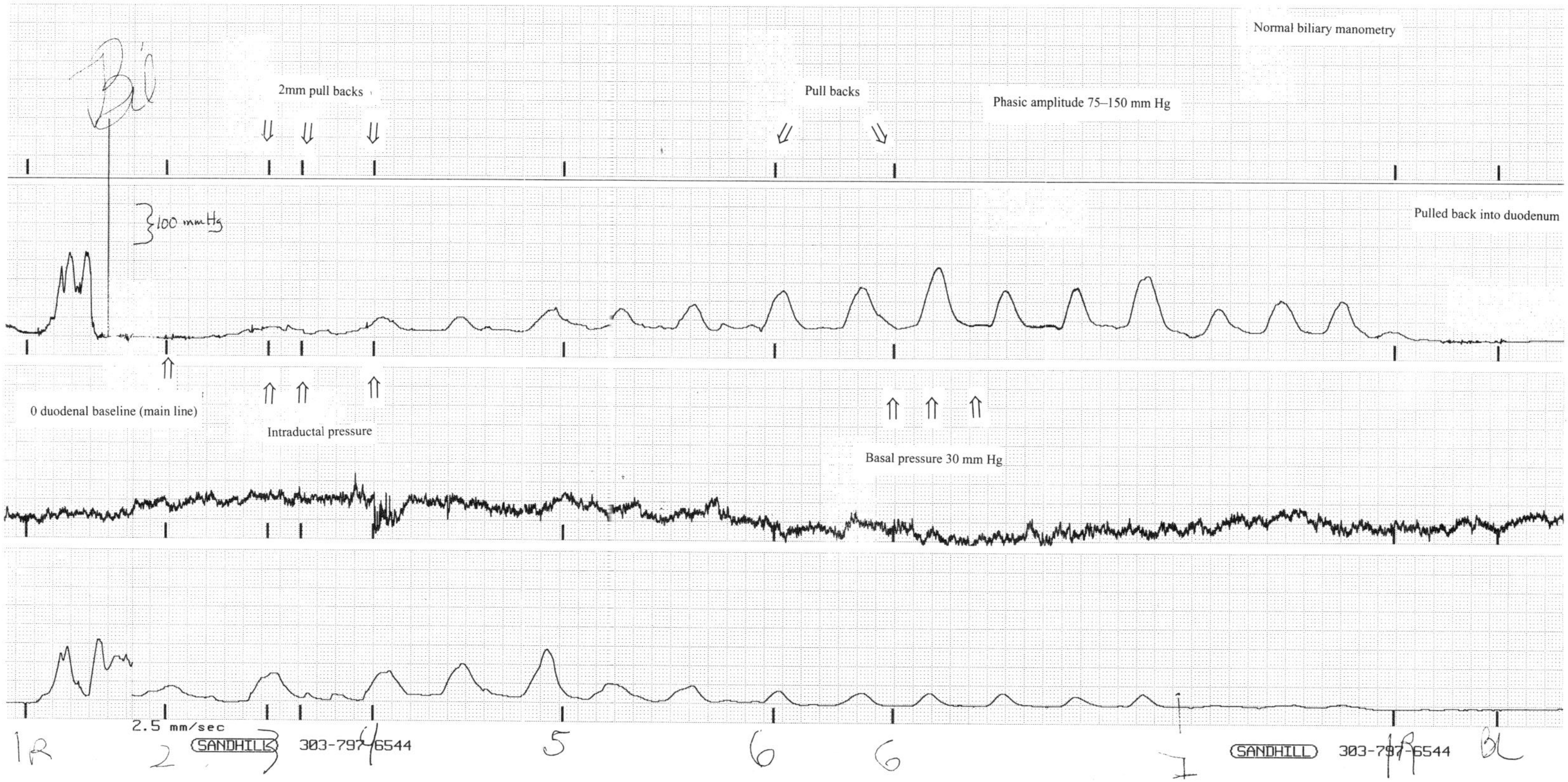

Figure 2. This tracing shows a station-pull-through technique through a normal sphincter of Oddi. As the sensor straddles the high pressure zone, the basal pressure rises (PR+40 mm Hg) there is superimposed phasic pressure waves [channel 2].

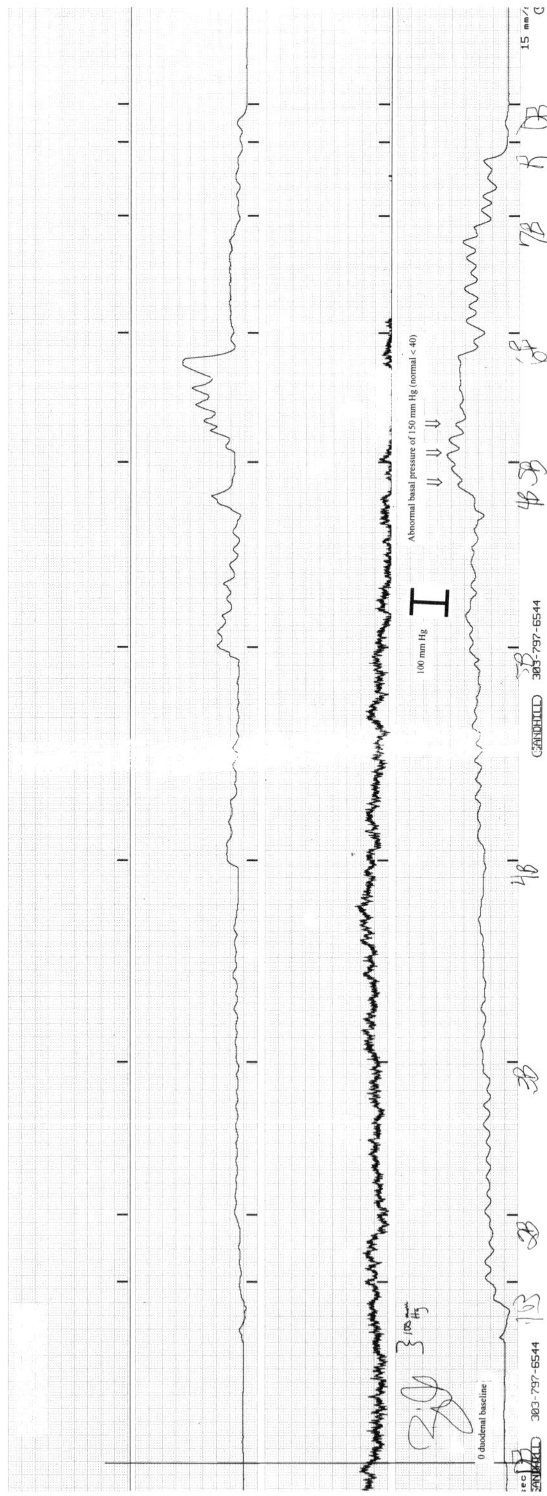

Figure 3. This is an example of sphincter of Oddi dysfunction with abnormally high basal sphincter pressure (> 150 mm Hg).

A variety of methods have been proposed to decrease the incidence of post-manometry pancreatitis. These include: a) use of an aspiration catheter; b) gravity drainage of the pancreatic duct after manometry; c) decrease the perfusion rate to 0.05 to 0.1 ml/lumen/minute; d) limit pancreatic duct manometry time to less than 2 minutes (or avoid pancreatic manometry); e) use the microtransducer (non-perfused) system. In a prospective randomized study, Sherman and colleagues (9) found that the aspirating catheter (this catheter allows for aspiration of the perfused fluid from the end hole and the side hole while accurately recording pressure from the two remaining sideports) reduced the frequency of pancreatic duct manometry-induced pancreatitis from 31% to 4% (9). The reduction in pancreatitis with the use of this type of catheter in the pancreatic duct and the very low incidence of pancreatitis after bile duct manometry has lend support to the notion that increased pancreatic duct hydrostatic pressure is a major cause of this complication. Thus, when examining the pancreatic duct sphincter, aspiration of pancreatic juice is strongly recommended.

CONCLUSIONS

In selected cases, SOM can provide invaluable information and facilitate optimal management of patients with right upper quadrant pain. However, the procedure requires specialist expertise. Before undertaking the procedure, its risk versus benefit should be carefully assessed.

REFERENCES

1. Oddi R. D'une disposition a sphincter speciale de l'ouverture du canal choledoque. Arch Ital de Biol 1887;8:317.
2. Lehman GA: Endoscopic sphincter of Oddi manometry—a clinical practice and research tool. Gastrointest Endosc 1991, 37:490–491.
3. Richards RD, Yeaton P, Shaffer HA, Pambianco DJ, Pruett TL, Stevenson WC, Mittal RK, McCallum RW: Human sphincter of Oddi motility and cholecystokinin response following liver transplantation. Dig Dis Sci 1993, 38:462–468.
4. Becker JM, Parodi JM: Basic control mechanisms of sphincter of Oddi motor function. Gastrointest Endos Clin N Am 1993, 3:41–66.
5. Anderson TM, Pitt HA, Longmire WP: Experience with sphincteroplasty and sphincterotomy in pancreatobiliary surgery. Ann Surg 1985, 201:399–406.
6. Lans JL, Parikh NP, and Geenen JE: Application of sphincter of Oddi manometry in routine clinical investigations. Endoscopy 1991, 23:139–143.
7. Sherman S, Gottlieb K, Uzer MF, Smith MT, Khusro QE, Earle DT, Brunelle RL, Hawes RH, Lehman GA: Effects of meperidine on the pancreatic and biliary sphincter. Gastrointest Endosc 1996, 44:239–242.
8. Geenen JE, Hogan WJ, Dodds WJ, Toouli J, Venu RP: The Efficacy of endoscopic sphincterotomy after cholecystectomy in patients with suspected sphincter of Oddi dysfunction. N Engl J Med 1989, 320:82–87.
9. Sherman S, Troiano FP, Hawes RH, Lehman GA: Sphincter of Oddi manometry: Decreased risk of clinical pancreatitis with the use of a modified aspirating catheter. Gastrointest Endosc 1990, 36:462–466.
10. Raddawi H, Geenen J, Hogan W, Dodds W, Venu R, Johnson G: Pressure measurements from biliary and pancreatic segments of Sphincter of Oddi—Comparison between patients with functional abdominal pain, biliary, or pancreatic disease. Dig Dis Sci 1991, 36:71–74.
11. Rolny P, Ärlebäck A, Funch-Jensen P, Kruse A, Jarnerot G: Clinical significance of manometric assessment of both pancreatic duct and bile duct sphincter in the same patient. Scand J Gastro 1989, 24:751–754.
12. Silverman WB, Ruffolo TA, Sherman S, Hawes RH, Lehman GA: Correlation of basal sphincter pressures measured from both the bile duct and pancreatic duct in patients with suspected sphincter of Oddi dysfunction. Gastrointest Endosc 1992, 38:440–443.

13. Eversman D, Sherman S, Bucksot L, Earle D, Rusche M, Gottlieb K, Lehman GA: Frequency of abnormal biliary and pancreatic basal sphincter pressure at sphincter of Oddi manometry (SOM) in 463 patients. Gastrointest Endosc 1997; 45:A131.
14. Kalloo AN, Tietjen TG, Pasricha PJ: Does intrabiliary pressure predict basal sphincter of Oddi pressure? study in patients with and without gallbladders. Gastrointest Endosc 1996, 44:696–699.
15. Guelrud M, Mendoza S, Rossiter G, Villegas MI: Sphincter of Oddi manometry in healthy volunteers. Dig Dis Sci 1990, 35:38–46.
16. Rolny P, Anderberg B, Ihse I, Lindstrom E, Olaison G, Arvill A: Pancreatitis after sphincter of Oddi manometry. Gut 1990, 31:821–824.

8

COLONIC TRANSIT AND ANORECTAL MANOMETRY

Satish S. C. Rao*

Department of Internal Medicine
University of Iowa College of Medicine

INTRODUCTION

Disorders of colonic and anorectal function affect 10–20% of the population (1). Over the last two decades the development of several new tests has facilitated better diagnostic approaches and treatment of these problems. This article provides an overview of the tests used for assessing colonic and anorectal function.

MANOMETRIC ASSESSMENT OF COLONIC FUNCTION

This includes an evaluation of a) colon transit and b) colon manometry. Because colonic manometry is primarily a research tool and is under evaluation, its role will not be discussed.

Colon Transit

Rationale. Patient recall of stool habit is often inaccurate (2). There is also a poor correlation between a patient's recall of stool frequency and their colon transit time.(3). Hence, assessment of colon transit not only provides an objective information regarding colonic dysfunction but also enables the physician to pursue a rational approach to further investigations or treatment. Several methods have been described (4–8).

Patient Preparation. It is important to ensure that patients refrain from using laxatives, enemas or suppositories or consume unusually high fiber diet during the study. They

* Phone: 319/353-6602, Fax: 319/353-6399, E-Mail: satish-rao@uiowa.edu

Gastrointestinal Motility, edited by Rao *et al.*
Kluwer Academic / Plenum Publishers, New York, 1999.

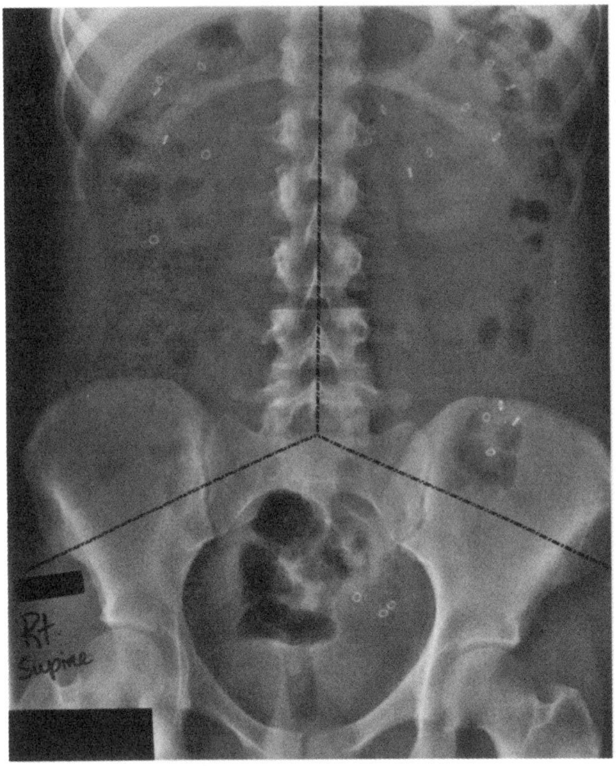

Figure 1. Colon transit study showing multiple radiopaque markers scattered throughout colon suggestive of slow transit constipation.

should be encouraged to stay on their routine diet. Some consideration should be given to concomitant medications, although in practice it is often difficult to stop many drugs. However, opiate analgesics, anticholinergics, antispasmodics and prokinetics must be discontinued. In a patient with stool impaction, it is advisable to clean the colon and wait for at least two weeks before performing the transit study.

Single Capsule Technique. This test consists of administering a single Sitzmarks capsule (Konsyl Pharmaceuticals, Fort Worth, Texas) containing 24 radio opaque markers on day 1, and obtaining a supine x-ray of the abdomen on day 6 (i.e., 120 hours later). If there are five or less markers on the x-ray, then colon transit is normal (5,9). Conversely, if more than five markers are present, the transit is slow and suggests colonic inertia or sluggish propulsion of stool (Figure 1). The x-ray is arbitrarily divided into three zones (8). Patients with obstructive defecation may also show delayed transit that is greater retention of markers. But unlike patients with slow transit constipation who have markers that are distributed throughout the colon, in the obstructive group, a greater number of markers are often found in the rectosigmoid segment. The advantages of this technique are its simplicity and the reduced radiation exposure. The disadvantages are that it does not accurately quantify transit time nor does it assess segmental transit.

Multiple Capsule Technique. In this technique, the patient swallows one Sitzmarks capsule daily, for three consecutive days. A supine x-ray of the abdomen is obtained on day 4 or 5 and again on day 7 (6). With this technique, segmental and total colonic transit time can be assessed and is calculated by the formula; mean transit=1.2 $(n_1 + n_2 + n_3 + \ldots +$

n_j), where n= number of markers present on each film and nj = number of x-rays. Segmental colonic transit is calculated by the product of 1.2 (nX), where nX is the number of markers in each colonic segment (right colon, left colon and rectosigmoid regions). The normal mean colon transit time is 35 hrs (95th centile = 68 hrs) and the mean segmental transit time is 12 hours (95 th centile =36 hrs) (6). The advantages of this technique are its ability to quantify segmental transit time and to improve sensitivity. But this kind of information is more useful for research studies than for routine clinical practice. Moreover, for most patients either taking multiple capsules or having multiple x-rays can be confusing. This may not only affect compliance but also adds cost and inconvenience. Also, a recent study has questioned the validity of the multiple capsule technique (10).

MANOMETRIC TESTS OF ANORECTAL FUNCTION

A list of the various tests that are commonly performed for evaluating anorectal function is shown in Table 1. The most commonly performed test is anorectal manometry and this has been discussed in detail below. A few, in particular videofluoromanometry (11), impedance planimetry (12), somato-sensory evoked potentials (13), and ambulatory manometry (14,15,16) are research techniques which in the future may prove clinically useful.

Anorectal Manometry

Anorectal manometry quantifies anal sphincter tone and assesses anorectal sensory responses, recto anal reflexes and rectal compliance i.e., it provides comprehensive information regarding anorectal function. The manometric system has two major components; the manometric probe and the pressure recording apparatus. Currently, several types of probes and pressure recorders are available. Each system has distinct advantages and disadvantages (9). The most commonly used probes and recording devices are reviewed here. The indications for anorectal manometry are shown in Table 2.

Water Perfused Catheter. This is the most commonly used system for anorectal manometry (4,17,18). The catheter has multiple side-holes through which water is perfused slowly using a pneumohydraulic pump [Arndorfer, Milwaukee, WI; MUI Scientific Ltd., Toronto, Canada]. Although the catheter is placed inside an organ, the pressure transducers are located outside the body and across the flow of water. When the lumen contracts, the

Table 1. Tests for assessing defecation disorders

• Anorectal manometry
• Balloon expulsion test
• Saline continence test
• Endosonography
• Pudendal nerve terminal motor latency
• Electromyography
• Defecography
• Rectoanal scintigraphy
• Somato sensory evoked potentials
• Videofluromanometry
• Impedance planimetry
• Ambulatory anorectal manometry

Table 2. Indications for anorectal manometry

• Fecal incontinence
• Chronic constipation
• Hirschsprung's/megarectum
• Biofeedback retraining
• Rectal prolapse
• Proctalgia
• Pre/post surgery (pouch)
• Solitary rectal ulcer

intraluminal pressures increases which occludes the flow of water through the side holes. This transient occlusion generates resistance to the flow of water, which is measured as the intraluminal pressure. The probe and the recording system are inexpensive, versatile and useful. A catheter with closely-spaced sensors may also help to discriminate between the external and internal sphincter activity (17).

Solid State Probe with Microtransducers. This system has pressure sensors (microtransducers) that are mounted directly on the probe (Figure 3). This allows a more accurate measurement of pressure activity, by placing the pressure sensors directly at the source, without complex plumbing. The transducers are true strain gauges, i.e. they consist of pressure sensitive diaphragms with semiconductor strain gauges that are mounted on its inner surface. Currently, this is the most accurate catheter system for performing manometry (10) and is particularly useful for anorectal manometry (9). It is user friendly and offers higher fidelity. It is free of the limitations imposed by the perfused systems. However, it is expensive and fragile, features that restrict its wider usage.

Pressure Amplifiers/Recorders. The pressure signals that are obtained from the transducers are amplified and recorded either on a chart recorder or on a digital recorder.

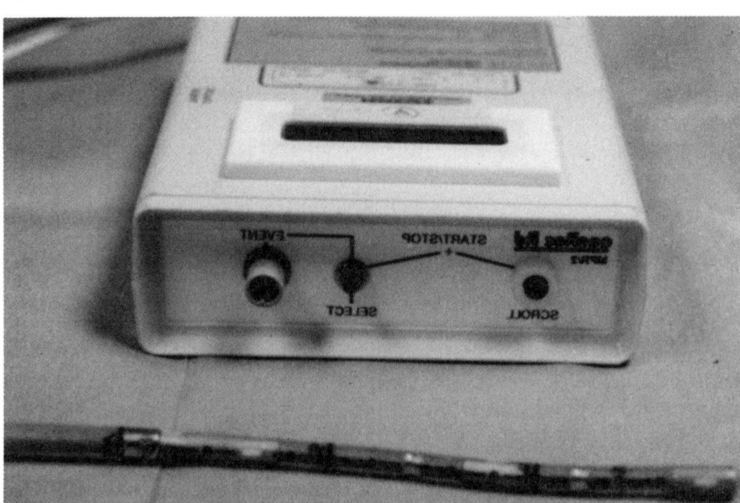

Figure 2. This shows a solid state manometry probe with multiple radially arrayed microtransducers and a latex balloon (Front), together with a solid state data logger (Back).

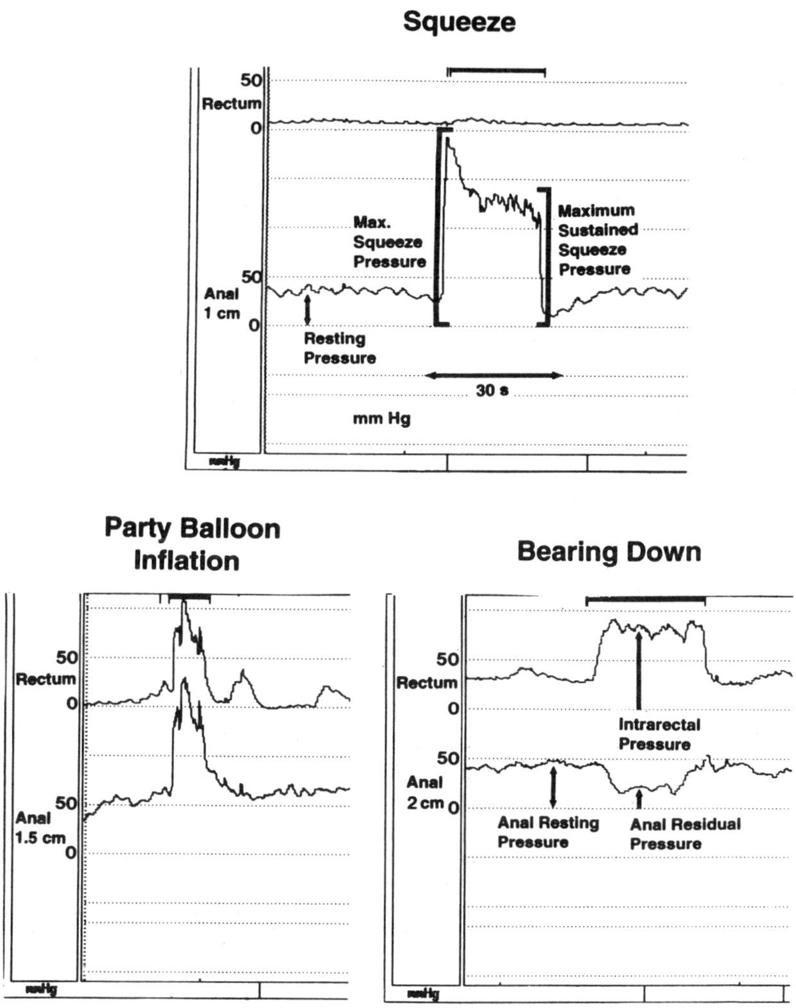

Figure 3. This is a graphic representation of the normal responses and the technique of measuring anal sphincter pressures at rest, during squeeze, during party balloon inflation and when bearing down. The recording was obtained by using a radially arrayed, solid state, multi-sensor pressure probe. The top trace shows pressure changes in the rectum and the lower trace shows pressure changes in the anal canal at 1 cm from the anal margin. To aid clarity, pressure traces from the other sensors have been omitted. As can be seen during squeeze, there is a rapid rise in anal sphincter pressure which slowly stabilizes and hence provides two separate measurements; maximum squeeze pressure and maximum sustained squeeze pressure. When blowing up a party balloon there is simultaneous rise in rectal and anal pressures with anal pressures exceeding rectal pressures. When bearing down there is rise in rectal pressure together with a fall in anal sphincter pressure. This technique obviates the need to perform a station pull through.

They can also be displayed on-line, on a monitor screen. Several analogue amplifiers/recorders are now available. But with the advent of computerized systems, several digital and portable recorders have now become available (Gaeltec 7-MPR, Isle of Skye, UK; Polygraph and DigiTrapper, Synetics Medical Inc., Irving, TX; Sandhill Scientific Inc., Highland Ranch, CO) that can also record data from many channels. They are small, compact, often portable and not only serve as amplifiers and recorders but also facilitate analysis of data and convenient storage on diskettes for future retrieval or for generating a data base.

Table 3. Normal values for anorectal manometry in 40 healthy subjects

	All (n=40)	Male (n=18)	Female (n=22)
Max anal rest Pr mm Hg	67(59–74)	71 (52–90)	64 (53–75)
Max squeeze Pr mm Hg	138 (124–152)	163 (126–200)*	117 (100–134)
Rectal Pr—straining mm Hg	61(52–71)	71 (57–85)	59 (50–68)
Anal residual pr-strain mm Hg	48(40–56)	41 (30–52)	45 (37–53)
Defecation index	1.7(1.2–2.2)	2.5 (2–2.9)*	1.4 (1.2–1.6)
Rectal compliance ml mm Hg	6.6(5.7–7.6)	6.9(5.3–8.5)	6.4(5.3–7.6)
First sensation cc	20(17–23)	20 (11–29)	20 (16–24)
Desire to defecate cc	107(91–123)	109 (78–140)	106 (85–127)
Urgency to defecate cc	181(162–200)	185 (139–231)	178 (156–200)
Max. tolerable vol. cc	241(223–259)	249 (113–305)	235 (214–259)
Balloon expulsion(sec)	57(23–91)	39 (36–42)	58 (9–107)
Vol. infused-Ist leak cc	632(553–711)	770(730–810)	516 (396–638)
% retention (800 cc)	90(83–97)	98 (96–100)*	82 (71–93)

* $p < 0.05$ (male vs. female).
Mean ± 95% confidence interval.

Essentials of Anorectal Manometry. A comprehensive assessment of anorectal manometry consists of measuring the following parameters; 1) anal sphincter function - maximum resting sphincter pressure, maximum squeeze sphincter pressure, and duration of squeeze, 2) anorectal pressure responses to increases in intraabdominal pressure, 3) changes in anal tone during straining, 4)length of anal sphincter, 5) rectal motor function, 6) rectal sensation, 7) rectal compliance, and 8) rectoanal reflexes. The normal values for anorectal manometry from our laboratory are shown in Table 3.

Anal Sphincter Function. The anal canal is 2–4 cm long. The internal anal sphincter is 0.15–0.5 cm thick. At least 70% of the resting sphincter pressure is due to tonic activity of the internal anal sphincter muscles. The external anal sphincter is a striated muscle and is under voluntary control. Its motor function is mediated by the pudendal nerve. Voluntary contraction of the anus increases pressure throughout the anal canal, but the pressure rise is maximal in the lower canal where the bulk of the EAS is situated. Squeeze sphincter pressures are higher in male subjects compared with female subjects (19, 20) and are perhaps lower in elderly compared to younger subjects (21).

Resting Sphincter Pressure. Currently, two methods are available for assessing this function. *Station pull through:* in this technique the most distal sensor of a multiport cathether assembly is placed 5 cm above the anal margin. At every 30 second intervals, the catheter is withdrawn by 0.5 cm either manually (20), or with a probe withdrawal device (22). As the sensors straddle the high pressure zone, there is a step up of pressure (Figure 3). The length and the highest resting pressure of the anal sphincter is then measured. Because pull through excites anal contraction and the individual is conscious and aware of these movements, the recorded pressure is higher (20). For the same reason, a rapid pull through is not an accurate method and is not advisable for measuring anal sphincter function. *An alternative method (Stationary)* is to use a radially arrayed multiport catheter (at least 3 sensors, 1 cm apart) that is placed in the anal sphincter zone, i.e., 0–3 cm from the anal verge (19,23). After allowing the traces to stabilize, the highest sphincter pressure that is observed at any level in the anal canal is taken as the maximum resting sphincter pressure (Figure 3)

Squeeze Sphincter Pressure. This can be measured with either the station pull through or the alternative technique. In the station pull through technique; after placing the multiport assembly as described above, at each level the subject is asked to squeeze (Figure 3) and to maintain the squeeze for as long as possible (at least 30 seconds). Alternatively with a multiport catheter in place, the subject is instructed to squeeze on three separate occasions, with a minutes' rest between each squeeze (Figure 3). The average of the three highest sphincter pressures recorded at any level in the anal canal is taken as the maximum anal squeeze pressure (19, 23).

Responses to Increases in Intra-Abdominal Pressure. An increase in intra-abdominal pressure, brought about by asking the subject to blow up a party balloon or by coughing is associated with a reflex increase in the activity of the EAS (19, 23). This reflex response causes the anal sphincter pressure to rise above that of the rectal pressure so that continence is maintained. The response may be triggered by receptors in the pelvic floor and mediated through a spinal reflex arc. In patients with supra conal spinal cord lesions, this reflex response is present but the voluntary squeeze may be absent whereas in patients with lesions of the cauda equina or of the sacral plexus, both the reflex response and the voluntary squeeze are absent.

Straining as if to Defecate. In this maneuver, the subject is asked to bear down, and simulate the act of defecation. The normal response consists of an increase in rectal pressure coordinated with a relaxation of the external anal sphincter (Figure 3). This maneuver is under voluntary control and is primarily a learned response (9). The inability to perform this coordinated maneuver may suggest a diagnosis of obstructive defecation (anismus), a common cause of constipation (19, 23, 24, 25).

Assessment of Rectal Function. The rectum is 12 to 15 cm long and has three flexures. Unlike the rest of the colon where the longitudinal muscle is arranged in three long bundles, the taeniae coli, at the rectosigmoid junction, these muscles fan out to form a continuous layer of longitudinal muscle that envelopes the rectum. This muscle arrangement facilitates shortening and emptying of the rectum.

Rectal Motor Function. Under resting conditions, the rectum has a basal pressure of 5 mm Hg. Usually, the arrival of fecal material or gas stimulates rectal contraction. Three types of contractile activities have been described: runs of simple contractions occurring at a frequency of 5–10 cycles/min., isolated contractions occurring randomly, and periodic rectal motor activity (PRMA) that consists of slow rhythmic contractions at 3 cycles/min (14–16). PRMA occurs more frequently at night, is predominately segmental or propagates retrogradely (14). The PRMA is not related to the phase III activity of the migrating motor complex (MMC) of the small intestine (26) or to REM sleep (15, 16). Recently, it has been shown that patients with constipation exhibit a higher incidence of PRMA (27).

Rectal Sensory Function. This can be assessed either by the intermittent rectal distension method or by the ramp inflation method.

Intermittent Rectal Distension. This method mimics the sudden arrival of stool or gas in the rectum. It is performed by inflating a balloon in the rectum using a hand-held syringe. After each inflation, the balloon is deflated completely and after a rest period is reinflated to the next volume. Initially, the subject may report a transient sensation of fullness, followed by a desire to defecate, an urgency to defecate, and ultimately pain or intolerable discomfort (4, 18, 19, 23).

Ramp Inflation. In this method, the rectum is progressively distended without deflation (18). This is performed by a) continuously inflating the balloon at a constant rate with a peristaltic pump or with a syringe using increasing volumes of air or fluid or b) in a stepwise fashion with a one minute interval between each incremental inflation of 10–30 cc (18). Subjects may describe sensory responses similar to those described above with the intermittent distension technique. A desire to defecate is associated with the onset of phasic rectal contractions and occurs when the rectum can no longer accommodate. Each sensation occurs at a lower volume when the infusion rate is slowed from 100 ml/min through 50 and 20 to 10 ml/min and this is associated with a shallower pressure-volume curve, indicating greater receptive relaxation or rectal (18).

Assessment of rectal sensation is useful in patients with fecal incontinence associated with neurogenic problems such as diabetes mellitus or multiple sclerosis (4, 28). It is also useful in patients with constipation; an impaired rectal sensation has been reported in 50–70% of patients with obstructive defecation or pelvic floor dysfunction (25). Importantly, when rectal sensation is impaired, neuromuscular conditioning using biofeedback techniques can be effective in improving this dysfunction (4, 25, 28).

Rectal Compliance. This is a measure of the rectal reservoir function. Rectal filling, induces phasic contractions initially followed by a decrease in intrarectal pressure, the "accommodation response" (4, 19). The rectum can hold more than 300 ml without significant increase in rectal pressure. The rectal compliance is usually measured by the ramp distention method, either by infusing air or water until a sensation of maximum tolerable volume is reported. Immediately after rectal distention, there is usually a rapid phasic contractile response, which decreases and reaches a steady state. The individual values for the steady state intrarectal pressure is plotted against the volume of fluid that was infused. This pressure-volume curve is usually reversed S-shaped. The slope of the steady-state intrarectal pressure (dP) and the volume used for rectal inflation (dV) is used to calculate rectal compliance (4, 19).

The phasic contractile response to rectal distension is reduced or absent in patients with lesions involving the low spinal cord, suggesting that this is mediated by a spinal reflex. Rectal tone is increased and compliance is reduced in patients with high spinal lesion, whereas in patients with low spinal lesions the rectal tone is reduced and compliance is increased (29). In many patients, particularly elderly, a poor rectal compliance may be the cause of incontinence rather than weak anal sphincter.

Rectoanal Inhibitory Reflex. This consists of reflex relaxation of the internal anal sphincter in response to rectal distension. As the volume of rectal distension is increased, the amplitude and duration of relaxation also increases (17). This reflex may facilitate sampling of rectal contents by the sensory receptors in the upper anal canal and may also help to discriminate flatus, from liquid or solid stools. This reflex is mediated by the myenteric plexus. In patients with Hirschsprung's disease and in those with a history of rectal transection this reflex is absent. However, in patients with spinal cord injury and in patients with transection of the hypogastric nerves or lesions of the sacral spinal cord, it is present (29).

BALLOON EXPULSION TEST

A test of simulated defecation is useful in the assessment of patients with constipation (19, 23, 30). This is performed either by placing either a 4 cm long balloon attached to a 3 mm diameter flexible tube or a silicon-filled deformable device-fecom (31) into the rectum. The balloon is filled with 50 cc of water at 37° C. The patient is asked to sit on a commode and to expel the simulated stool, in privacy. If the subject is unable to expel this artificial stool after 5 minutes of straining, the balloon is deflated and removed. The time taken for expelling the balloon is recorded. In case of any difficulty, gentle traction is applied to the balloon while instructing the patient to relax the pelvic floor. Most normal subjects can expel the device within three minutes (19, 23, 30), whereas patients with obstructive defecation may fail to expel this even after straining for five minutes (24, 25).

SALINE CONTINENCE TEST

This test assesses the overall capacity of the defecation unit to maintain continence during conditions that simulate diarrhea (19,32,33). With the patient lying on the bed, a 2 mm plastic tube is introduced approximately 10 cm into the rectum and taped in position. Next, the patient is transferred to a commode. The tube is connected to an infusion pump and 800 cc of warm saline (37° C) is infused into the rectum at a rate of 60 cc/minute. The patient is instructed to hold the liquid for as long as possible. The volume of saline infused at the onset of first leak (defined as a leak of at least 15 cc), and the total volume retained at the end of infusion are recorded. In their original study, 1500 ml of saline was infused (32). However, the sensitivity of this test can be improved by infusions lesser volume (19, 33). Most normal subjects should retain 800 ml without leakage (19), whereas patients with fecal incontinence (33, 34), or patients with impaired rectal compliance such as ulcerative colitis (35) leak at much lower volumes. The test is also useful in assessing objective improvement of fecal incontinence after biofeedback therapy (36).

NORMATIVE DATA

There is a dearth of well controlled, comprehensive information regarding normal values of anorectal physiology. Unpublished data from our center in a large group of healthy normal subjects using solid state technology is shown in Table 3.

ELECTROPHYSIOLOGICAL ASSESSMENT OF ANORECTAL FUNCTION

Pudendal Nerve Terminal Motor Latency [PNTML]

This test measures the neuromuscular integrity between the terminal portion of the pudendal nerve and the anal sphincter muscles (37). The nerve latency quantifies the degree of nerve damage. During its course through the levator ani muscles, the pudendal nerve is vulnerable to traumatic injury particularly during vaginal delivery and during application of forceps (38). Nerve injury can lead to sphincter weakness, loss of muscle control and fecal incontinence. The PNTML is measured using a disposable electrode (St. Marks pudendal electrode, 13640, Dantec electronics, Bristol, UK) and a recorder (Cadwell quantum Q8U). The electrode mounted on a gloved finger is inserted into the rectum (37). The ischial spine is palpated bilaterally and as the finger is swept medially electrical impulses (5–15MA) are delivered. When the nerve is stimulated, a prompt response is seen on the monitor. The time interval between the stimulus artifact and the onset of myoelectrical response is measured as the nerve latency time (37). The normal latency is 2.2 ms (19,37). A prolonged nerve latency is suggestive of pudendal neuropathy. This may occur following obstetric trauma (19, 38) and occasionally following surgical trauma and has also been reported in patients with idiopathic fecal incontinence (39). It has been postulated that excessive perineal descent, over many years, may lead to pudendal neuropathy and cause fecal incontinence (37). However, a recent prospective study has shown that pudendal neuropathy and excessive perineal descent may each occur in approximately 33% of patients with defecation disorders, but an association between the two problems was seen in only 8% of patients (40). Thus, these two conditions are independent variables.

CLINICAL UTILITY OF ANORECTAL FUNCTION TESTS

In a recent prospective study, it was shown that a combined assessment of anorectal function not only confirmed a clinical suspicion but also provided new information that was not detected clinically (23). Furthermore, the information obtained from these tests influenced the management and outcome of patients with defecation disorders (Table 4). These findings have been further confirmed by another recent prospective study, which showed that colorectal physiological tests provided definitive diagnoses in 75% of patients with constipation, 66% of patients with incontinence and 42% of patients with intractable anorectal pain (41).

CONCLUSION

In a patient with constipation that is refractory to fiber supplementation, a colon transit study may reveal slow transit or retention of markers in the rectosigmoid segment confirming slow colon transit. However, 50% of patients with either obstructive defecation or pelvic floor dysfunction may also have abnormal transit. In these patients, anorectal manometry together with balloon expulsion test can provide objective evidence for an underlying dysfunction. In a patient with fecal incontinence, manometry not only quantifies the degree of sphincter weakness but also provides information regarding rectal sensation rectal compliance and the thresholds for anal relaxation. Assessment of pudendal nerve latency provides a pathophysiological basis for a weak anal sphincter pressure. Because no single test of anorectal function can provide complete information, one or more tests are required. Based on the clinical findings, a clinician should request one or more tests. Undoubtedly, they provide useful information that can guide management.

Table 4. ARM —clinical usefulness

	Constipation (n=69)	Incontinence (n=56)	Pre-op (n=10)	Other (n=8)	Total (n=143)
Confirmed diagnosis	75%	98%	—	74%	—
New information	81%	98%	80%	87%	88%
Influenced treatment	67%	84%	100%	74%	76%
Normal study	19%	2%	20%	13%	12%
Not helpful	14%	14%	0%	13%	12%

With permission from Rao & Patel (Ref 23).

ACKNOWLEDGMENTS

The authors wish to acknowledge the excellent secretarial support of Ms. Susie McConnell.

REFERENCES

1. Drossman, D.A., Zhiming, L., Andruzzi, E., Temple, R.D., Talley, N.J., Thompson, W.G., Whitehead, W.E., Janssens, J., Funch-Jensen, P., Corazziari, E., Richter, J.E. and Koch, G.G. US householder survey of functional gastrointestinal disorders: Prevalence, sociodemography, and health impact. *Dig Dis Sci,* 1993;38:1569–1580.
2. Manning AP, Wyman JB, Heaton KW. How trustworthy are bowel histories? Comparison of recalled and recorded information. *BMJ* 1976; 2:213–214.
3. Ashraf W, Park F, Quigley EMM, Lof J. An examination of the reliability of reported stool frequency in the diagnosis of idiopathic constipation. *Am J Gastroenterol* 1996;91:26–32.
4. Wald A. Colonic and anorectal motility testing in clinical practice. *Am J Gastroenterol* 1994;89:2109–2115.
5. Evans RC, Kamm MA, Hinton JM, Lennard-Jones JE. The normal range and a simple diagram for recording whole gut transit time. *Int J Colorectal Dis* 1992;7:15–17.
6. Metcalf AM, Phillips SF, Zinsmeister AR, McCarty RL, Beart BW, Wolfe BG. Simplified asessment of segmental colonic transit. *Gastroenterology* 1987; 92: 40–47.
7. Rao SSC, Read NW, Brown C, Bruce C, and Holdsworth CD: Studies on the mechanism of bowel disturbance in ulcerative colitis. *Gastroenterology* 1987; 93:934–940.
8. Arhan P, Devroede G, Jehannin B, et al. Segmental Colonic transit time. *Dis Colon Rectum* 1981;24:625–9.
9. Rao SSC. Manometric evaluation of constipation-Part 1. *The Gastroenterologist* 1996; 4: 145–154.
10. Sasaki Y, Munakata A. High viscous luminal content causes different manometric outcomes between methodologies. *Gastroenterology* 1995;108:A683 (abstract).
11. Pelsang, R.E., Rao, S.S.C. and Welcher, K. Video flurodefecometry: A new technique for studying defacatory disorders. *Gastroenterology* 1996;110:A733.
12. Rao, S.S.C., Hayek, B. and Summers, R.W., Impedance planimetry: An integrated approach for assessing sensory, active and passive biomechanical properties of the esophagus. *Am. J Gastroenterol* 1995;90:431–438.
13. Loening Bauchke, V. and Yamada, T. Is the aberent pathway from the rectum impaired in children with chronic constipation and encopresis? *Gastroenterology* 1995;109:397–403.
14. Rao, S.S.C. and Welcher, K. Periodic rectal motor activity: The intrinsic colonic gatekeeper. *Am. J Gastroenterol* 1996;91:890–897.
15. Kumar, D., Williams, N.S., Waldron, D., Wingate, D.L. Prolonged manometric recording of anorectal motor activity in ambulant human subjects: Evidence of periodic activity. *Gut* 1989;30:1007–1011.
16. Orkin, B.A., Hanson, R.B., Kelly, K.A., Phillips, S.F., and Dent, J. Human anal motility while fasting, after feeding, and during sleep. *Gastroenterology* 1991;100:1016–1023.
17. Sun, W.M., Donnelly,T.C. and Read, N.W. Anorectal function in normal human subjects: Effect of gender. *Int. J Colorect Dis* 1989;4:188–196.
18. Sun, W.M., Read, N.W., Prior, A., Daly, J.A., Cheah, S.K. and Grundy, D. Sensory and motor responses to rectal distention vary according to rate and pattern of balloon inflation. *Gastroenterology* 1990;99:1008–1015.
19. Rao, S.S.C., Hatfield, R., Soffer ,E., Rao S, Leistikow J, Conklin, F.J.. Manometric tests of anorectal function in healthy adults. *Am J Gastroenterol* 1999;94:773–783.
20. McHugh, S.M. and Diamant, N.E. The effects of age, gender and parity on anal canal pressure: Contribution of impaired anal sphincter function to fecal incontinence. *Dig Di.s Sci* 1987;32:726–736.
21. Bannister, J.J., Abouzekry ,L. and Read, N.W. Effect of aging on anorectal function. *Gut* 1987;37:183–186.
22. Coller, J.A. 1990, Computerized anal sphincter manometry performance and analysis; in Smith LE (ed): Practical Guide to Anorectal Testing. New York, Igaku-Shoin, pp65–111.
23. Rao SSC , Patel RS. How useful are manometric tests of anorectal function in the management of defecation disorders. *Am J Gastroenterol* 1997;92: 469–475.
24. Preston DM, Lennard-Jones JE. Anismus in chronic constipation. *Dig Dis Sci* 1985;30:413–418.

25. Rao SSC, Welcer KD, Leistikow J. Obstructive Defecation (O.D.): A failure of recto-anal coordination. *Am J Gastroenterol* 1998;93:1042–1050.
26. Prior A, Fearn UJ, Read NW. Intermittent rectal motor activity: a rectal motor complex? *Gut* 1991;32:1360–1363.
27. Rao SSC, Welcher KD, Leistikow J. Altered periodic rectal motor activity. A mechanism for slow transit constipation. *Am J Gastroenterol* 1997; 92: A1686.
28. Miner, Jr., P.B. Fecal incontinence: Issues, assessment, and management. *Gastrointestinal Diseases Today* 1993;2:10–18.
29. MacDonagh, R., Sun,W.M., Thomas, D.G., Smallwood, R., Read, N.W. Anorectal function in patients with complete supraconal spinal cord lesions. *Gut* 1992;33:1532–1538.
30. Bannister, J.J., Davison, P., Timms, J.M., Gibbons, C. and Read, N.W. Effect of stool size and consistency on defecation. *Gut* 1987;28:1246–1250.
31. Rao, S.S.C., Rao, S., Hatfield, R. and Chamberlain, M. Is the stimulated defecation test influenced by position, presence of artificial stool and its consistency? *Gastroenterology* 1996;110:A741.
32. Haynes, W.G. and Read, N.W. Ano-rectal activity in man during rectal infusion of saline: A dynamic assessment of the anal continence mechanism. *J Physiol.* 1982;330:45–56.
33. Allen ML, Orr WC, Robinson MG. Anorectal functioning in fecal incontinence. *Dig Dis Sci* 1988; 33: 36–40.
34. Sun WM, Donnelly TC, Read NW. Utility of a combined test of anorectal manometry, electromyography, and sensation in determining the mechanism of idiopathic fecal incontinence. *Gut* 1992;33:807–813.
35. Rao SCC, Read NW, Davison P, Bannister JJ and Holdsworth CD: Anorectal sensitivity and responses to rectal distension in patients with ulcerative colities. *Gastroenterology* 1987; 93:1020–1026.
36. Rao SSC, Welcher KD, Happel J. Can biofeedback therapy improve anorectal function in fecal incontinenece? *Am J Gastroenterol* 1996; 91:2360–2366.
37. Henry, M.M., Parks, A.G. and Swash, M. Investigation of disorders of the anorectum and colon. *Ann. Roy. Col.l Surg. (Engl)* 1992;33:807–813.
38. Sultan, A.H., Kamm, M.A., Hudson, C.N., Bartram, C.I. Third degree obstetric anal sphincter tears: Risk factors and outcome of primary repair. *BMJ* 1994;308:887–891.
39. Kiff, E.S. and Swash, M. Normal proximal and delayed distal conduction in the pudendal nerves of patients with idiopathic (neurogenic) fecal incontinence. *J Neurol Neurosurg Psychiatry* 1984;47:1269–1273.
40. Jorge, J.M.N., Wexner, S.D., Ehrenpreis, E.D., Nogueras, J.J. and Jagelman, D.G. Does perineal descent correlate with pudendal neuropathy? *Dis Colon Rectum* 1993;36:475–483.
41. Wexner, S.D. and Jorge, J.M.N. Colorectal physiological tests: Use or abuse of technology? *Eur. J Surg* 1994'160:167–174.

9

DEFECOGRAPHY AND ANAL ENDOSONOGRAPHY

Retta E. Pelsang*

The University of Iowa College of Medicine
Department of Radiology

INTRODUCTION

Disorders of anorectal function can be assessed by many modalities. Imaging studies can complement physiologic studies in order to best diagnose and treat patients with defecatory disorders. Imaging studies provide anatomic delineation and if dynamic in nature can provide functional information. Here, I discuss the two most popular tests to evaluate pelvic floor function, namely defecography and anal endosonography.

Defecography

Rationale. Defecography is a dynamic imaging technique for assessing anorectal anatomy, morphology, and function (1,2). The anatomical outline and orientation of these structures within the pelvis can be directly seen while the location and function of the levator ani complex muscles can be inferred. Morphologic changes such as rectocele(s), mucosal intussusception, or prolapse can be demonstrated. The function of the anorectal unit can be assessed during the dynamic portion of the examination establishing a coordination of sphincter activity and rectal emptying. Defecography is most useful in patients with obstructive symptoms. Obstructive defecation includes constipation as well as the sensation of incomplete emptying (3). Defecography has limited value in evaluation of anal incontinence, unless associated with obstructive symptoms (3).

Technique. No patient preparation is required before the rectum is filled with a thick barium paste. Commercially available pastes include, Anatrast (E-Z-EM, Westberry, NY) or Evacupaste (Lafayette Pharmaceutical, Lafayette, IN) which requires a cocking gun for rectal filling or Intropaste (Lafayette Pharmaceutical, Lafayett, IN) which can be injected

* Phone: 319-353-8729, Fax: 319-356-2220, E-Mail: retta-pelsang@uiowa.edu

Gastrointestinal Motility, edited by Rao *et al.*
Kluwer Academic / Plenum Publishers, New York, 1999.

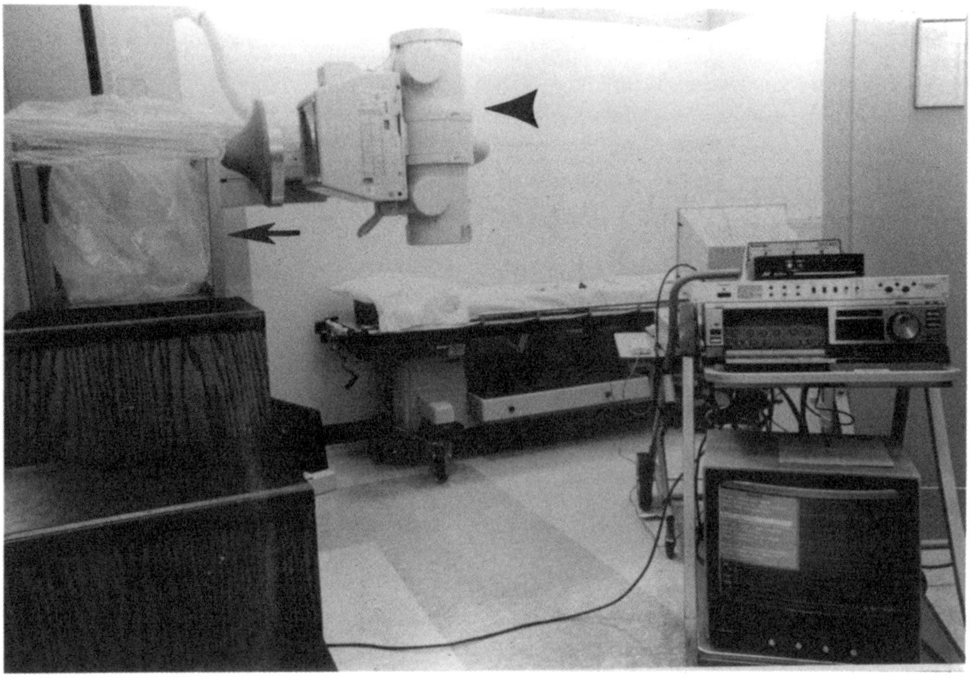

Figure 1. Room set up for defecography with a modified commode (black arrow) and remote fluoroscopic tower. (arrowhead).

with a hand held syringe. With the subject in left lateral decubitus position, approximately 150 to 250cc of barium paste is placed into the patient's rectum. Patient is then seated on a radiographic commode and imaged in the lateral position using a fluoroscopic tower (Figure 1).

Once the patient is seated and no longer needs further assistance, the patient is left in the room alone for privacy. The examination should be performed by remote control, all monitors should be turned off inside the room, and the light should be dimmed. This will facilitate a more physiologic examination without embarrassing the patient or distorting the examination. The patient is first imaged at rest (Figure 2), then asked to "squeeze" to assess activity of the levator ani muscles, then asked to cough or blow up a party balloon to assess for any incontinence, and finally the patient is asked to bear down to assess evacuation (Figure 3). These dynamic maneuvers are recorded on videotape. Additionally, static images can be obtained by taking spot x-rays or from the videotape recording.

Radiation Exposure. Radiation is calculated by taking into consideration the technique, scatter, and body parts that are being exposed. Since the patient is imaged in the lateral position, there is significant scatter of radiation. Furthermore, because a majority of patients are women, the ovaries are often exposed to radiation. Ovaries are weighted heavier than other parts of the body for calculating radiation exposure. Therefore, although short in duration, it is radiation intensive. For example, the whole body radiation dose during defecography is 0.47 rad while an abdominal film is 0.06 rad and a barium enema is 0.41 rad. Dose to the ovaries themselves is 1.45 rad and to the testes 0.09 rad.

Normal Subjects. At rest, the anorectal angle has a wide range from 70–135° (4, 5). During squeezing, the levator ani muscles are activated which produces an indentation on

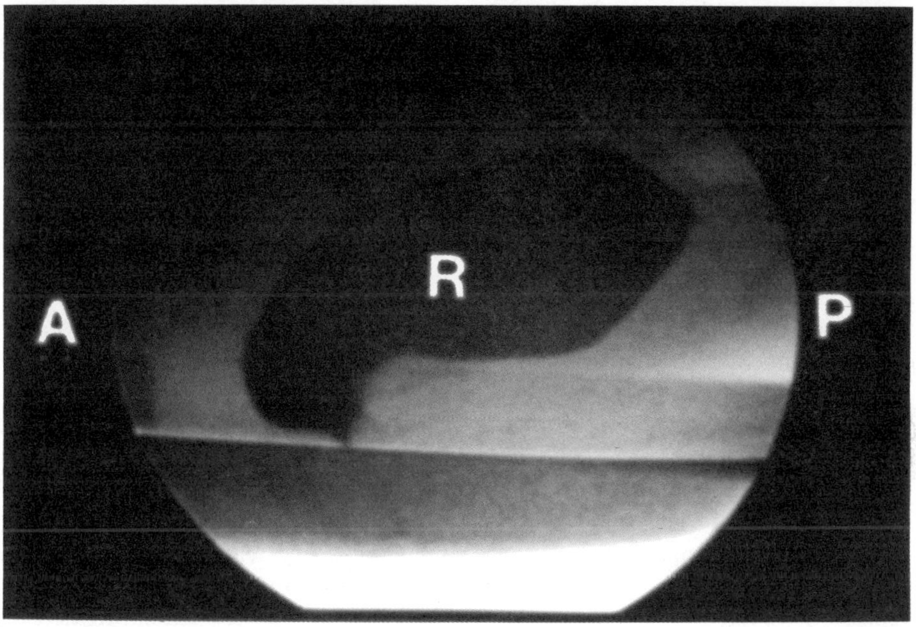

Figure 2. In the lateral position, this patient is imaged at rest after filling the rectum with 150 cc of barium paste. A=anterior, P=posterior, R=rectum.

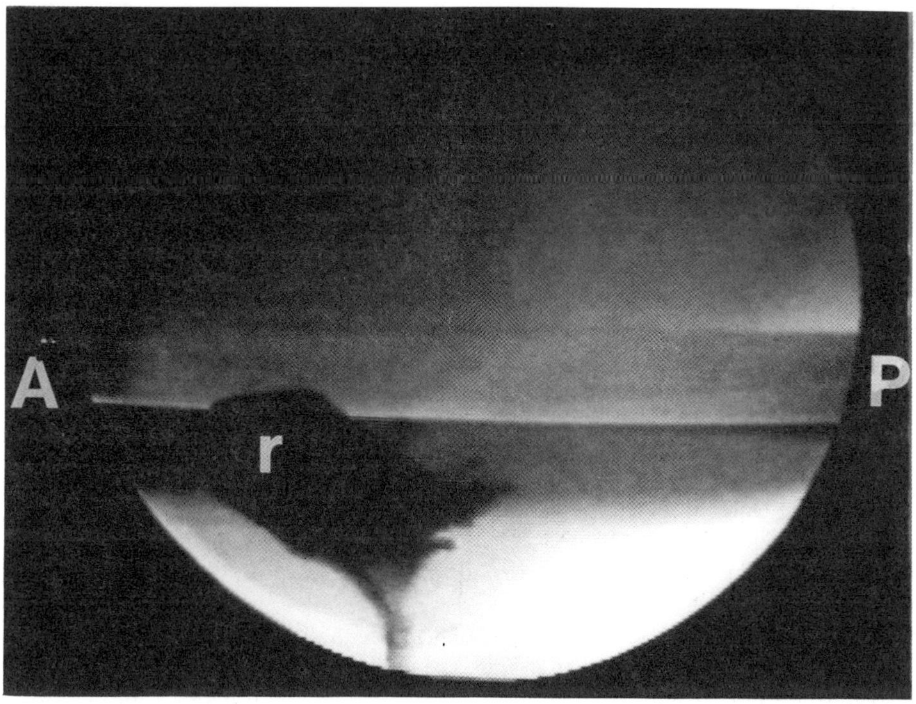

Figure 3. At the end of defecation, the anorectum is almost completely empty with some residual barium in the anterior rectocele. A=anterior, P=posterior, r=rectocele.

the posterior wall of the rectum, elevates the posterior rectal wall, and decreases the anorectal angle. (4, 5) Evacuation involves many morphological changes including; 1) relaxation of the pelvic floor muscles resulting in an increase in anorectal angle and a descent of the anorectum into the pelvis, 2) coordinated opening of the anal sphincter, and 3) a stripping wave generated from the rectum. The degree of rectal emptying is variable and total emptying is seen in only half of normal subjects. (1) Various morphologic changes can be seen during defecation. Mild degree of intussusception of mucosa can be seen normally (4, 5). Similarly, 80% of asymptomatic women have small anterior rectoceles usually less than 2cm in depth (4, 5).

Symptomatic Patients. Rectoceles. Rectoceles are thought to result from weakening of the rectovaginal septum due to trauma or chronic straining. Rectoceles although common in asymptomatic women are almost always present with pelvic floor weakness. (6) Defecography can identify those women with rectoceles of moderate (2–4cm) to large (>4cm) sizes. More importantly, defecography may help to distinguish those rectoceles that retain barium paste and may be responsible for dyschezia with those that do not retain barium and are therefore asymptomatic (7).

A proportion of women with rectoceles complain of persistent straining and feeling of incomplete evacuation. Some of the patients report the necessity to apply digital pressure to the vagina or perineum to assist in evacuation. This maneuver performed during defecography can demonstrate further clearing of the barium "trapped" in the rectocele. After surgical repair of rectoceles, lack of relief of obstructive symptoms suggests underlying defecatory disorder. Thus, it is not surprising that 71% of patients with symptomatic rectoceles have evidence of spastic pelvic floor (8). This will be discussed later in the chapter.

Enterocele and Sigmoidocele. Enterocele is a herniation of the posterior peritoneal cul-de-cac into the rectovaginal space and generally contain small bowel. Enterocele (18%) and sigmoidocele (5%) have been identified in patients with pelvic floor weakness (7). Many enteroceles are not recognized at surgery because the patient is supine and relaxed. (9) Identifying an enterocele may be more important prior to surgery for genuine urinary stress incontinence or cystocele repair than for obstructive defecation. Enteroceles are identified in 64% of patients with hysterectomy and in 27% of patients with cystopexy. (10) An enterocele is easier to identify if the small bowel is opacified. This can be performed by giving the patient 300–500cc of liquid barium by mouth, approximately 2 hours before defecography. Enteroceles are best seen at the end of defecation. Repeated straining after evacuation may be necessary to recognize an enterocele (11). Enteroceles are frequently associated with rectoceles, and both may result from chronic straining (7). Sigmoidoceles are rare and can be seen as the separation of the rectum from the sacrum during defecography.

Obstructive Defecation or Spastic Pelvic Floor. This condition is known by many names; anismus, paradoxical sphincter reaction, non-relaxing puborectalis syndrome, and rectal dyschezia. Lack of relaxation of levator ani complex muscles and poor opening of the anorectal junction or anal canal leads to a functional disorder involving inappropriate contraction of the voluntary pelvic floor muscles during defecation. By electromyography, during defecation, there is a paradoxical increase in activity of the puborectal and external sphincter muscles (12). It is important to identify this behavioral disorder prior to surgery to improve surgical outcome. This behavioral disorder can be successfully treated with biofeedback techniques (13–15).

Spastic pelvic floor may be associated with rectocele, rectal intussusception, solitary rectal ulcer syndrome, and slow transit constipation (16, 17). This is believed to be secondary to chronic straining and poor coordination of the anorectal muscles and pelvic floor (17).

Rectal Intussusception and Rectal Prolapse. Rectal intussusception consists of mucosal infoldings of the mid rectal folds at approximately 6–8cm inside the rectum. Commonly, the anterior rectal wall is involved and rarely it can be circumferential (18). Varying degrees of intussusception can be seen extending into the rectal ampulla (intrarectal), into the anal canal (intra-anal), or passing through the anal canal resulting in external rectal prolapse (18). Radiographically, abnormal loosening of the rectum and sigmoid colons from the sacrum can be seen. As the intussusception occurs the rectum pulls the anterior peritoneal covering downward, resulting in a deep pocket anteriorly that an enterocele can occur. Enteroceles are seen in 24% of patients with rectal intussusception. (15) At the end of defecation, the intussusception may sequester barium paste into the anterior rectocele keeping it from being evacuated. This material may be involuntarily expelled after relaxation and reduction of the prolapse leading to incontinence. (15) Rectal intussusception maybe associated with solitary rectal ulcer syndrome and spastic pelvic floor. (19, 20)

Fecal Incontinence. Direct trauma or denervation injury of the puborectalis and external sphincters can result in fecal incontinence (21). Internal sphincter weakness is suggested when leakage of barium paste is seen during defecation. Inability to narrow an open anal canal during squeezing suggests a loss of external sphincter. A weakness of the levator ani complex muscles can be seen by lack of anorectal angle reduction during squeeze. There is a sub-group of patients with both fecal incontinence and obstructive symptoms which may benefit from defecography. Non-emptying rectoceles, spastic pelvic floor, and rectal intussusception may coexist in these patients (14, 22).

Anorectal angle is often wide in patients with fecal incontinence however the measurement is highly variable. Anal endosonography may prove more useful as an imaging technique and this will be discussed later in the chapter.

Conclusion. Radiographic findings on defecography overlap between normal individuals and symptomatic patients. Hence these findings must be in interpreted in the sight of clinical correlation. Identification of an underlying functional obstruction may improve surgical repair of pelvic floor defects. The ability of the dynamic defecography to demonstrate anorectal dysfunction in combination with cystography and small bowel or sigmoid enhancement may help to identify multiple compartment pelvic floor defects. Patients are often bothered by defects in one or more compartments and/or functional abnormalities that may be observed by defecography.

ANAL ENDOSONOGRAPHY

Introduction

Anal endosonography is an imaging modality that can complement physiologic studies particularly in patients with fecal incontinence (23–25).

Rationale. Ultrasound is an inexpensive method of imaging the anal sphincters. Identifying a defect in the anal sphincters may direct surgical planning in patients with fecal incontinence. Anal endosonography is more accurate than clinical evaluation and conventional physiologic methods (anorectalmanometry/EMG concentric needle mapping) for morphologic evaluation of the anal sphincters (26).

Technique. A special ultrasound transducer is required for anal endosonography. A high frequency 7 MHz transducer is required for spatial resolution of the near field and a 360° axial rotating image is necessary to view the circular sphincter structures. This requires a dedicated transducer that is not currently available on standard ultrasound machine, such as you might use for liver imaging. Currently, the required rotating endo probe must be purchased with a dedicated ultrasound unit, for example Bruel and Kjaer, (Naerum, Denmark) sonographic scanner (model 3535) 7 mHz rotating endoprobe (8539, focus length 2–5cm). The probe itself is best imaged with a hard plastic cone cover and the patient is imaged in either the prone or left lateral decubitus positions. Probe sizes range from 1–2cm in diameter. Images are recorded on a video recorder and static images can be obtained on a paper recorder. The examination is brief and causes little patient discomfort. No patient preparation is required.

Anatomy. Five layers can be seen corresponding to the mucosa, submucosa, internal anal sphincter, intersphincteric plane, and external sphincter (27). The internal anal sphincter is smooth muscle oriented in oblique bundles in contiguity with the muscle of the rectum (28). Sonographically it is a hypoechoic (dark) uninterrupted circle ranging from 1–3mm in width (25, 27, 29, 30). High in the anal canal the internal anal sphincter anteriorly in women is thin or ill-defined. The termination of the internal anal sphincter is at the junction of the superficial and subcutaneous parts of the external sphincter. The internal anal sphincter extends from the anal rectal junction to just below the level of the anal valves (27) (Figure 4).

The external anal sphincter is composed of striated muscle and consists of either 2 or 3 segments. The deep portion is a circular muscle merging with the puborectalis posteriorly and a superficial component more elliptical with fibers desiccating anteriorly and a circular subcutaneous component (31). Sonographically there is no anatomic demarcation of these distinct components nor is there distinct cleavage between the puborectalis muscle and the deep external sphincter (32). The external anal sphincter is variable in its echo

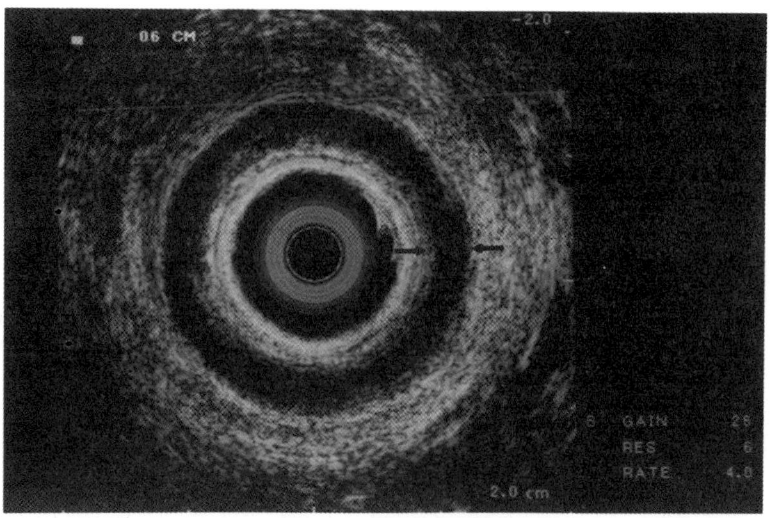

Figure 4. Normal intact internal anal sphincter. Completely circular, black band of smooth muscle (between arrows).

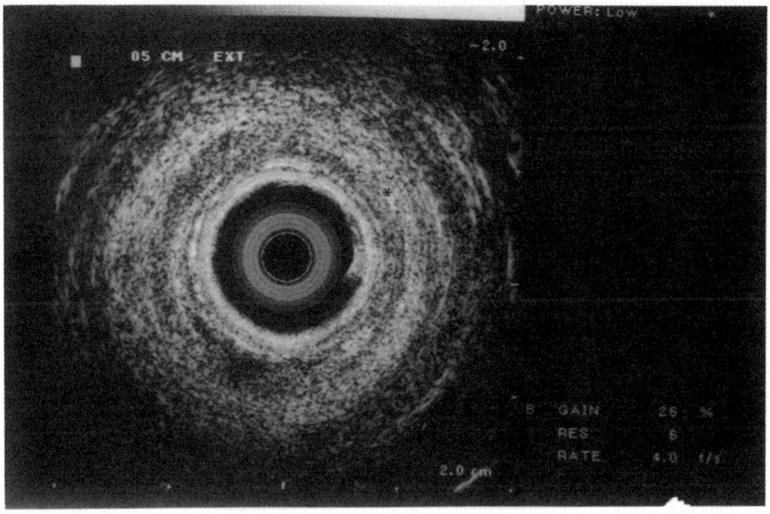

Figure 5. Normal external anal sphincter. Heterogeneous striated muscle (asterisk).

texture, generally hyperechoic (bright) with mixed signal ranging from 7–8mm in thickness (29,30,32). The external anal sphincter can be traced from the puborectalis component of the levator ani muscle to its cutaneous insertion (Figure 5).

Effects of Probe Size. Larger probes can "thin" the appearance of the internal anal sphincter. A decrease of 30% in width of the sphincter from a 1cm-diameter probe to a 2cm-diameter probe has been reported (33).

Effects of Age. With aging, internal anal sphincter muscle is replaced with fibroelastic tissue . This increases the thickness of the muscle. Between 19 and 65 years-of-age, internal anal sphincter thickness of 1.9 to 2.3 mm is seen (31.5 median age) and over 50 years of age increases to 2–3mm in thickness (28).

Gender Difference. The external anal sphincter is thicker in men secondary to body mass rather than to gender. The internal anal sphincter shows no difference in thickness between men and women except anteriorly, where it is thin in women (28).

Internal Anal Sphincter Abnormalities. Thinning or Tears. Atrophy of smooth muscle which occurs in conditions such as progressive systemic sclerosis can decrease the thickness of the internal anal sphincter. Similarly, in idiopathic fecal incontinence the thickness of internal anal sphincter may vary. Marked thinning of the internal anal sphincter is a sign of trauma. Any break in the continuity of the sphincter is abnormal and implies trauma (Figure 6). Obstetric tears are usually situated anteriorly. Isolated tear of internal anal sphincters is probably not related to obstetric trauma. Sphincterotomy may involve the whole thickness of the sphincter leading to incontinence since the anal canal is shorter in women. Anal stress procedures may damage the internal anal sphincter and external anal sphincter. A range of defects from abnormal patchy thinning to single defect to multiple defects where the sphincter has been fragmented may be seen.

Thickening. This can be seen with age as previously described.

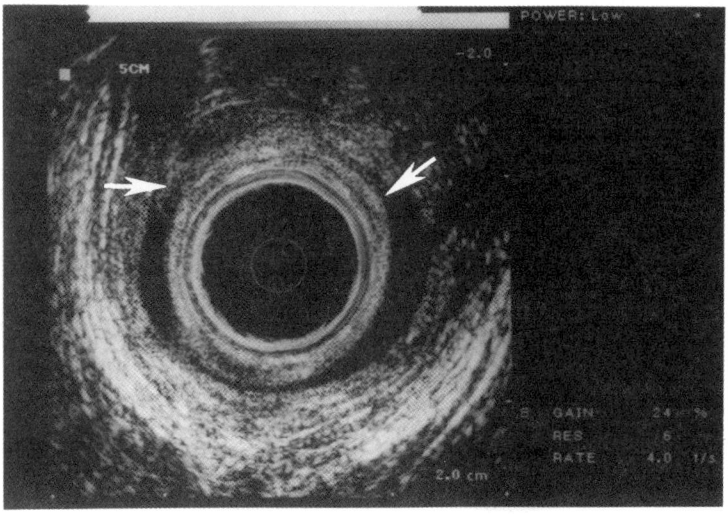

Figure 6. Anterior sphincter defect from 11:00 to 1:00 in the internal anal sphincter in a patient with fecal incontinence. (Defect between the arrows.)

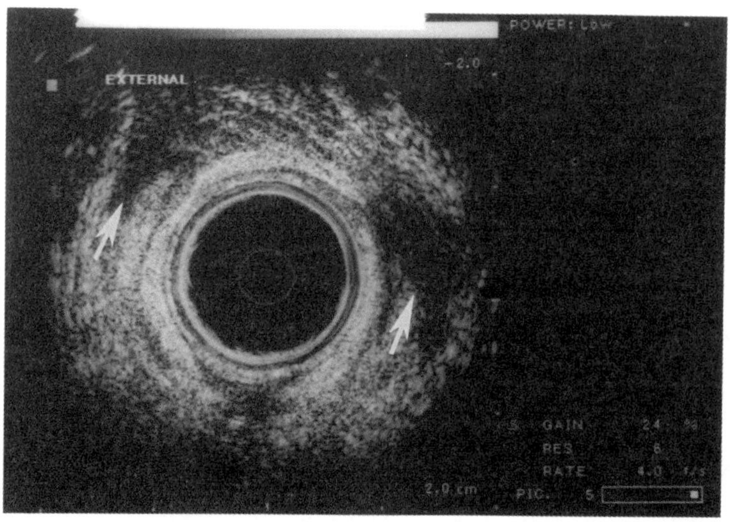

Figure 7. Defects can be seen at 11:00 and 3:00 in the external anal sphincter in a patient with fecal incontinence. (arrows point to defects).

Neurogenic incontinence can lead to a loss of smooth muscle cells leading to an increase in collagen fibrils content in the internal anal sphincter which may lead to an increase in diameter of the internal anal sphincter (24).

External Sphincter Abnormalities. Thinning. With neurogenic incontinence there is a thinning of the external anal sphincter in comparison to controls (23). With idiopathic fecal incontinence, there is no difference in the external anal sphincter between patients and controls (25).

Tears. Due to its complex structure, tears may be difficult to recognize and evaluate fully. The commonest causes of external anal sphincter trauma is childbirth and usually does not involve the puborectalis muscle (Figure 7). Other causes of trauma such as motor vehicle accidents or penetrating injuries may involve the puborectalis muscle.

CONCLUSION

Because anal sphincter defects can be seen in asymptomatic patients, sonographic changes must be correlated with clinical condition Anal endosonography is a simple and useful imaging method for evaluating anal sphincters in patients with fecal incontinence. This information may have a direct impact on reconstructive surgery.

REFERENCES

1. Mahieu P, Pringot J, Bodart P. Defecography 1. Description of a new procedure and results in normal patients. Gastrointest Radiol 1984;9:247–51.
2. Kelvin FM, Stevenson GW. Radiologic investigation. The anorectum and vagina. In: Benson JT ed. Female pelvic floor disorders. Investigation and Management. New York: WW Norton, 1992:70–88.
3. Kelvin FM, Maglinte DDT, Benson JT. Evacuation Proctography (Defecography): An aid to the investigation of pelvic floor disorders. Obstetrics & Gynecology 1994;83(2):307–13.
4. Bartram CI, Turnball GK, Lennard-Jones LE. Evacuation proctography: An investigation of rectal expulsion in 20 subjects without defecatory disturbance. Gastrointest Radiol 1988;13:72–80.
5. Shorvon PF, McHugh S, Diamant NE, Somers S, Stevenson GW. Defecography in normal volunteers: Results and implications. Gut 1989;30:1737–49.
6. Richardson AC. The rectovaginal septum revisited its relationship to rectocele and its importance in rectocele repair. Clin Obstet Gynecol 1993;36:976–83.
7. Kelvin FM, Maginte DDT, Hornback JA, Benson JT. Pelvis prolapse: Assessment with evacuation proctography (defecography). Radiology 1992;184:547–51.
8. Johanson C, Nilsson BY, Holmstom B, Dolk A, Mellgren A. Association between rectocele and paradoxical sphincter response. Dis Colon Rectum 1992;35:503–509.
9. Nichols DH. Enterocele, In: Nichols Dh, Randall CL eds. Vaginal Surgery. Baltimore; Williams and Wilkins 1989:313–27.
10. Hock D, Lombard R, Jehaes C, etal. Colpocystodefecography. Dis Colon Rectum 1993;36:1015–21.
11. Ekberg O, Nylander G, Fork FT. Defecography. Radiology 1985;155:45–48.
12. Kuijpero HC, Bleyinberg G. The spastic pelvic floor: a cause of constipation. Dis Colon Rectum 1985;28:669–72.
13. Rao SSC, Welcher KD, Pelsang RE. Effects of biofeedback therapy on anorectal function in obstructive defecation. Dig Dis Science 1997;42(11):2197–2205.
14. Fleshman JW, Dreznick Z, Meyer K, Fry RD, Carney R, Kodner JJ. Outpatient protocol for biofeed back therapy of pelvic floor outlet obstruction. Dis Colon Rectum 1992;35:1–7.
15. Johansson C, Ihre T, Ahlback SO. Disturbance in the defecation mechanism with special reference to intussusception of the rectum (internal procidentia) Dis Colon Rectum 1985;28:920–24.

16. Halligan S, Bartam CI, Park HJ, Kamm MA. Proctography features of anismus. Radiology 1995;197:679–82.
17. Halligan S. Commentary: Imaging of the anorectal function. Br J Radiology 1996;69:985–88.
18. Mahieu P. Pringot J, Bodart P. Defecography II contribution to the diagnosis of defecation disorder. Gastroent Radiol 1984;9:253–61.
19. Goei R, Baeten C. Rectal intussusception and rectal prolapse: detection and postoperative evaluation with defecography. Radiology 1990;174:124–126.
20. Halligan S, Nicholls RJ, Bartrum CI. Evacuation proctography in patients with solitary rectal ulcer syndrome: anatomic abnormalities and frequency of impaired emptying and prolapse. AJR 1995;164:91–5.
21. Penninckx FM. Fecal incontinence. Int J Colorectal Dis 1987;2:173–86.
22. Rex DK, Lappas JC. Combined anorectal manometry and defecography in 50 consecutive adults with fecal incontinence. Dis Colon Rectum 1992;35:1040–45.
23. Emblem R, Dhaenens G, Stein R, Morkrid L, Aasen AO, Bergan A. The importance of anal endosonography in the evaluation of idiopathic fecal incontinence. Dis Colon Rect 1994;37(1):42–8.
24. Swash M, Bray A, Lubowski DZ, Nicolls RJ. Ultrastructural changes in internal anal sphincter in neurogenic facal incontinence. Gut 1988;29:1692–8.
25. Eckardt VF, Jung B, Fisher B, Lierse W. Anal endosonography in health subjects and patients with idiopathic fecal incontinence. Dis Colon Rect. 1994;37(3):235–42.
26. Sultan AH, Kamron MA, Talbot IC, Nicholls RJ, Bartam CI. Anal endosonography for identifying external sphincter defects confirmed histologically. Br J Surgery 1994;81:463–65.
27. Law PF, Bartram CI. Anal endosonography: technique and normal anatomy. Gastrointest Radiol 1989;14:349–53.
28. Bartam CI, Frudinga A. Handbook of anal endosonography. Wrightson Biomedical Publishing LTD. Petersfield UK 1997.
29. Nielsen MB, Pedersen JF, Hauge L, Rasmussen O, Christiansen J. Endosonography of the anal sphincter: findings in healthy volunteers. AJR 1991;157:1199–1202.
30. Gantke E, Schafer A, Emck P, Lubke H. Sonographic, manometric and myographic evaluation of the anal sphincters morphology and function. Dis Colon Rect 1993;36(11):1037–41.
31. Milligan ETC, Morgan CN. Surgical anatomy of the anal canal. Lancet 1937;2:1150–56, 1213–17.
32. Sultan AH, Kamm MA, Hudson CN, Nicholls JR, Bartam CI. Endosonography of the anal sphincters: normal anatomy and comparison with manomety. Clin Radiol 1994;49:368–374.
33. Papachrysostomou M, Pye SD, Wild SR, Smith AN. Anal endosonography: which endoprobe? Br J Radiol 1992;65(776)715–17.

10

OROPHARYNGEAL DYSPHAGIA

Konrad Schulze-Delrieu* and Bruce Brown†

INTRODUCTION

History and physical examination are important tools in the diagnostic assessment of swallowing disorders. They help to define treatment strategies, to select appropriate tests or to determine the need for consultation with other disciplines. The history proceeds from a definition of the swallowing problem proper to associated and ancillary symptoms. The physical examination proceeds from a general assessment of the patient's physical and mental health to an assessment of swallowing structures and functions. Both history and physical examination seek to establish the adequacy and safety of oral food intake, and to identify remediable causes of abnormal swallowing (1, 2).

Oropharyngeal swallowing is often impaired in neurologic disorders (3, 4, 5), after interventions to treat head and neck neoplasms, in esophagitis and esophageal tumors. Neurologic diseases commonly associated with abnormal swallowing include cerebrovascular accidents (CVA), head trauma, cerebral palsy, mental retardation and dementias, and dyskinesias including Parkinson's disease, myasthenias, muscular dystrophies, degenerative disorders including multiple sclerosis (MS) and amyotrophic lateral sclerosis (ALS), and sleep apnea (3).

Most patients seek help because they have recognized some malfunction of their swallowing. Hospitalized or institutionalized patients may be unable to recognize or to describe dysphagia. Accounts of swallowing problems must then be elicited from attendants who witnessed them. The bedside encounter should screen for the likelihood and seriousness of a swallowing impairment.

HISTORY

The swallowing history should determine whether the difficulty involves the oral cavity, the pharynx, the esophagus, or all three (Table 1). Is the dysphagia caused by inflamma-

* Phone: 319/356-2793, fax 319/353-6399

† Phone: 319/356-4374, E-mail: bruce-brown@uiowa.edu

Gastrointestinal Motility, edited by Rao *et al.*
Kluwer Academic / Plenum Publishers, New York, 1999.

Table 1. Sites of dysphagia

• **Oral**: problems with chewing, bolus gathering or initiation of swallow, may compensate by suckling, using straw, pushing bolus into pharynx, cough may precede swallow
• **Pharyngeal**: problems immediately after swallow, coughing or sensation of choking during swallow, sensation of phlegm, need to clear throat or to swallow repeatedly; *both oral and pharyngeal problems typically worsen with liquids (particularly if carbonated) than with viscous/pasty materials*
• **Esophageal**: sensation of sticking seconds after swallow particularly if projected onto lower sternum, motor dysfunction similar with liquids and solids

tion, dysfunction, or tumor localized in the swallowing structures, or is it part of a systemic disease? Is the patient at risk for respiratory or nutritional complications (1, 2, 3)?

Dysphagia, Odynophagia, and Globus Sensation

Oral dysphagia typically presents itself as a difficulty with initiating a swallow. There is prolonged struggling with the bolus, but the involuntary pharyngeal and esophageal stages of swallowing may proceed normally once the reflex is triggered. Patients may compensate for failure of their tongue to push the bolus by drinking through a straw or by using their fingers. Oral dysphagia is characteristic of dementia, pseudobulbar palsy, and cerebral palsy with oral motor incoordination.

Pharyngeal dysphagia may present with coughing or choking during swallowing. Patients may report that water and other thin liquids like coffee or tea present the greatest problem. Patients may try to compensate by tilting or turning their head, or by pressing against one side of their neck while swallowing. Patients may experience accumulation of phlegm in the throat due to pharyngeal stasis. In severe cases, tracheal aspiration and nasal regurgitation will be evident. Impaired phonation and speech may be reported.

The consistency of the foods which produce the dysphagia provide diagnostic clues (8). With neuromuscular dysfunction of the pharynx, liquids frequently cause greater problems with bolus gathering, formation and clearance than more viscous materials. In the esophagus, mechanical obstruction is characterized by greater ease in swallowing liquids than solids, at least early in the disease. Esophageal motility disorders frequently cause almost as many problems with liquids.

Odynophagia

Pain during swallowing, or *odynophagia*, indicates the disruption of mucosal integrity. Subacute or chronic disruption is produced by infections (e.g. candidiasis, cytomegalovirus or herpes infections). Caustic injury may be caused by ingesting acidic or lye products or swallowing batteries. It may also result from the ingestion of corrosive drugs including antibiotics (particularly tetracycline and derivatives), non-steroidal antiinflammatory drugs (aspirin, indomethacin, ibuprofen, etc.), iron, anticholinergics, quinidine and procainamide (9). Malignancies ulcerate as they enlarge, and so may also cause odynophagia (6). Any condition leading to odynophagia may also lead to a continuous pain at rest, with a sore throat and substernal chest pain.

Globus Sensation

Sometimes, the sensation of a chronic fullness, tightness or heaviness in the oropharynx or substernal area is assumed to relate to a problem in swallowing mechanisms

Table 2. Speech and voice

• **Hoarse speech, weak cough**: vocal cord paralysis.
• **Wet, gurgly voice**: overflow of secretions retained in pharyngeal recesses (pharyngeal paralysis)
• **Slurred, mumbly speech**: weakness or incoordination of muscles needed for articulation as well as for swallowing

and structures. If this sensation is centered in the lower neck and not affected by swallowing, it is called globus sensation (10, 11, 12).

Associated Symptoms

Symptoms which coexist with the dysphagia provide information on its origin, symptoms which are absent improve the prognostic outlook. Key issues involve dehydration and weight loss. Weight loss implies a serious swallowing problem. Patients may maintain nutrition by adjusting their diet. Those with strictures may compensate with high caloric, semi-liquid diets (6). Those with poor pharyngeal coordination may restrict the intake of liquids. Dehydration leads then to chronic fatigue, weakness or faintness.

Speech and Phonation

The speech may be slurred or the voice raspy. Hoarseness points to primary laryngeal disease, particularly if speech is otherwise articulate. Table 2 lists conditions in which swallowing is likely to be impaired.

Poor pharyngeal clearance can lead to a wet quality of the voice or to a cough from overflow aspiration after multiple swallows. It can be a sign of cricopharyngeal spasm accompanying esophageal disease, particularly gastroesophageal reflux. In this setting, the onset of the swallowing problem will usually coincide with the onset of problems with voice, speech, and other neurologic deficiencies.

Aspiration

One must determine to what extent the patient is at risk from aspiration (5,8). Recurrent respiratory tract infections should raise the suspicion of previous aspiration. Also, are cough or dyspnea during meals may be caused by aspiration?

The mechanism underlying the aspiration can be deduced from its timing. Aspiration which occurs while the bolus is gathered in the mouth implies a problem with initiating the swallow reflex. Aspiration during swallowing implies poor pharyngeal control. Aspiration as a consequence of reflux and regurgitation is unrelated to swallowing, and particularly common after large meals, during recumbency and sleep. Absence of a cough or other respiratory symptoms does not exclude the possibility of serious aspiration. Silent aspiration (Table 3) occurs with laryngeal anesthesia, chronic debilitation and in dementia

Table 3. The gag and the cough reflex

• Gag highly variable, poor predictor for swallowing, abnormal if asymmetric
• Cough needed for clearance of airway test adequacy of cough at will
• **Problem**: silent aspiration in laryngeal denervation or anesthesia (for instance after intubations)

(8). Conditions which increase the risk for aspiration include depressed consciousness, a flat body position (particularly when combined with an inability to bend the neck) and intubation of the trachea or the esophagus.

Problems with Salivation

A decrease in salivary secretions impairs the ability to masticate and to form a lubricated bolus (13). In extreme cases, patients may, upon chewing a dry cracker for 5 minutes, be found to have the crumbled pieces stuck to their tongue, palate and gums. The mucous membranes are dry even when patients are fully hydrated, and the tongue may become shriveled and fissured. This may be the result of radiation or of an autoimmune process as in Sjögrens or "sicca syndrome". More commonly it is a side effect of drugs, (including diuretics, anticholinergics and many others.)

Gastroesophageal Reflux Disease (GERD)

Common complaints in gastroesophageal reflux disease include regurgitation of sour or bitter fluid, postprandial fullness and belching, food intolerance and heartburn. Heartburn or pyrosis is an unpleasant sensation of a substernal burning which can become painful. It is particularly common after acidic, spicy, sweet or greasy meals. Patients may consume large amounts of antacids, including baking soda or may have been found to have a hiatus hernia, esophagitis or Barrett's esophagus (a replacement of squamous mucosa by columnar epithelium).

Tongue, Larynx, and Other Structures Mediating Swallowing

The tongue collects the bolus, triggers the swallow by contact with the fauces, and drives the bolus into the esophagus in a piston like action (8). With loss of tongue function, the bolus is misdirected, and there are repetitive rolling movements and false starts. As the bolus escapes from the tongue, no swallow is triggered. Parts of the bolus may run over the base of the tongue and penetrate the fully exposed larynx. The delivery of the bolus from the pharynx into the esophagus will be less forceful.

Nasopharyngeal closure depends on the elevation of the soft palate. In its absence, nasal regurgitation will occur and the voice assumes a nasal quality. Airway protection occurs by the anterior and superior movement of the larynx, tilting of the arytenoids, and inversion of the epiglottis. In addition, there is adduction of the vocal cords. Impaired laryngeal elevation and protection is likely to result in aspiration during swallowing. Widening of the esophageal inlet depends on relaxation of the UES and the anterior traction exerted by the hyoid and laryngeal movement. If sphincter compliance is impaired because of muscular fibrosis, there will be insufficient widening for rapid bolus passage. Pharyngeal pressure will raise abnormally during swallowing and aspiration. This may be associated with globus sensation, and may eventually be accompanied by formation of a hypopharyngeal or Zenker's diverticulum (14, 15). Large diverticula will give rise to regurgitation of putrid, undigested food between meals, or upon manipulation of the neck. Pharyngeal peristalsis (a sequential contraction of the superior, middle and inferior pharyngeal constrictors occurring against the background of longitudinal pharyngeal contraction and shortening) leads to eversion of the pharyngeal recesses and their clearance of any residue. Poor pharyngeal clearance leads to accumulation of residue in the valleculae and pyriform sinuses, and leads to the risk of overflow penetration or aspiration (16, 17).

Neuromuscular Diseases

Many inherited and acquired neuromuscular diseases may render individuals susceptible to problems with swallowing (3). Oculopharyngeal dystrophy is particularly common among French-Canadians and leads to ptosis with its characteristic facial expression followed by pharyngeal dysphagia. Myasthenia gravis may cause similar symptoms, but displays fatiguing, with symptoms aggravated by muscle use and exercise. Patients with lower motor neuron disease like amyotrophic lateral sclerosis experience early dysphagia as they have atrophy of their tongue and pharyngeal muscles. The risk of aspiration and malnutrition is high. Patients with Parkinson's disease are prone to oral, pharyngeal and esophageal dysfunction with swallowing. Rigidity of the tongue may interfere with bolus collection and propulsion, and of pharyngeal musculature to inadequate pharyngeal clearance. Esophageal contractions are often of excessive amplitude and duration, causing poor bolus transit and chest pain. A slowly progressive muscular atrophy even of musculature not affected by the acute attack occurs in the wake of poliomyelitis. Palatal, pharyngeal and laryngeal weakness may result in dysphagia and aspiration. Cerebrovascular accidents are a common cause of acute and chronic oropharyngeal dysphagia. Dysphagia may be part of the syndromes of bulbar or pseudobulbar palsy. Stroke victims are more likely to succumb to pneumonia, dehydration, and stroke if they experience dysphagia than if they do not. Cortical strokes are less likely to cause significant and persistent dysphagia than brain stem strokes but aspiration is a recognized complication of cortical strokes even when there is no alteration in consciousness.

Drugs

The drug history (2, 9) should focus on agents which depress the level of consciousness and those agents which may interfere with salivary secretions. These include particularly diuretics and anticholinergic agents. Anticholinergic agents also can aggravate gastroesophageal reflux. Additionally, attention should be paid to agents which may anesthetize the oropharynx, such as dental adhesives and lozenges.

PHYSICAL EXAMINATION

One observes nutritional status, alertness, social independence, and physical mobility. From there, one proceeds to an assessment of the structural integrity of the head, neck and oropharynx, to testing of normal function of the nerves and muscles involved in swallowing reflexes and to a systematic examination of other body parts.

Mental Status and Social Function

Patients with a depressed level of consciousness, from a disease process or from sedative drugs, might not initiate their own oral intake and are at increased risk to of aspiration. Patients with severe muscle weakness or spatial incoordination may be unable to handle their intake, and patients with cognitive deficits because of delirium or dementia may not recognize food for what it is. If these conditions are severe, an assessment of swallowing function and rehabilitation might be irrelevant until functions have been restored. Recovery might occur in conditions which were caused by an accident, acute illness, or drug treatment.

Intubation

Indwelling cannulas interfere with swallowing, increase the risk of aspiration, and interfere with effective clearance of aspirated material (18). By increasing oropharyngeal secretions, they increase potential substrate for aspiration. Tracheal cannulas interfere with the ease of laryngeal ascent, thereby hampering the opening of the UES and the protection of the airway under the base of the tongue. By diverting the air stream, they also interfere with the protective closure of the glottis, which occurs when material other than air flows through it. Effective clearance of airways during coughing depends on the pressure build up in the airways against the closed glottis. The puncture of the airway at the tracheostomy site interferes with the pressure build up and hence with airway clearance. Inflated cuffs can obstruct the clearance of material out of the lung and entrance of material into the esophagus. Even remote, mechanical impediments from tracheostomies may leave local fistulas, scars and deviations, that may cause functional impairment.

Nutritional and Hydration Status

Temporal wasting characterizes generalized malnutrition and muscular atrophy, as may hollow cheeks and other features of emaciation. Sunken eyeballs, dry mucous membranes, shriveling of the tongue, decreased skin turgor, hypotension and elevated temperature may all result from dehydration.

Head and Neck Structures

The Tongue. This organ collects the bolus and propels it through the pharynx into the esophagus (15). Atrophy and fasciculation of the tongue are often pronounced in lower motor neuron diseases (such as ALS), and interfere with oral feeding in disease progression. Upper motor neuron disease may lead to spastic weakness of the tongue and/or its inability to perform its sequential functions (apraxia). Mobility and strength of the tongue are assessed by having patients protrude their tongue or push it into their cheek against the examiner's index finger. Deviation of the tongue from the midline is a feature of unilateral paralysis. Immobility of the tongue may also be caused by infiltrative processes, including tumors of the floor of the mouth and scarring (Table 4).

Examination of the Neck. One inspects and palpates the neck for the presence of scars, tumors, and lymph nodes. Scars may indicate a resection (larynx, strap muscle), which are involved in swallowing, or they may tether these structures and be associated with muscular contractures or atrophy.

Table 4. Examination of head and neck

- **Tongue**: needed for chewing, gathering of bolus, forcing bolus into and through pharynx assess bulk, mobility, force, mucosal integrity
- **Salivation**: lubrication of bolus, oral hygiene decreased by many drugs, sicca syndrome apparent increase with pharyngeal stasis
- **Laryngeal ascent**: save position, inversion of epiglottis and opening of UES
- **Vocal cords**: closure for airway protection, air way clearance by cough

Table 5. Bedside testing of swallowing

• Only if no serious impairment is obvious
• Start with dry swallows, ice chips, advance to semi-liquids, crackers
• Observe for ease of swallowing and of breathing
• Watch out for cough, clearing of throat, wet voice
• Inspect oropharynx for residue
• Palpate laryngeal ascent

Laryngeal ascent protects the larynx and pulls the upper esophageal sphincter (UES) open. To assess laryngeal ascent, the index finger is placed between the hyoid bone and notch of the thyroid cartilage with slight pressure. The larynx should move forcefully a couple of centimeters up on a voluntary swallow. Poor laryngeal ascent may be the result of resection or paralysis of the musculature producing the elevation. Induration of the skin and subcutaneous structures after operations or irradiation also may impair elevation. Particularly in elderly patients, the larynx may be located in an abnormally low position (laryngoptosis) so that it cannot move into optimal position for airway protection during swallowing.

BEDSIDE TESTING OF SWALLOWING

Challenging the patient to take food, to chew and to swallow can provide useful diagnostic information. Clinicians may discover discrepancies between the patient's own observations and theirs, or may press the patient to clarify their complaints with regard to the timing, location or the nature of the problem. Through auscultation and inspection, they may also be able to judge the effectiveness of chewing and salivation, of the movements of the tongue, the pharynx and the esophagus (Table 5).

Contraindications

The preceding history or physical examination might have indicated problems which will make bedside testing unnecessary or unsafe. This includes cues for recent and continuous aspiration, the presence of severe dysarthria, impairment of mental status and respiratory compromise. The findings of food residue or excessive saliva in the recesses of the mouth, are also relative contraindications to the performance of swallowing tests.

Cough is the final protective mechanism against aspiration. The cough reflex may be suppressed by the presence of a tracheostomy. Patients with a tracheostomy may not sense, and thereby not protect the upper airway when laryngeal penetration occurs.

Performance

One observes whether patients are able to recognize food components and to select, prepare, deliver and hold a "mouthful" in their oral cavity. Does the patient initiate appropriate chewing movements, and is speech and respiration arrested on attempts to swallow? Is the swallow effortless, or does the patient try repeatedly and laboriously? Is one swallow enough to clear the mouth and the pharynx, and can the patient take repeatedly swallows in a rapid sequence without fatiguing, coughing, clearing of throat or increased respiratory effort or voice change? Does the patient need to make extraneous movements to initiate a swallow? If one auscultates the side of the neck, is there the characteristic swish and click produced by the epiglottis and the rush through the esophageal inlet, and is the phonation of an "e" as crisp as before the swallow? Also, on auscultation over the epigastrium, does a single major splash occur within 10 s of the laryngeal ascent or is this delayed with several minor arrivals barely enhancing the baseline gurgling of the stomach and intestines? This simple assessment of the effectiveness of primary esophageal peristalsis should be performed with the patient supine so as to eliminate gravity.

Diagnosis and Prognosis

The bedside encounter should lead to a formulation of the likely cause for the dysphagia, its prognosis and its management. The indications for specific tests, for the cessation of oral feeding or for specific therapeutic measures should be elaborated.

If the bedside assessment points to a local mechanical process, diagnostic studies to assess the integrity of the swallowing structures are a primary choice. If one suspects lesions, pharyngoscopy and laryngoscopy are indicated. If it is unclear whether the pharynx or the esophagus are primarily involved, a barium contrast study is helpful. If neuromuscular dysfunction of the oropharynx is suspected, this can be best assessed by a dynamic radiographic contrast study "cookie swallow" (8).

RADIOGRAPHIC EVALUATION OF OROPHARYNGEAL DYSPHAGIA

To understand the radiographic evaluation of oropharyngeal dysphagia, one must understand the basic anatomy of the region (Chap. 3) as well as the mechanics of a normal swallow.

During the oral-preparatory phase, a bolus is formed in the central groove of the tongue. It is formed anteriorly by closure of the lips and apposition of the tip of the tongue and hard palate. Posteriorly it is formed by the base of the tongue, the soft palate, and the posterior oropharyngeal musculature. As the swallow begins, the tip of the tongue is raised, and the posterior oropharyngeal musculature relaxes. As the bolus passes through the posterior oropharynx, the swallow reflex is initiated (see Figure 1A-D).

While the oral-preparatory phase of swallowing is controlled primarily by the voluntary movements of the intrinisic muscles of the tongue, the swallow reflex is an involuntary, highly-organized, combination of sensory and motor events requiring sensory afferent input from touch, taste and pressure receptors supplied by cranial nerves IX and X which are integrated in the nucleus solitarius of the brainstem. The result of this input activates a stylized, coordinated motor neuron output from cranial nerves V, VII, IX, X, XI, XII and the first three cervical nerves. The muscular activity produced, elevates the soft palate to seal the vocal cords.

The larynx is mobile and suspended anterior to the esophagus in the neck by striated muscles which attach to the hyoid and laryngeal cartilage. Contraction of these muscles during the swallow reflex retracts the airway upward and anteriorly under the base of the tongue, momentarily tipping the epiglottis to cover the larynx and pulling open the relaxed cricopharyngeus just in time to allow the bolus to be propelled into the upper esophagus by the peristaltic motion of the superior, middle, and inferior constrictors.

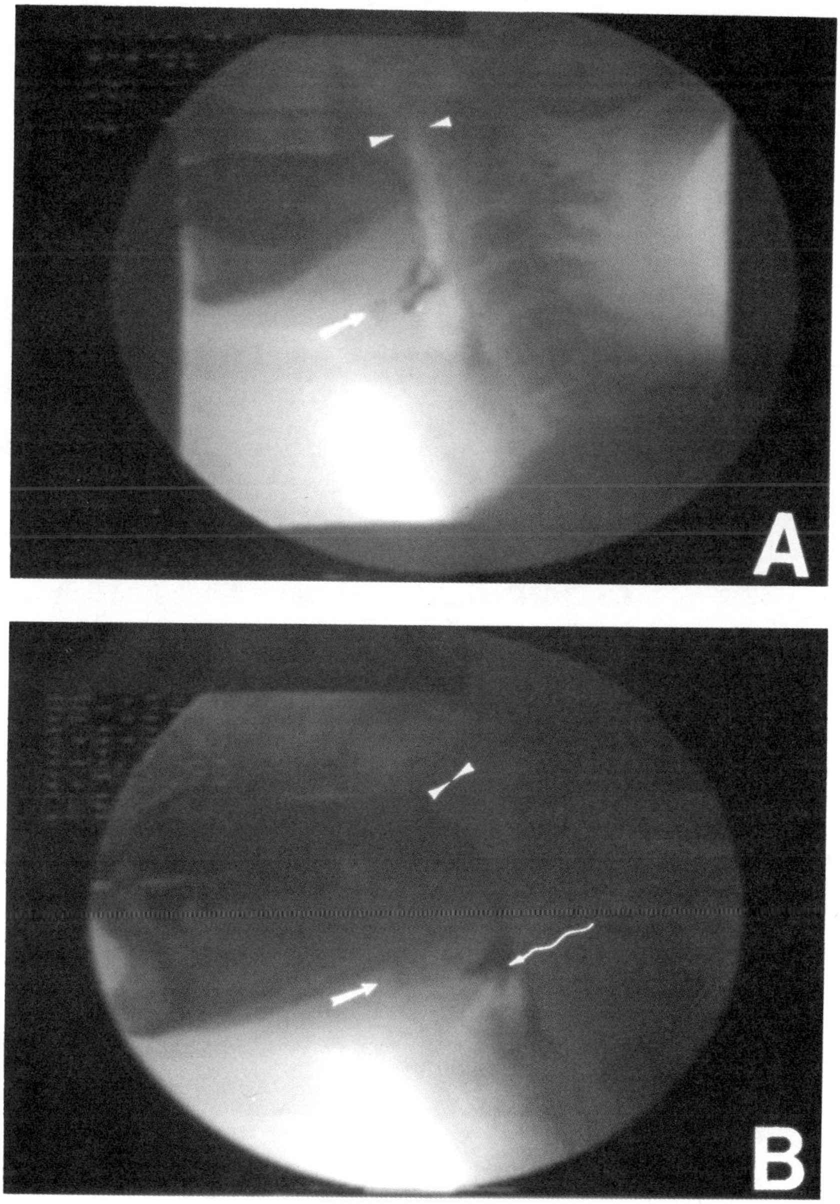

Figure 1. A. The bolus is sealed anteriorly by the lips and posteriorly by the soft palate and the musculature of the posterior oropharynx. Note the hyoid (arrow) is well below the angle of the mandible and that the nasopharynx is patent (small arrowheads). **B.** As the tip of the tongue is raised, the bolus is transported in the central grove of the tongue, through posterior oropharynx into the hypopharynx. This is the beginning of the swallow reflex. Note that the soft palate rises and seals the posterior nasopharynx (small arrowheads). The arrow shows that the hyoid has risen to just below the angle of the mandible. Contrast is passing over the back of the tongue and filling the vallecula. The epiglottis (wavy arrow) is beginning to tip as it is pulled anteriorly and superiorly under the base of the tongue.

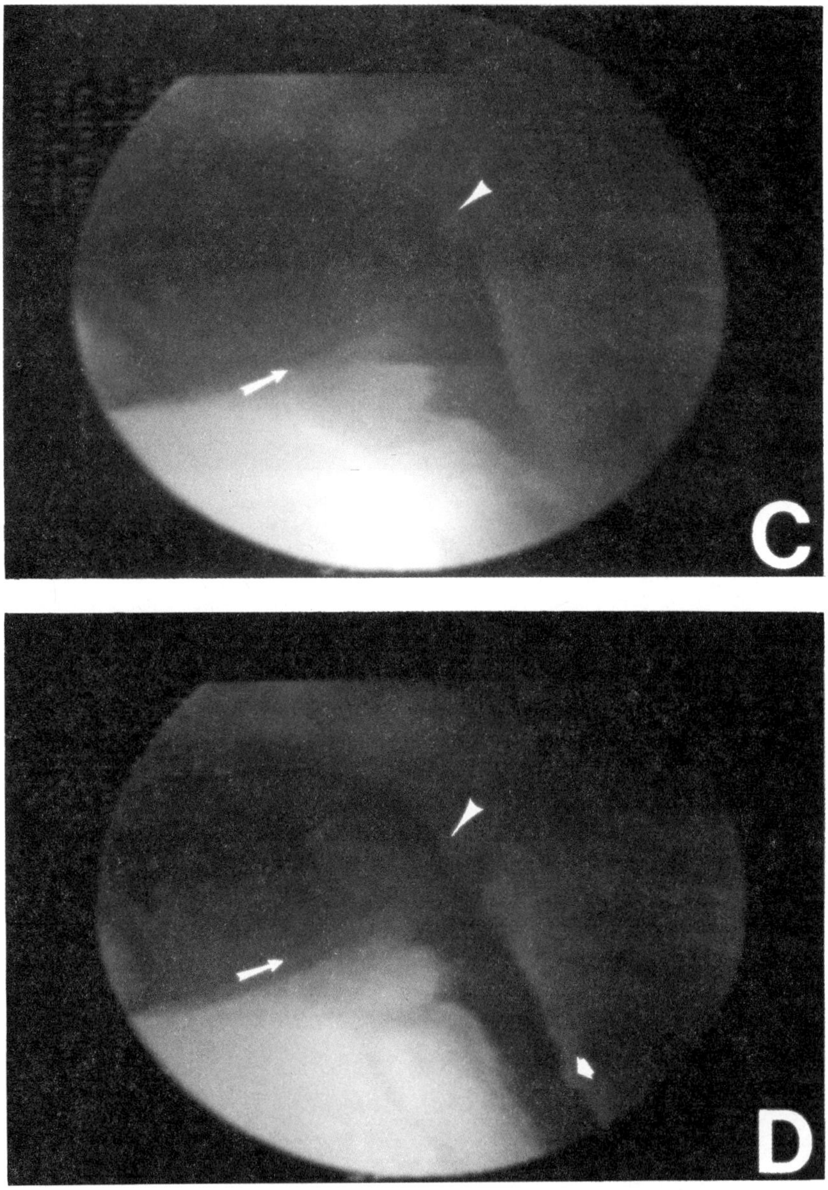

Figure 1 (continued) **C.** The hyoid has reached the level of the angle of the mandible, and contrast courses over the sides of the epiglottis such that we can no longer see it in the barium path. The small arrowhead demonstrates the beginning of the coordinated downward peristaltic movement of the superior constrictor. Contrast has reached a level just above the cricopharyngeus. **D.** The hyoid (arrow) remains at the level of the angle of the mandible. The contraction of the superior constrictor (large arrowhead) courses caudally. The cricopharyngeus (small arrowhead) is open and contrast passes into the proximal esophagus.

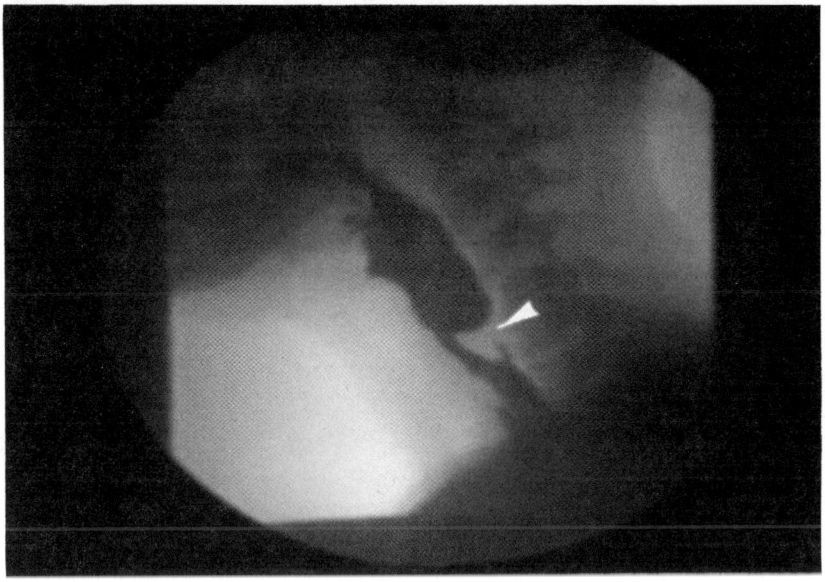

Figure 2.

Video-Radiographic Case Studies in Patients with Oropharyngeal Dysphagia

1. A 47-year-old patient with a long history of gastro-esophageal reflux presents with a feeling of fullness and food sticking in the mid-neck.
A cine-radiography was performed and is shown in Figure 2. What is your diagnosis?

   a. Stricture secondary to GERD.
   b. Esophageal Cancer.
   c. Cricopharyngeal Bar
   d. Plummer-Vinson Syndrome.

2. A 67-year-old white female presents with a long history of being able to eat only small amounts of liquids and no solids. Over the years, she has steadily lost some weight.

Cline-radiographic films are shown in Figure 3, Frame A and Frame B. What is the underlying abnormality?

   a. Cricopharyngeal Achalasia.
   b. Esophageal stricture with penetration.
   c. Pulmonary aspiration.
   d. Cricopharyngeal web with aspiration.
   e. Esophageal cancer.

3. A 67 year-old smoker has had previous partial layngectomy and tracheostomy for squamous cell cancer of the neck. He now presents with progressive fullness in the mid neck.

A cine-radiolography was performed and is shown in Figure 4. What is your diagnosis?

   a. Post-surgical stricture.
   b. Recurrent squamous cell carcinoma.
   c. Bolus impaction.
   d. Spasm of neck muscles.

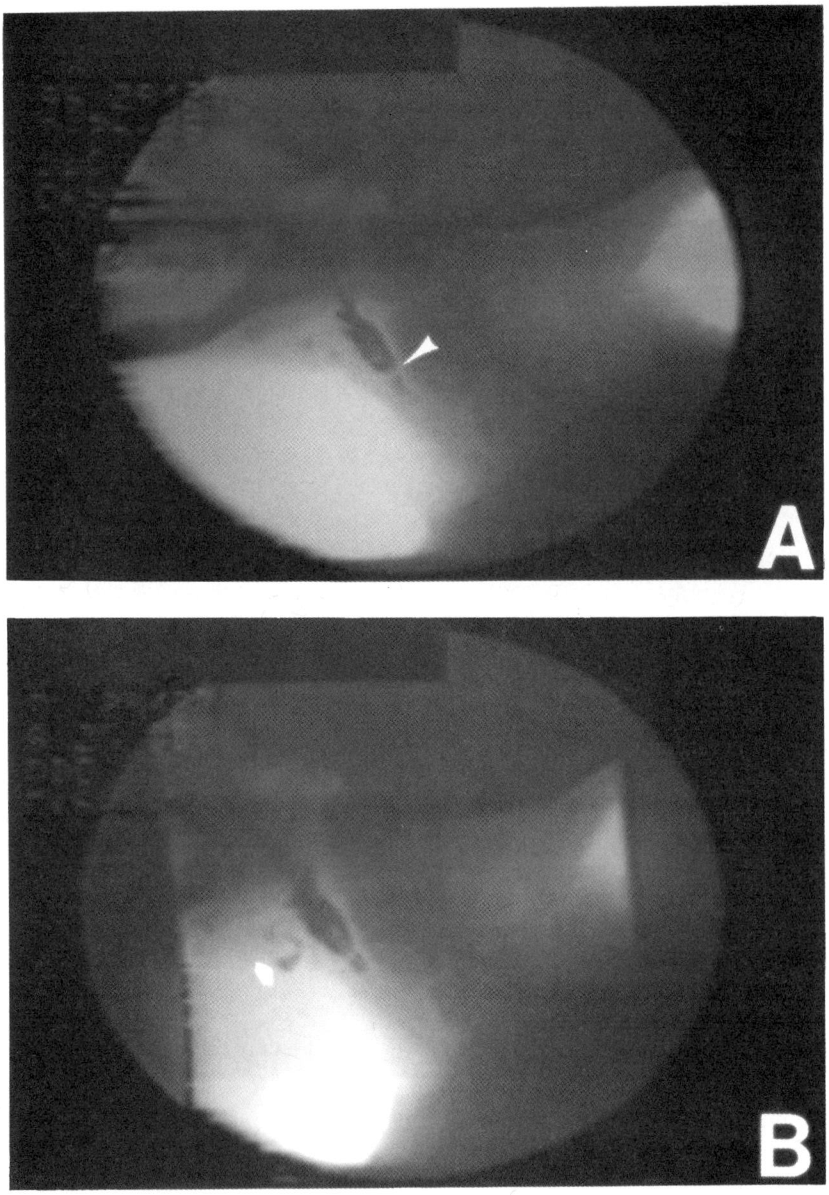

Figure 3.

ANSWER KEY AND EXPLANATIONS

1. Cine-radiology from the lateral position (Figure 2) shows transient protrusion of the cicopharyngeus (arrowhead) into the barium column with delay of liquid contrast distally.

Although the cricopharyngeus protrudes somewhat into the barium column during a normal swallow, in this patient, the protuberance is abnormally prominent and there is delay of contrast passage distally. This is typical of a cricopharyngeal bar. The patient did well with a cricopharyngeal myotomy. The best option is **C**.

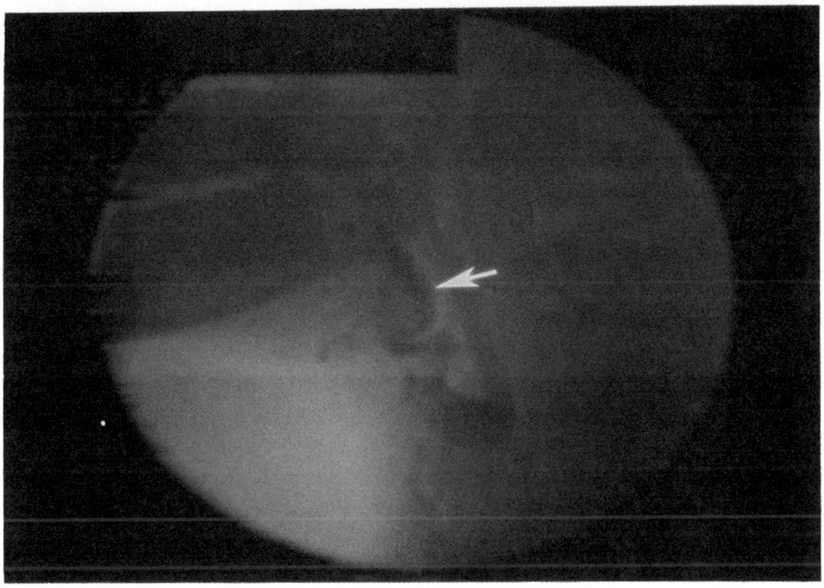

Figure 4.

2. The cine-radiography shows a thin membrane-like stricture (Frame A - arrow head) which allows only small amounts of liquid to pass distally. Because of the extreme narrowing, there is significant retention of fluid proximal to the stricture which on repeated swallows, passes into the airway to the level of the cords (Frame B - arrow). Thus, because of the stricture, she has penetration and is at risk for aspiration. Such high strictures can be seen in patients with skin disorders such as epidermolysis bullosa, and pemphigus vulgaris. The best option is **b**.

3. Cine-radiography shows a lobulated mass at the base of the tongue which narrows the hypopharynx (arrow). This is most likely due to recurrent squamous cell carcinoma. The best option is **b**.

REFERENCES

1. Schulze-Delrieu K and Miller R pp125–152. Clinical Assessment of Dypshagia. In Deglutition and its disorders, ed by A Perlman and K Schulze-Delrieu, Singular 1997
2. Castell DO, Donner MD: Evaluation of dysphagia: a careful history is crucial. Dysphagia, 2:65–71, 1987.
3. Jones B, Donner MW: How I do it: examination of the patient with dysphagia. Dysphagia, 4:162–172, 1989.
4. Buchholz D: Neurologic evaluation of dysphagia. Dysphagia, 1:187–192, 1987.
5. Logemann JA: Evaluation and treatment of swallowing disorders. College Hill Press, 1983.
6. Truszkowksi J and Schulze-Delrieu K. Dysphagia and esophageal obstruction. pp459–464 in Conn's Current Therapy ed by R Rakel, Saunders, 1997
7. Ergun GA and Kahrilas P. Medical and Surgical Treatment interventions in deglutitive dysfunction, in Deglutition and its disorders, ed by A Perlman and K Schulze-Delrieu, Singular, 1997.
8. Perlman A, Lu C, and Jones B. Radiographic contrast examination of the mouth, pharynx and esophagus, in Deglutition and its disorders, ed by A Perlman and K Schulze-Delrieu, Singular, 1997.
9. Christensen J: Effects of drugs on esophageal motility. Arch Intern Med, 136:532–537, 1976.
10. Cook IJ, Dent J, Collins SM: Upper esophageal sphincter tone and reactivity to stress in patients with a history of globus sensation. Dig Dis Sci 34:672–676, 1989.

11. Moser G, Vacariu-Granser G, Schneider C, Thalia-Anthi A, Pokieser P, Stacher-Jonotta G, Gaupmann G, Weber U, Wenzel T, Roden M, Stacher G: High incidence of esophageal motor disorders in consecutive patients with globus sensation. Gastroenterology, 101:1512–1521, 1991.
12. Ravich WJ, Wilson RS, Jones B, Donner MW: Psychogenic dysphagia and globus: Reevalaution of 23 patients. Dysphagia, 4:35–38, 1989.
13. Hughes CV, Baum BJ, Fox PC, Marmary Y, Yeh CK, Sonies BC: Oral-pharyngeal dysphagia: A common sequela of salivary gland dysfunction. Dysphagia, 1:173–177, 1987.
14. Cook J. Cricopharyngeal function and dysfunction. Dysphagia, 8, 244–251, 1993
15. GA Ergun and P Kahrilas. Oropharyngeal dysfunction in the elderly. Practical Gastroenterology, 17, 9–16, 1993
16. Stein M, William AJ, Grossman F, et al.: Cricopharyngeal dysfunction in chronic obstructive pulmonary disease. Chest, 97:347–352, 1990.
17. Olson NR: Laryngopharyngeal manifestations of gastroesophageal reflux disease. Voice Disorders, 24:1201–1213, 1991.
18. Sasaki CT, Suzuki M, Horiuchi M, et al.: The effect of tracheostomy on the laryngeal closure reflex. Laryngoscope, 87:1428–1433, 1977.

11

ESOPHAGEAL DYSPHAGIA

Jeffrey L. Conklin*

Department of Internal Medicine
University of Iowa College of Medicine, and
VA Medical Center
Iowa City, Iowa 52242

INTRODUCTION

The organs involved in swallowing; that is, the pharynx, the esophagus and their associated sphincters, transport from the mouth to the stomach what is swallowed, prevent the reflux of gastric contents into the esophagus, pharynx and respiratory tract, and clear from the esophagus any gastric contents that are refluxed. Peristaltic contractions of the musculature of these organs transport what is swallowed from the mouth to the stomach. The disrupted, uncoordinated or failed peristaltic contractions may disturb the propulsion of a swallowed bolus and lead to the symptoms of *dysphagia* (difficult swallowing) or *odynophagia* (painful swallowing).

CASE STUDY

R.E. is a 35 year old man who works as a postman. He suffered from dysphagia since his middle teen years. He describes the dysphagia as a sensation of something sticking in the lower substernal region. After the food gets stuck, he often has chest pain that goes away with resolution of the dysphagia. He says that for many years his dysphagia could be overcome by drinking a large volume of fluid to wash down what was stuck. Over the last year, his dysphagia became more prominent and drinking fluids was less likely to resolve the problem. He lost 25 pounds of weight in the last 6 months because eating is difficult, and he is awakened at night with coughing and the presence of undigested food in his mouth. He does not have heartburn, nausea, vomiting, hematemesis,

* Phone: (319) 356-2131, Email: jeffrey-conklin@uiowa.edu

Gastrointestinal Motility, edited by Rao *et al.*
Kluwer Academic / Plenum Publishers, New York, 1999.

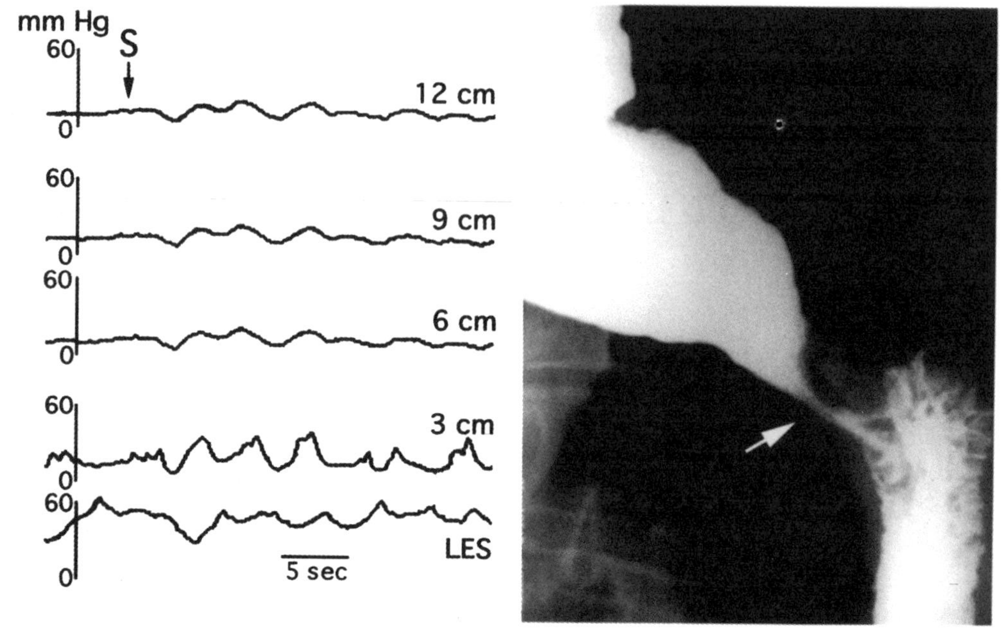

Figure 1. Radiographic and manometric changes in a patient with dysphagia. See case history and answer key for explanations.

melena or hematochezia. His primary care physician obtained a barium swallow to evaluate his dysphagia: it is seen in Figure 1.

1. The most likely diagnosis is:

   A. Esophageal spasm
   B. Peptic esophageal stricture
   C. Achalasia
   D. Scleroderma
   E. Adenocarcinoma of the esophagus

2. What will be the most useful next step?

   A. 24 hour ambulatory pH study
   B. Esophageal manometry
   C. Hurst dilation
   D. Esophageal scintigraphy

3. The esophageal manometry performed on this patient is seen in Figure 1.

   A. It confirms the diagnosis of achalasia
   B. Is consistent with the diagnosis of scleroderma
   C. Documents esophageal spasm
   D. Demonstrates only a nonspecific esophageal motor abnormality

4. Prior to treating this patient which other study is indicated and why?

   A. Esophageal manometry
   B. 24 hour ambulatory pH study

C. Barium swallow with Cine study
D. CAT scan of abdomen
E. Upper endoscopy

STRUCTURE AND FUNCTION OF THE SWALLOWING ORGANS

The pharynx and esophagus are a tubular conduit composed of neuromuscular elements that are functionally integrated to perform the job of swallowing. The funnel-shaped pharynx connects the mouth and nasopharynx to the esophagus and trachea, making it part of the respiratory and digestive systems. Three overlapping sheets of striated muscle, the pharyngeal constrictors, make up the muscular wall of the pharynx. The inferior pharyngeal constrictor, the most caudal of these muscles, thickens to form a distinct band of striated muscle called the cricopharyngeus. The cricopharyngeus is situated at the pharyngo-esophageal junction, and it functions as the upper esophageal sphincter (1, 2).

The esophagus forms a continuous muscular tube that runs from its junction with the pharynx to the stomach. Its musculature consists of inner circular, and outer longitudinal layers that are named according to the axial orientation of their constituent muscle cells. The esophageal musculature is striated down to about the level of the aortic arch (22–24 cm from the incisor teeth). There is a transition zone of several centimeters over which the striated muscle blends with and is progressively replaced by smooth muscle. Smooth muscle makes up the distal half of the esophagus. The circular muscle layer thickens at the esophago-gastric junction to form a muscular ring called the lower esophageal sphincter (LES). The LES also receives muscle fibers from the third, inner muscle layer of the stomach.

When not in use, the pharyngeal musculature generates a weak, tonic contraction that stiffens the pharyngeal wall without occluding its lumen. The upper esophageal sphincter produces a powerful tonic contraction. This contraction, along with the viscoelastic properties of surrounding structures in the neck, occludes the lumen at the pharyngo-esophageal junction. Tonic contraction of the lower esophageal sphincter occludes the lumen at the esophago-gastric junction, making the resting esophagus a tube that is closed at both ends (1, 2).

Swallowing begins as a preparatory phase, which includes mastication of the food and formation of the bolus to be swallowed. The tongue takes on a bowl shape to hold the bolus in the mouth. With initiation of the swallow, the tongue elevates against the palate in a sequential manner from anterior to posterior to squeeze the bolus into the pharynx. The soft palate elevates to make contact with the posterior pharyngeal wall as the bolus is propelled into the pharynx; thereby, sealing the oropharynx from the nasopharynx so that nasopharyngeal reflux does not occur. The larynx elevates and the arytinoids move forward to contact the epiglottis. This closes the laryngeal opening to protect against aspiration. As the swallow reflex is initiated, the pharynx shortens in its long axis. This decreases the distance the bolus must travel and it eliminates the recesses formed by the pyriform sinuses, valleculae and the laryngeal vestibule so that none of the bolus is caught in them. A peristaltic contraction of the pharyngeal musculature sweeps down the pharyngeal lumen to help propel to the esophagus what was swallowed. The upper esophageal sphincter (UES) relaxes as it is approached by the pharyngeal peristaltic contraction and it is actively pulled open as the larynx elevates. The upper esophageal sphincter contracts powerfully as the pharyngeal peristaltic sequence passes, and then rapidly regain its basal tone. The peristaltic contraction advances uninterrupted across the UES, into and along the esophageal body. Long before the peristaltic contraction reaches the esophago-gastric junction, the tonically contracted LES relaxes. Its relaxation persists until the peristaltic contraction strips the bolus into the stomach. The peristaltic contraction closes the lower esophageal sphincter with a transitory, powerful contraction that abates to the tonic contraction of the resting sphincter.

NEUROMUSCULAR CONTROL OF SWALLOWING

The motor behaviors of the swallowing organs comprised of striated muscle; the pharynx, upper esophageal sphincter, and proximal esophageal body, are controlled by somatic nerves that originate in the nuclei of the glossopharyngeal (IX) and vagus (X) nerves. Axons of neurons arising in these nuclei pass via cranial nerves, without synaptic interruption, to terminate in motor end plates on the striated muscle cells. Tone is maintained in these striated muscles by the continuous discharge of their innervation: their relaxation results from inhibition of this tonic neural discharge. The peristaltic contraction that propagates along these organs results from a vigorous discharge of their somatic innervation. Orderly peristaltic contractions are generated because the motor units; i.e., the single motor neurons and striated muscle cells that they supply, are activated in a fixed, craniocaudal sequence (1–3).

The smooth muscle esophagus is controlled by neuromuscular mechanisms that are quite different from those of the striated muscle organs. The dorsal motor nucleus of the vagus gives rise to axons that travel via the vagus nerve to synapse with neurons in the myenteric plexus of the esophagus. Axons from nerve cell bodies in the myenteric plexus supply the terminal motor innervation of the smooth muscle of the esophagus.

In the smooth muscle segment of the esophagus, the neural systems appear to be inactive at rest. The resting contraction of the lower esophageal sphincter results primarily from myogenic tone, but the tone may be modulated by myenteric neurons supplying the muscle. Relaxation of the lower esophageal sphincter is the first swallow-induced event to occur in the smooth muscle esophagus. It results from the activation of inhibitory myenteric nerves that release nitric oxide at the neuromuscular junction (1–3).

Propagation of the peristalsis in the smooth muscle esophagus, unlike peristalsis in the striated muscle esophagus, is not programmed in the CNS: it is controlled by neuromuscular mechanisms that are intrinsic to the esophagus. Myenteric neurons that supply the circular smooth muscle of the esophagus release nitric oxide when stimulated. These nerves are activated nearly simultaneously along the length of the esophagus when a swallow occurs, to produce a nearly contemporaneous inhibition at all levels of the smooth muscle esophagus. The duration of this inhibition gets progressively longer the more distal the site in the esophagus. Excitation that follows the period of inhibition is responsible for the progressive nature of the peristaltic contraction (1–3).

DYSPHAGIA—DIFFICULTY SWALLOWING

Dysphagia is the perception of difficulty with swallowing. It is normal to be cognizant of the oral and most of the pharyngeal phases of swallowing; that is, chewing motions, the formation of a bolus in the mouth, and transport of the bolus into the oropharynx. Beyond this most of us have little perception of the swallowing process. Dysphagia may be only a feeling that something has gone awry after the initiation of swal-

Table 1. Causes of dysphagia

- **Dysphagia from the pharynx and striated muscle esophagus**
- Neuromuscular diseases
  - Diseases of the central nervous system
    - Cerebral vascular accident
    - Multiple sclerosis
    - Amyotrophic lateral sclerosis
    - Parkinson's disease
    - Tumors of the brainstem
  - Diseases of the peripheral nervous system
    - Poliomyelitis
    - Myasthenia gravis
    - Peripheral neuropathies (diabetes mellitus)
  - Myopathic diseases
    - Muscular dystrophies
    - Polymyositis, dermatomyositis
    - Metabolic myopathic processes
      (myxedema ,thyrotoxicosis, steroid-induced myopathy)
- Structural abnormalities
  - Neoplasms (squamous cell carcinoma, lymphoid tumors)
  - Infectious processes
  - Compression form external lesions (enlarged thyroid, lymph nodes, cervical vertebral osteophytes)
  - Esophageal rings or webs (Plummer-Vinson syndrome, congenital webs)
- **Esophageal dysphagia**
  - Mechanical obstruction
    - Esophageal neoplasms (mucosal or intramural tumors)
    - Peptic stricture
    - Foreign bodies
    - Paraesophageal hernia
    - Mediastinal tumors
  - Esophagitis
  - Myopathic diseases as above
  - Peripheral neuropathies (diabetes)
  - Primary motor abnormalities
    - Achalasia
    - Diffuse esophageal spasm

lowing, it may be a fleeting sensation of the food slowing in its passage to the stomach, or at its worst, it may be the perception of food getting stuck. These symptoms occur when either a mechanical obstruction, or weak or disordered peristalsis impairs transport of a swallowed bolus along the pharyngo-esophageal conduit. Dysphagia can be classified as either oropharyngeal or esophageal. This distinction is useful because the neuromuscular mechanisms controlling the function of these regions (see above) and the pathological processes that give rise to disordered swallowing in them are different (Table 1).

EVALUATION OF THE PATIENT WITH DYSPHAGIA

Historical Features—Characteristics of Dysphagia (Figure 2)

Eliciting the characteristics of a patient's dysphagia and the symptoms associated with it is essential for directing the work-up. The patient will usually be able to indicate

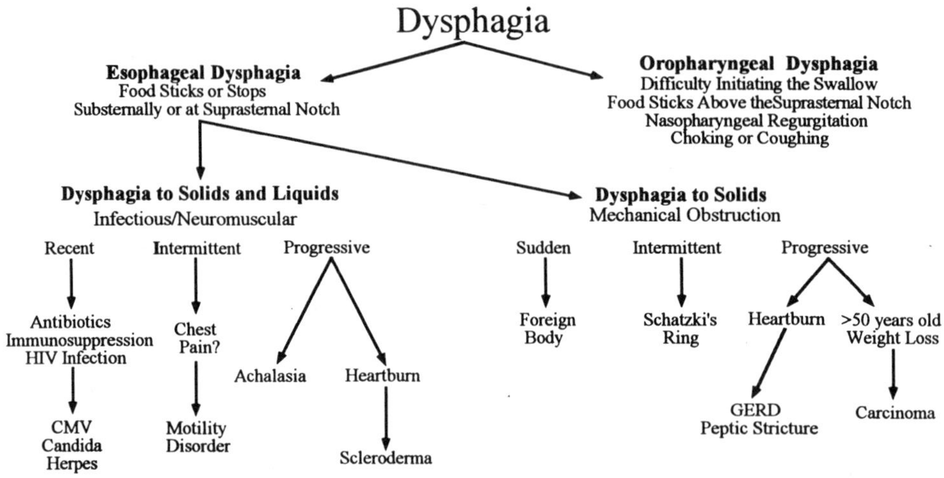

Figure 2. Use of the history to determine the cause of dysphagia.

where the dysphagia is sensed and *when* it occurs in relationship to swallowing. When the patient says that things stick anterior in the throat above the suprasternal notch the dysphagia is usually oropharyngeal in origin. Esophageal dysphagia is most often sensed at the suprasternal notch, substernally or on the anterior chest wall, but the location of the sensation does not predict the level of the lesion. In most cases, the sensation of dysphagia is above the lesion. Dysphagia that occurs promptly with the initiation of swallowing is likely to be oropharyngeal, while dysphagia that follows swallowing by several seconds is characteristic of esophageal dysphagia.

Exacerbating and Relieving Factors

Asking the patient what exacerbates or improves the dysphagia and what provokes it often affords valuable hints as to the origins of the problem. Patients who have more difficulty swallowing liquids than solids are likely to have oropharyngeal dysphagia. They often learn that eating and swallowing carefully or positioning the head in certain ways gives some relief. Patients with mechanical lesions obstructing the lumen of the esophagus complain of dysphagia that is worse with solids than with liquids. The foods that are most bothersome are things that are not chewed well; meats, breads and peanut butter. Patients with mechanical lesions may get relief by "washing it down" with liquids. Dysphagia produced by both solids and liquids is indicative of either a near complete obstruction of the esophagus or disordered esophageal motor function. Dysphagia that is associated with chest pain like angina pectoris, or dysphagia that is brought on or worsened by drinking very cold liquids or eating ice cream is characteristic of esophageal motility problems. These symptoms are suggestive of esophageal spasm.

Chronology of Symptoms

Esophageal dysphagia often starts as only mild, intermittent dysphagia and worsens over time. Intermittent dysphagia that is not progressive or only slowly progressive over a long period of time suggests a benign cause; i.e., a stable stricture or a motility disorder. Long asymptomatic periods punctuated by brief episodes of relatively severe dysphagia suggest the presence of a thin esophageal mucosal ring, a Schatzki's ring. The dysphagia is intermittent in nature because the ring is ruptured by the bolus as it is transiently lodged and passes the ring. Dysphagia that progresses from occurring only with solid boluses to occurring with both solids and liquids suggest an esophageal neoplasm. The sudden onset of dysphagia is uncommon and should trigger the thought of either an infectious cause; or foreign body, like a chicken bone or pill, that can lodge in the esophagus.

Associated Symptoms or Illnesses

Asking the patient about concurrent illnesses or other symptoms associated with their dysphagia often aids in ascertaining the etiology of the dysphagia. Patients with oropharyngeal dysphagia have difficulty initiating the swallowing process, nasopharyngeal reflux of swallowed liquids and choking or coughing with swallowing. They may describe other symptoms of the neuromuscular disease that underlies their swallowing disorder (Table 1).

Dysphagia accompanying heartburn suggests that inflammation, usually chronic gastroesophageal reflux disease (GERD), is responsible for the dysphagia. Chronic esophagitis may precipitate an abnormality of esophageal motor function that perturbs esophageal peristalsis and inhibits esophageal transit, or it may contribute to the genesis of an esophageal stricture. Odynophagia (painful swallowing) suggests a more severe inflammatory process like esophageal ulcers arising from GERD, pills lodging in the esophagus, infections or neoplasms.

Immunocompromised patients; i.e. those being treated with cancer chemotherapeutic agents, transplant patients and those suffering from AIDS, may have dysphagia arising from infectious processes. Cytomegalovirus, Herpes simplex virus or Candida albicans are the most common infectious agents identified in biopsies of esophageal ulcers. Patients with AIDS can also have dysphagia from Kaposi's sarcoma involving the esophagus. Patients with systemic diseases like collagen vascular diseases (especially scleroderma), diabetes mellitus and hypothyroidism may also present with dysphagia. The dysphagia associated with these diseases may arise from primary abnormalities of neuromuscular function or from complications of gastroesophageal reflux disease.

RADIOGRAPHIC EVALUATION OF THE PATIENT WITH DYSPHAGIA

Historically, the first diagnostic test to be undertaken in the patient with dysphagia is a barium x-ray of the swallowing organs. Now, with the widespread use of videofluoroscopy, radiologist can provide insights into a number of pathogenic processes producing dysphagia. Videofluoroscopy is an essential component of the evaluation of oropharyngeal dysphagia. This radiographic technique involves the videotaping of fluoroscopic images of barium boluses as they are formed in the patient's oral cavity and subsequently swallowed. The video images are reviewed in slow motion or frame-by frame to assess carefully the complex muscular actions of the oropharynx during swallowing (see Structure and Function of the Swallowing Organs). Neuromuscular dysfunction of any component of the oropharyngeal swallowing mechanism can produce discrete motor defects: uncoordinated motions of the tongue disrupt formation of the bolus and impair its passage along the pharynx; regurgitation of the bolus into the nasopharynx results from dysfunction of the soft palate; failed or uncoordinated closure of the larynx leads to aspiration; and feeble or uncoordinated contraction of the pharyngeal musculature will allow part of the bolus to be left behind as a residue in the valleculae or pyriform sinuses after the swallow. A crico-pharyngeal bar (a thickened cricopharyngeus muscle) and/or a Zenker's (pharyngo-esophageal) diverticulum, both of which indicate failure of opening of the upper esophageal sphincter, are only identified radiographically.

The barium swallow or video esophagram also help to identify the etiology of esophageal dysphagia. Many endoscopists maintain the bias that endoscopy is superior to barium studies in the evaluation of esophageal dysphagia because endoscopy is required to determine the histopathological identity of esophageal lesions and to treat lesions obstructing the esophageal lumen. In fact, the barium swallow is a very effective way to rule out neoplasms or strictures of the esophagus as the cause of dysphagia. It is superior to endoscopy in identifying upper esophageal webs or Schatzki's rings, and it is a safer way to identify pharyngo-esophageal (Zenker's) diverticula. Barium studies are more likely than endoscopy to identify intramural lesions that do not involve the esophageal mucosa and esophageal compression by extraesophageal lesions like mediastinal tumors or calcified excrescence on cervical vertebra.

The video esophagram may also provide insights into potential motor abnormalities underlying the symptom of dysphagia because it provides a very clear representation of bolus transport, making it possible to determine how effectively the esophagus moves what is swallowed to the stomach. The video esophagram effectively identifies the failure of lower esophageal sphincter relaxation and aperistalsis that characterize achalasia. It also demonstrates the spontaneous, repetitive and uncoordinated esophageal contractions that define esophageal spasm and presbyesophagus. The upper GI x-ray shows failed esophageal peristalsis and bolus escape. It may also provide some evidence for the presence of gastroesophageal reflux.

ENDOSCOPIC EVALUATION OF THE PATIENT WITH DYSPHAGIA (Figure 3)

Although radiographic studies identify many causes of dysphagia, there is no doubt upper GI endoscopy is the best way to determine the etiology of a lesion obstructing the lumen of the esophagus. In many cases, endoscopic techniques can be used to treat obstructing lesions. Endoscopy is particularly useful in the patient suffering from dysphagia and heartburn. When these symptoms co-exist endoscopy should be done without resorting to barium studies for two primary reasons. First, endoscopy with biopsy is a more sensitive way to identify esophagitis: endoscopy is more sensitive than barium studies in identifying esophageal erosions, and biopsies of the esophageal mucosa may demonstrate inflammation that is not apparent on endoscopic or radiographic examination of the mucosa. Second, complications of GERD (peptic strictures, or Barrett's esophagus) can be treated endoscopically. Esophageal strictures can be dilated by distending balloons of various sizes in the stricture or passing dilators over a guide wire placed across the stricture at the time of endoscopy. Barrett's epithelium is now being treated by endoscopic ablation (4). While this approach reconstitutes the squamous mucosa of the distal esophagus, there is some evidence that some specialized epithelium may remain beneath the squamous epithelium. Thus, it is unclear whether ablation therapy alters the natural history of the disease process and decreases the risk of developing esophageal cancer. Biopsies and cytological brushings should always be obtained as part of the endoscopic evaluation of Barrett's esophagus, esophageal ulcers or intraluminal esophageal lesions obstructing the

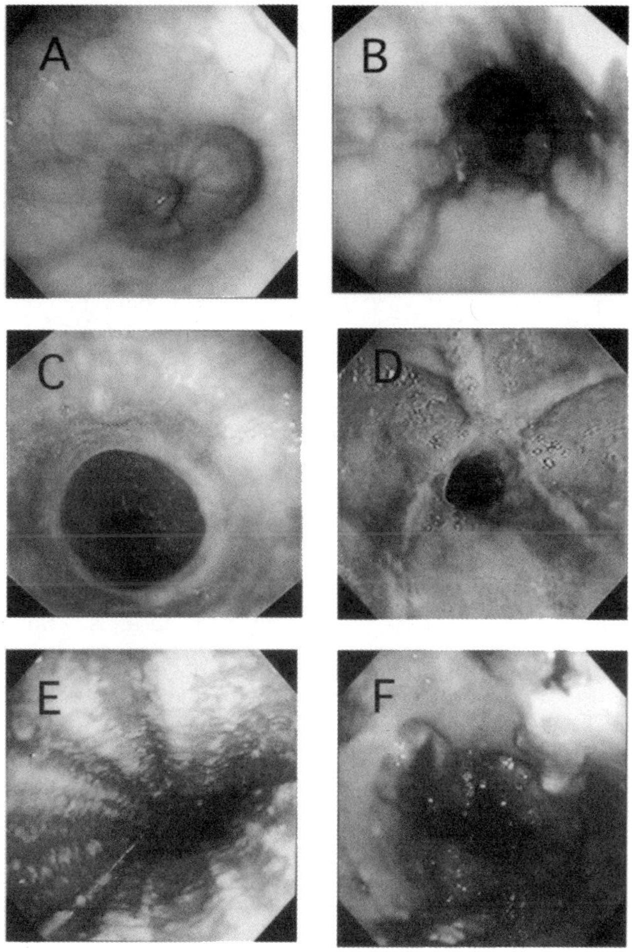

Figure 3. Endoscopic lesions seen in patients with dysphagia: (A) normal esophagus, (B) peptic esophagitis, (C) Schatzki's ring, (D) peptic esophageal stricture, (E) Candida albicans esophagitis, and (F) esophageal cancer.

lumen of the esophagus. Endoscopy should be part of the routine work-up of patients suspected of having achalasia to rule out pseudoachalasia, a tumor of the cardia that mimics achalasia. Endoscopy is warranted for several reasons in immunosuppressed patients: radiographic studies may not identify the subtle mucosal changes induced by infectious agents, and biopsies for histopathological evaluation and cultures are required to target therapies to treat specific infectious agents.

MANOMETRY IN THE EVALUATION OF THE PATIENT WITH DYSPHAGIA

Manometric Techniques (6,7)

The water-perfused catheter system and solid state pressure sensors are used for esophageal manometry. The perfused catheter is the oldest and most widely used method. The catheter consists of a bundle of thin tubes, each of which has a lateral opening. Each opening is at a different level equidistant from its neighbor so that pressure can be measured simultaneously at several sites along the catheter. A low compliance pump slowly perfuses the tubes with bubble-free water, and each tube is attached to a pressure transducer so that changes in pressure can be monitored. Manometry is performed by introducing the tube into the pharynx or esophagus and placing it at selected locations to measure the motor events induced by swallowing. The recorded pressure rises when the flow of water through the lateral opening is occluded by contraction of the esophagus. The amplitude and timing of the pressure changes reflect the force and timing of circular muscle contraction. The other commonly utilized manometric method uses solid state pressure sensors placed at varying lengths along a flexible tube. The timing and force of circular muscle contraction are sensed reliably by both methods.

Manometry in the Evaluation of Oropharyngeal Dysphagia

Manometric techniques in common clinical use are of limited value for evaluating oropharyngeal dysphagia because they cannot record motor events in the pharynx and upper esophageal sphincter with precision. These structures are not radially symmetrical, and the pressure profile generated when they contract is irregular. Thus, the radial orientation of the sensor will greatly influence the recorded pressure: what is seen as a powerful contraction in one axis may appear to be feeble in another. In addition, the recording device stays at a fixed length while the pharynx shortens during swallowing. This is particularly troublesome when trying to evaluate the upper esophageal sphincter because the recording port may slip out of the sphincter as the sphincter is pulled upward. This produces an artifactual drop in pressure. Finally, perfused catheter systems do not respond rapidly enough to adequately follow the motor events produced in the striated muscle regions of the pharynx and UES. Some recent literature, however, suggests that recording intrabolus pressures during the pharyngeal phase of swallowing predict which patients with oropharyngeal dysphagia may respond well to crycopharyngeal myotomy (8). Myotomy seems to work best when the intrabolus pressure is high. Intrabolus pressures can be measured accurately with either type of recording device because this pressure does not depend upon the speed of the recording system or the radial position of the recording port. None-the-less, it is still best to consider manometry of the oropharynx a research tool at this time.

Manometry in the Evaluation of Esophageal Dysphagia

Esophageal manometry should be considered after careful radiographic studies and/or endoscopy fail to identify either a lesion obstructing the esophageal lumen or esophagitis. Esophageal manometry is widely used in hopes of identifying any abnormalities of esophageal motor function that may cause dysphagia. However, prior to embarking upon a manometric work-up of a patient with dysphagia the clinician should understand the relatively low probability of finding a motor abnormality that will clearly explain the symptom of dysphagia (6). Of all patients complaining of dysphagia without other symptoms of esophageal disease and no obstructing lesion, somewhere in the range of 19% will have achalasia and fewer (7%) will have diffuse esophageal spasm (9). If the patient complains of chest pain as a predominant symptom, the yield drops dramatically. Most investigators agree that both of these disorders cause dysphagia and abnormal bolus transit. As many as 27% to 50% of patients with dysphagia and no mechanical obstruction of the esophagus will have what are called nonspecific motor abnormalities and 5% will have what is called nutcracker esophagus (9). In most cases these nonspecific abnormalities do not appear to affect bolus transit in the esophagus, making their relationship to the symptom of dysphagia unclear.

Esophageal Scintigraphy in the Evaluation of Dysphagia

Scintigraphic techniques have been explored as a way to evaluate esophageal motor function. These techniques can tell whether the patient has gastroesophageal reflux and how well the esophagus is able to clear its intraluminal contents. Its clinical applications and clinical value remain undetermined, and it adds little to our understanding of esophageal motor function or dysfunction.

PRIMARY ESOPHAGEAL MOTOR ABNORMALITIES THAT CAUSE DYSPHAGIA

Achalasia

Is the most well defined and best understood of the primary motor abnormalities of the esophagus. It is defined pathophysiogically by a loss of inhibitor myenteric neurons (10), and physiological studies suggest that it results form a loss of nitric oxide producing nerves (11). A barium swallow in a patient with achalasia reveals a dilated, flaccid esophagus and a smooth tapering at the level of the gastroesophageal junction that is said to look like a *bird's beak* (Figure 1). The lower esophageal sphincter fails to relax normally with swallows. Achalasia is characterized manometrically by aperistalsis of the smooth muscle esophagus, and either failed or incomplete relaxation of the lower esophageal sphincter (Figure 1). While lower esophageal sphincter pressure is often abnormally high, this is not a criterion for diagnosing achalasia. Early in the course of the disease, there may be high amplitude, spontaneous and repetitive contractions that mimic esophageal spasm. This is a variant called *vigorous achalasia*. The functions of the upper esophageal sphincter and striated muscle esophagus remain normal in achalasia.

Endoscopy is absolutely necessary when the diagnosis of achalasia is entertained. Pseudo-achalasia is an achalasia-like dysfunction of the esophagus that is produced by tumors of the gastric cardia that do not obstruct the esophageal lumen (12). Tumors of this type are thought to produce achalasia by altering the function of the myenteric plexus; either by infiltration of malignant cells into the plexus or by the production of a substance that alters neuromuscular function. On occasion, lesions that occlude the esophageal lumen at the level of the esophago-gastric junction can mimic achalasia both radiologically and manometrically.

Provocative tests, while rarely needed to make the diagnosis of achalasia can be helpful when the diagnosis is in question. Two pharmacological agents, *Mecholyl* and *CCK*, are of value in this setting (13, 14). Mecholyl, a muscarinic cholinergic agonist, produces vigorous spasm of the esophagus in patients with achalasia. While diagnostic, this response produces significant discomfort for the patient. Cholecystokinin at a dose of 40 ng/kg relaxes the normal lower esophageal sphincter. In about 80% of patients with achalasia, CCK causes a paradoxical contraction of the sphincter.

Diffuse Esophageal Spasm

This is often suspected when simultaneous and repetitive contractions of the smooth muscle esophagus seen on a barium swallow produce what has been termed a *"corkscrew esophagus* (Figure 4)." Although the manometric criteria used to diagnose diffuse esophageal spasm vary depending upon the author, some common features are fairly well agreed

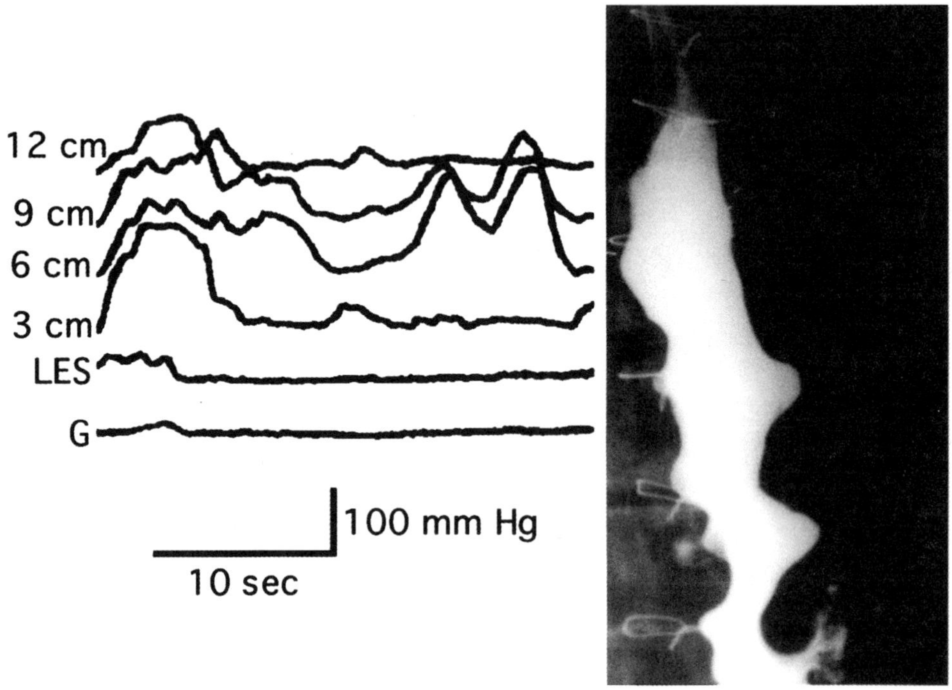

Figure 4. Radiographic and manometric identification of esophageal spasm. On the right is a barium swallow from a patient with diffuse esophageal spasm. Note the irregular corkscrew-like appearance of the esophagus. On the left is an esophageal manometry from a patient with esophageal spasm. G is intragastric pressure, LES is the pressure in the lower esophageal sphincter and the tracings above it are at 3 cm intervals above the sphincter. Note the spontaneous, simultaneous and repetitive contractions.

upon. The frequency of *simultaneous esophageal contractions* following swallowing appears to be the most sensitive way to identify esophageal spasm, but there is debate about what percentage of swallows must produce simultaneous contractions before the diagnosis can be made (14). Other manometric abnormalities seen with spasm include *spontaneous esophageal contractions* occurring in the absence of swallowing and *repetitive esophageal contractions* that occur either after swallows or spontaneously. There usually is normal relaxation of the lower esophageal sphincter. Unfortunately none of these abnormalities is specific for esophageal spasm.

The value of provocative tests in the diagnosis of diffuse esophageal spasm is limited, and these tests are of little use at present. They are of limited value because they can produce spasm-like motor patterns in people who do not symptoms of esophageal disease.

Nutcracker Esophagus

Is characterized manometrically by very high-pressure peristaltic contractions (>200 mm Hg) that propagate normally along the esophagus. The lower esophageal sphincter relaxes normally. Radiographic studies are usually normal in these patients and there is no alteration in bolus transport. As with most other esophageal motor disorders, it has been difficult to correlate this manometric abnormality with symptoms, and provocative test are of limited usefulness.

Nonspecific Esophageal Motor Abnormalities

These disorders encompass all of the as yet undefined abnormalities seen during esophageal manometry. These abnormalities may be seen in up to 40%-50% of patient with dysphagia (9). They may accompany *reflux esophagitis*. The motor changes seen with these disorders, while often characteristic, are not diagnostic, and provocative testing does not help define them.

Esophageal Motility Abnormalities Associated with Systemic Diseases

The majority of patients with *scleroderma* and its attendant skin changes have symptoms of esophageal motor dysfunction (15). Early in the course of the disease, peristaltic contractions of the esophagus become feeble. Later, there will be aperistalsis and hypotension of the lower esophageal sphincter. These changes result from a combination of myenteric neuropathy, and smooth muscle atrophy and fibrosis.

While patients with diabetes mellitus do not frequently report dysphagia, diabetics with a peripheral or autonomic neuropathy often have abnormalities of esophageal motor function. These abnormalities are not specific and may include features of either spasm of the esophagus or hypomotility with feeble peristalsis and hypotension of the lower esophageal sphincter.

A decrease in the amplitudes of peristaltic contractions in the smooth muscle esophagus is seen in a number of other systemic diseases: *amyloidosis, hypothyroidism, alcoholism and idiopathic intestinal pseudo-obstruction*. Such changes in motor function are by no means specific for these diseases and do not usually produce dysphagia.

ANSWER KEY AND EXPLANATIONS

1. On the right is a barium swallow from a patient with achalasia. Note the smooth, tapered narrowing at the esophago-gastric junction (arrow), and the dilated esophagus above it. This is typical for achalasia of the esophagus. The best option is **C.**
2. On the left is an esophageal manometry from a patient with achalasia. The bottom tracing is a recording from the lower esophageal sphincter and the tracings above it are at 3 cm intervals above the sphincter. S indicates the timing of a swallow and the vertical bars indicate luminal pressure. Time is indicated by the horizontal bar. Note that the lower esophageal sphincter does not relax with swallowing and there is no esophageal peristalsis. A 24 hr pH study is unhelpful. Hurst dilatations will not relieve dysphagia in this patient. Scintigraphy may be useful to quantify delayed transit through the esophagus but does not provide any other additional information. Hence it is not indicated. The best option is **B.**
3. In patients with scleroderma there is normal peristalsis in the the striated muscle portion of the esophagus and either weak or absent peristalsis in the smooth muscle portion. In spasm the patient shows simultaneous contractions. Neither of these features are seen here. Nonspecific dysmotility is a poorly defined entity and comprises of one or more manometric abnormalities without precise delineation. The best option is **A.**
4. An upper endoscopy is essential to exclude the possibility of a carcinoma at the cardia as it may mimic both the manometric and radiographic appearances.

Sometimes CAT scan may provide information regarding an extrinsic tumor but this is extremely rare. A pH study may not be diagnostically useful. The best option is **E.**

REFERENCES

1. Conklin, J.L., and Christensen, J. Motor Functions of the Pharynx and Esophagus. In: *Physiology of the Gastrointestinal Tract*, L.R. Johnson (Ed.), Raven Press, New York, NY1994, pp. 903–928.
2. Miller, A., Bieger, M.D., and Conklin, J.L. Functional Controls of Deglutition. In: *Deglutition and its Disorders: Anatomy, Physiology, Clinical Diagnosis and Management*, A. Perlman, K. Schulze-Delrieu (Eds.), Singular Publishing Group, Inc., San Diego, CA, 1996;p. 43–97.
3. Conklin, J.L.: Nitric Oxide: Mediator of Esophageal Motor Function. J. Lab. Clin. Med.,Sampliner, R.E. Ablative therapies for the columnar-lined esophagus. Gastroenterol. Clin. N. Amer. 1997;26: 685–694.
4. Bremner, C.G., Demeester, T.R. Proceedings from an international conference on ablation therapy for Barrett's mucosa. Diseases of the Esophagus 1998;11: 1–27.
5. Kahrilas, P.J., Clouse, R.E., Hogan, W. American Gastroenterological Association Technical Review on the clinical Usefulness of Esophageal Manometry. Gastroenterology 1994;107: 1865–1884.
6. Orlando, R. Esophageal manometry. In: *Manual of Gastrointestinal Procedures*. Drossman, D.A. (Ed.) Raven Press, New York. pp. 36–49.
7. Ali, G.N., Wallace, K.L., Lauundl, T.M., Hunt, D.R., DeCarle, D.J., Cook, I.J. Predictions of outcomes following cricopharyngeal disruption for pharyngeal dysphagia. Dysphagia 1997;12: 133–139.
8. Katz, P.O., Dalton, C.B., Richter, J.E., Wu, W.C., Castell,D.O. Esophageal testing of patients with noncardiac chest pain or dysphagia. Ann. Int. Med. 1987;106: 593–597.
9. Csendes, A., Smok, G., Braghotto, I. et al. Gastroesophageal sphincter pressure and histological changes in distal esophagus in patients with achalasia of the esophagus. Dig. Dis. Sci. 1985;30: 941– .
10. Murray, J., Ledlow, A., Launspach, J., Evans, D., Loveday, M., and Conklin, J.L.: The Effects of Recombinant Human Hemoglobin on Esophageal Motor Function in Humans. Gastroenterology 1995;109:1241–1248.
11. Kahrilas, P.J., Kishle, S.M., Helm, K.F. Comparison of pseudoachalasia and achalasia. Am. J. Med1987;82:439–.
12. Heitmann, P., Espinoza, S.P., Csendes, A. Physiology of the distal esophagus in achalasia. Scand. J. Gastroenterol1969.; 4: 1–11.
13. Dodds, W.J., Dent, J.,Hogan, W.J. et al. Paradoxical lower esophageal sphincter contraction induced by cholecystokinin -octapeptide in patients with achalasia. Gastroenterology 1981; 80: 327–333.
14. Richter, J.E., Castell,D.. Diffuse esophageal spasm: A reappraisal. Ann. Int Med. 1984;100:242–245.
15. Rodnan, G.P., Medsger, T.A. Jr., Buckingham, R.B. Progressive systemic sclerosis-CREST syndrome. Observations on natural history and late complications in 90 patients. Arthritis Rheum. 1975;18:423.

12

TYPICAL AND ATYPICAL MANIFESTATIONS OF GASTROESOPHAGEAL REFLUX DISEASE

Donald O. Castell*

Kimbel Professor and Chairman
Department of Medicine
Allegheny University Hospitals, Graduate
Philadelphia, Pennsylvania

INTRODUCTION

Gastroesophageal reflux disease (GERD) usually presents with heartburn or regurgitation. These symptoms occur most commonly within two hours after eating or while in the recumbent position. In addition to this typical presentation, many so-called "atypical" forms of GERD also occur. This includes angina-like chest pain, asthma, chronic cough and chronic hoarseness of voice. Although some regard symptoms of GERD as mere an annoyance, in many this may have an important effect on the quality of life (1). Furthermore, up to 20% of patients have more serious complications such as ulcerative esophagitis, peptic stricture, or epithelial metaplasia (Barrett's esophagus) with its premalignant potential (2). In most individuals, particularly those who present to a gastroenterologist, GERD is often a chronic relapsing condition. Therefore, long-term treatment is the essence of managing this condition.

CASE STUDY

A 42-year old automobile mechanic is referred for evaluation of GERD. He reports a 4-year history of retrosternal burning sensation both at work and post-prandially. Treatment with ranitidine, 300 mg bid and cisapride, 10 mg at bed time, has improved symptoms. However, he has frequent relapses and the referring physician is concerned about

* Phone: 215/893-7128, Fax: 215/893-2472

Gastrointestinal Motility, edited by Rao *et al.*
Kluwer Academic / Plenum Publishers, New York, 1999.

the long term management. Physical examination reveals a moderately obese male with otherwise normal physical findings.

1. What would be your initial approach to this patient's problem?

   A. Discuss lifestyle modifications (Phase 1) therapy.
   B. Perform a Bernstein test.
   C. Perform endoscopy.
   D. Perform intraesophageal pH monitoring study.
   E. Order a GE scintiscan.

2. Endoscopic evaluation reveals linear erosions extending 6 cm above the GE junction. What is your next recommendation?

   A. Obtain intraesophageal pH monitoring study.
   B. Increase the dose of ranitidine.
   C. Cisapride 10–20 mg qid.
   D. Combination of ranitidine and cisapride.
   E. Begin treatment with a proton pump inhibitor.

3. On omeprazole, 20 mg/day the patient's symptoms recurred within three months. What is the preferred approach?

   A. Obtain intraesophageal and intragastric pH monitoring study while on therapy.
   B. Obtain a surgical consult.
   C. Increase omeprazole dose to 20 mg BID.
   D. Combine omeprazole and cisapride.
   E. Obtain esophageal manometry.

Prevalence of GERD

Epidemiologic studies have shown that daily heartburn occurs in about 10% of the US adult population and that up to 40% have occasional heartburn (3). These percentages only relate to the more typical manifestations of heartburn and regurgitation. Thus, the addition of patients with the "atypical" manifestations would make the estimated frequency of GERD even higher. The recent commercial attention to OTC medications for the treatment of heartburn and the greater than one billion dollars in sales of these products annually in the United States underscore the prevalence of this condition.

One of the important questions is whether Gastroesophageal reflux disease (GERD) is primarily a motility disorder. What is the evidence?

PATHOPHYSIOLOGY OF GERD

Lower Esophageal Sphincter (LES) Incompetence

An early report by Cohen and Harris compared radiographic evidence of hiatal hernias with lower esophageal sphincter pressures in patients with or without GERD symptoms and came up with the concept that the critical factor in anti-reflux competence was a resting LES pressure greater than 10 mmHg (4). Although this concept was attractive, subsequent studies have failed to confirm this observation. In one such study, Kahrilas and colleagues compared resting LES pressures in healthy volunteers with that of reflux patients without esophagitis, reflux patients with mild esophagitis, and reflux patients with severe esophagitis (5). Most individuals, whether healthy or with severe esophagitis, had a resting LES pressure in the normal range. Even in the severe esophagitis group, only an occasional patient had an LES pressure less than 10 mmHg. In fact, there were four patients with hypertensive LES pressures in the reflux groups. When originally published, these findings were somewhat surprising. In the 1990's we know that the explanation lies in the elegant studies performed by Dodds and colleagues on transient relaxations of the sphincter (6). These intermittent spontaneous relaxations, originally called "inappropriate," have become universally termed transient relaxations and are now generally considered to be the major mechanism by which most episodes of gastroesophageal reflux occur. We now know that excessive gastric distention pressure of fat in the stomach, and posture, particularly lying on the right side with the gastric pool against the cardia, will all trigger more of these relaxations and the associated reflux. Thus, in addition to those (few) patients with weak sphincter, most individuals with reflux disease have a normal sphincter but reflux because of intermittent relaxations. This concept better defines the motility disorder associated with reflux. In addition, it helps to focus pharmacologic development on seeking drugs that might interfere with these frequent relaxations rather than to alter the resting LES tone.

Esophageal Clearance

But, there is an even more interesting aspect of manometry in patients with reflux. Kahrilas and colleagues commented on this in their paper a decade ago (5). When evaluating the peristaltic sweep in response to water swallows, frequently low amplitude peristaltic contractions or normal amplitude contractions that are not transmitted all the way down the distal esophagus were observed. In another study of simultaneous videofluoroscopy and manometry Kahrilas and colleagues showed that when peristaltic pressures were below 35 mmHg in the distal esophagus, barium would escape proximally (7). They called these "ineffective contractions."

We have followed up on that concept manometrically and now use the term ineffective esophageal motility (IEM). In our laboratory, if 30% of the standard 10 wet swallows show amplitudes in the distal esophagus of less than 30 mmHg, we define these pressure waves as IEM (8).The rationale is in the study that we did ten years ago with 95 normal volunteers (9), none of the normals had more than two, (20%) low amplitude or nontransmitted peristalsis. In addition, the overall peristaltic contraction amplitude is low in patients classified as IEM. Comparing the amplitude of the effective contractions, that is those of greater than 30 mmHg in the distal esophagus, versus the number of ineffective swallows there is an excellent inverse relationship (r=0.92)

If we take this a step further and look at pH data in patients with IEM they are characterized by having median recumbent acid exposure time of 4.5%. The only group of patients that compared with them were those with scleroderma esophagus, and they were the same (Figure 2). Thus, the IEM group had reflux exposure at night that was similar to scleroderma. When we looked at recumbent acid clearing patterns, again it was only the scleroderma and the IEM patients that were very similar, i.e. they exhibited markedly prolonged acid clearance (Figure 3). We believe that IEM is an important dysmotility in the distal body of the esophagus in patients with reflux. We have since analyzed data from 100 consecutive patients who had manometry and abnormal pH. We found that the traditional hypotensive LES (<10 mmHg) was only present in 4% and ineffective esophageal motility was found in 34%. This observation suggests that IEM is a more frequent manometric marker of reflux than low resting LES pressure.

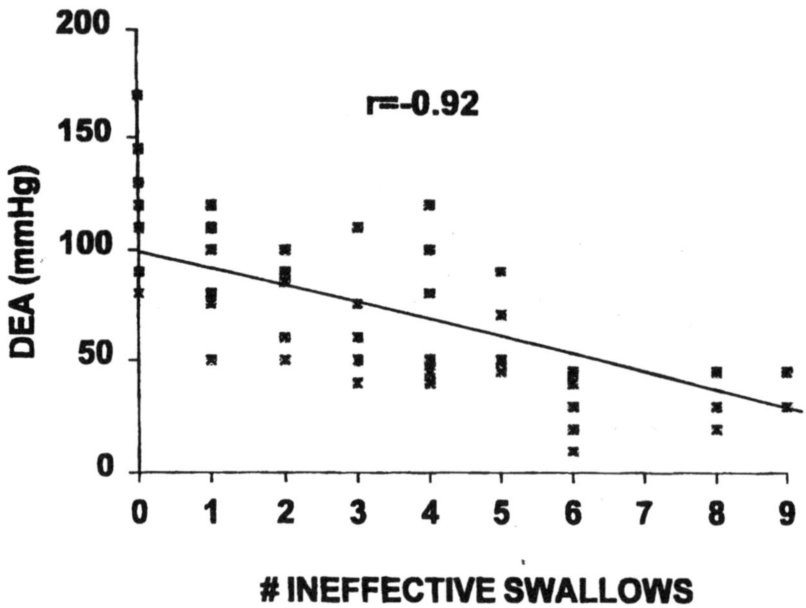

Figure 1. Relationship between the distal esophageal peristaltic amplitudes (DEA) of the effective contractions to the number of ineffective contractions from a total of 10 swallows for 45 patients. Each point represents the respective data from one individual patient in this study.

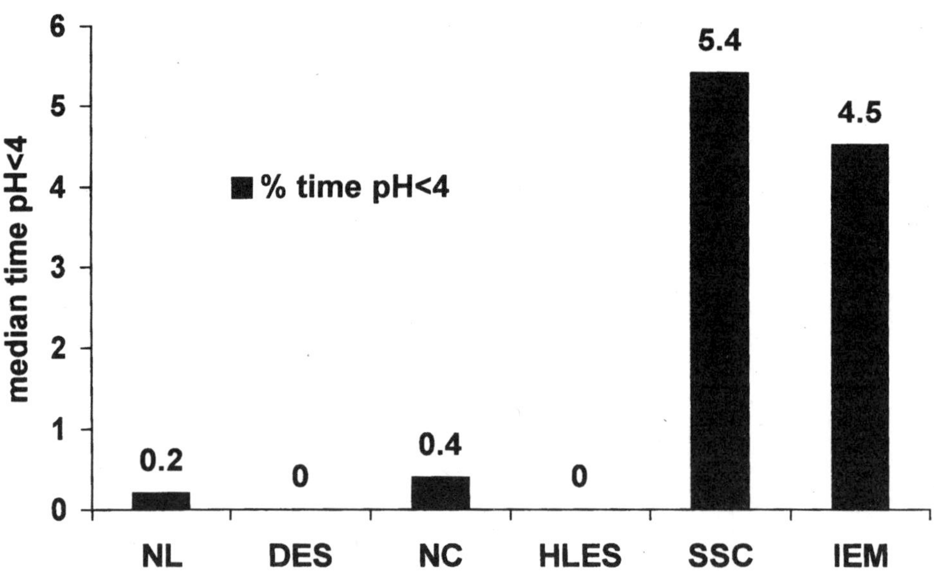

Figure 2. Median recumbent distal esophageal acid exposure, expressed as a percentage of total recumbent period for all patient groups. NL = normal motility, DES = diffuse esophageal spasm, NUT = nutcracker esophagus, HLES = hypertensive LES, SSc = scleroderma, IEM = ineffective esophageal motility. % time pH < 4 is significantly greater in patients with IEM compared to all other groups except SSc ($p<0.001$).

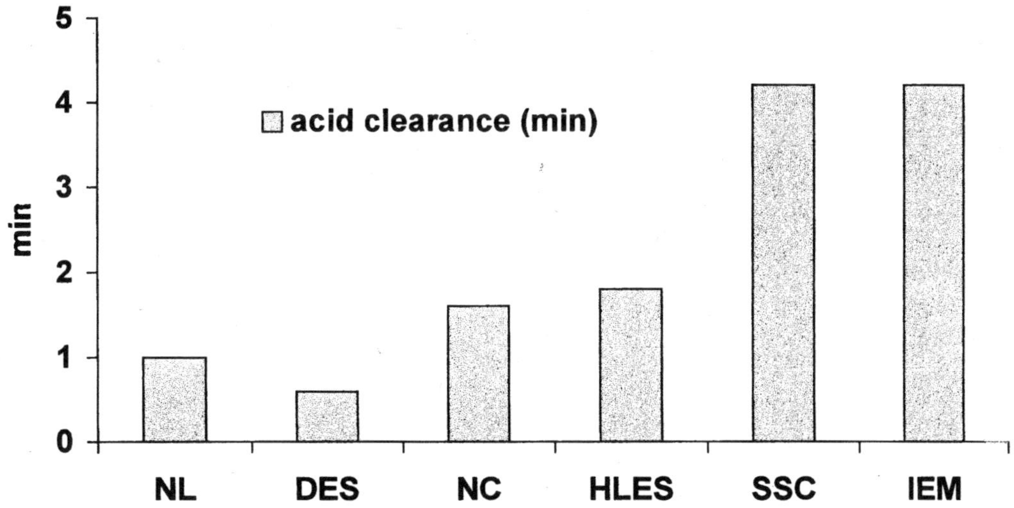

Figure 3. Median acid clearance, measured in minutes per episode, for all patient groups, NL = normal motility, DES = diffuse esophageal spasm, NUT = nutcracker esophagus, HLES = hypertensive LES, SSc = scleroderma, IEM = ineffective esophageal motility. Patients with IEM have slower nocturnal acid clearance than any group except SC ($p<0.01$).

PATHOPHYSIOLOGY OF ATYPICAL MANIFESTATIONS OF GERD

Most patients with reflux have the typical symptoms of heartburn and regurgitation. But, what about the so-called atypical symptoms including chest pain, hoarseness, asthma, and chronic cough? Are they truly manifestations of GERD, how often do they occur, and how do we establish the diagnosis? The asthma/cough relationship to GERD is really an interesting one, in which there are two proposed mechanisms (10). Certainly, if the patient aspirates acidic gastric contents into the airway it can produce an intense reaction. It does not take very much, possibly as little as 0.1 ml to produce bronchial spasm. It is also proposed that exposing the distal esophagus to acidic gastric contents can cause bronchiospasm through a neural reflex (11). So there are probably at least two mechanisms by which reflux may produce pulmonary symptoms. It is unclear whether both mechanisms occur in patients with laryngeal abnormalities. To study this, we modified our standard pH probe so that there were two sensors, one placed 5 cm above the LES, and a second electrode was placed 20 cm above the LES. We adopted this so that we could at least standardize the distance in each patient. We, then, established normal values for acid exposure as being less than 1.3% at the 20 cm site in the upright position and zero in the recumbent position (12). In patients with pulmonary symptoms, the pH monitoring study may identify proximal reflux followed by a cough. Additional data from our laboratory on dysmotility in these patients reveals a high prevalence of IEM in patients with abnormal reflux by pH-metry and asthma or laryngitis compared to GERD patient controls with heartburn only (13). Those with IEM also had prolonged acid clearance recumbent acid clearance, possibly contributing greater proximal esophageal acid exposure.

Sonnenberg, et al have attempted to determine if there really is an association between GERD and pulmonary or laryngeal symptoms (14). They used the VA data system and the CD codes to compare diagnoses in 100,000 patients with reflux, esophagitis, or stricture over a 14-year period. An identical number of controls without a diagnosis of esophagitis were used for comparison. Calculation of odds ratios for associated pulmonary and laryngeal symptoms revealed that asthma, laryngitis, pharyngitis, and aphonia had high odds ratios. Their conclusion was that in a patient with otherwise unexplained pulmonary or laryngeal symptoms, one should suspect GERD.

We found abnormal pH metry in 78% of patients who were referred to us with pulmonary symptoms and possible reflux (15). In addition, we found evidence of abnormal proximal reflux in 2/3rds of patients. We reviewed their records and found that most of them had improvement in their pulmonary symptoms when they were treated for reflux. Also abnormal proximal reflux appeared to predict those whose pulmonary symptoms may improve with therapy.

A similar conclusion was made by Harding, et al when they studied the effect of omeprazole on thirty patients with asthma and reflux that was documented by pH monitoring (16). They gave omeprazole starting with 20 mg and then each month titrated the dose with esophageal pH-metry to ascertain if a larger dose was needed to control reflux. Six patients required 40 mg and two needed 60 mg/day. In 73% of patients, asthma improved after treatment with omeprazole. Similar to our study, abnormal proximal reflux was shown to be a good predictor of response.

CLINICAL APPROACH TO ATYPICAL REFLUX

Figure 4 suggests an algorithm for patients with atypical GERD symptoms. Symptoms should initially be evaluated by the appropriate specialist; i.e. if the symptom is hoarseness an ENT physician should be consulted. If it is cough or asthma, a pulmonologist or allergist should be consulted. If they find no explanation for these symptoms, or if reflux is suspected, a trial of therapy is acceptable, although a pH study is preferred to define that there is actual abnormal reflux. Alternatively, if a trial of treatment fails, a pH study, off-therapy is important. If this is positive, then more aggressive anti-reflux therapy is appropriate. After a diagnostic pH study, if aggressive therapy fails, it is important to repeat pH metry on treatment to ensure that there is good acid control. If so, reflux is probably the wrong diagnosis. If not, advancing therapy is indicated.

TREATMENT MODALITIES FOR REFLUX

A brief review of the important concepts about the treatment of reflux is appropriate. What is the role of life-style modifications in 1999? I believe that they are important based on the concept that GERD is a chronic disease, much as recommending life-style changes to a patient with hypertension or diabetes. Sleeping with the head elevated should definitely be recommended. The diet modifications that are important are to decrease the volume and fat content because both of those will trigger transient LES relaxations and to instruct the patient to avoid recumbency for three hours after eating. Stopping smoking and use of OTC medications as needed are part of the patient's instructions.

When prescribing other medical therapies, remember that GERD is a chronic disease. Therefore, long-term treatment is essential. The excellent multicenter study by Vignieri et al, compared essentially all of the medical therapies that are used in the treatment of reflux (17). They included 175 patients with esophagitis and treated them all acutely with omeprazole until healing. Then the patients were randomized into five different treatment arms, 35 patients each. The percent that remained healed after a year in each

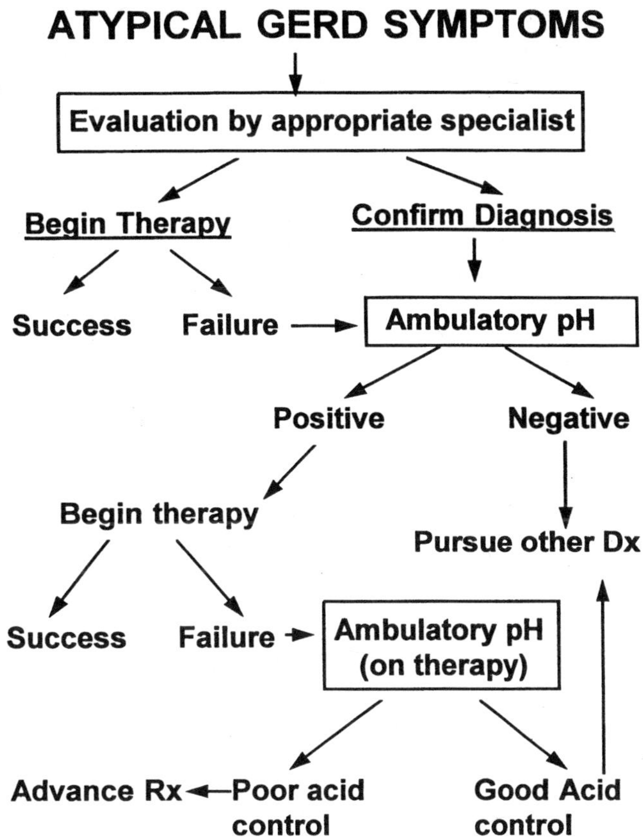

Figure 4. Algorithm for a patient presenting with atypical symptoms of gastroesophageal reflux disease (GERD). Based on the presenting symptom, the patient should be evaluated by the appropriate specialist. Specific therapies are not included in this scheme, although the preferred initial therapy would be a proton pump inhibitor given twice daily before meals.

treatment arm was as follows: ranitidine alone, 50%; cisapride alone, 54% (not significantly different); combination of ranitidine and cisapride, 66% (significantly greater than ranitidine alone); omeprazole alone, 80% (significantly better than the other three treatments); combination of cisapride with omeprazole resulted in a value (89%) numerically greater than omeprazole alone. These data summarize nicely what we see in practice when managing patients with long-term reflux disease. There is a role for each of the aforementioned therapies; but a precise treatment or a combination depends on the patient and on what they require for symptom control. The authors also make the point that patients with less severe, non-erosive disease are more likely to respond to an H_2 receptor-blocker or cisapride than those with erosive or complicated GERD who are more likely to require a proton pump inhibitor or anti-reflux surgery.

What about the comparison of medicine and surgery? The two year follow-up study from 10 VA hospitals by Spechler, et al was a randomized comparison of medical therapy versus anti-reflux surgery (18). Combination medical therapy with ranitidine and metoclopramide was less effective than surgery over the two year follow-up. However, new data from a three year follow-up comparing fundoplication with omeprazole has found identical remission rates from both approaches (19).

Long Term Management

The remarkable long-term follow-up by Klinkenberg, et al on five years of continuous treatment with omeprazole shows that approximately 50% of patients with erosive esophagitis remain healed and under symptom control with 20 mg of omeprazole/day, another 40% required 40 mg/day and the final 10% required 60 m/day (20). All of these patients, that is 100%, maintained healing for five years by adjusting the dose based on severity of symptoms. A very recent abstract reports similar results with 11 years of continuous omeprazole therapy (21). These observations suggest that today, a failure of medical treatment should be very unusual.

CONCLUSIONS

The following is a suggested philosophy for the long term management of GERD. I believe that it is possible to eliminate all acid reflux. If symptoms continue, it is more likely that they are not due to reflux. For the patient with definite GERD-related symptoms which is controlled by medical therapy, it is useful to advise them that continued medical therapy or surgery are both reasonable options for the long term management. In 1999, the indications for surgery should be poor compliance, patient choice, or price of medications - not medical failures because this is most unlikely. I conclude with the following therapeutic principle; The thoughtful gastroenterologist, and hopefully the thoughtful surgeon, does not recommend antireflux surgery until they are certain that the symptoms are due to GERD; i.e. they respond to appropriate medical therapy. Surgery is not a good diagnostic test for GERD! It is likely to result in both an unhappy patient and an unhappy surgeon.

ANSWER KEY AND EXPLANATIONS

1. Because of refractory symptoms in a 40 year old man it is important to perform an endoscopy now to document the severity of reflux, the possible presence of a Barret's esophagus, or to exclude other coexisting or sinister pathology. pH testing may also be appropriate but only in cases where endoscopy is negative. Gastric emptying study is inappropriate at this stage. The best option is **C.**
2. When there is endoscopic evidence for GERD, there is little immediate need for pH testing. pH may be useful in patients not responding to PPI therapy or to examine drug efficacy or compliance with treatment. A higher dose ranitidine may be useful but is not cost-effective, similarly cisapride may not be very effective in this situation. Combination of ranitidine and cisapride may be superior to either drug but the best choice is PPI therapy. The best option is **E.**
3. All of these options may be considered and the correct answer depends on the individual situation. The best response would be to double the dose of PPI therapy and if there is no response to perform pH study on treatment. But the patient has had symptoms for many years has refractory erosive disease and may be reluctant to take pills life-long. If so surgery would be appropriate. Esophageal manometry would be required prior to pH study and prior to surgery particularly to identify achalasia or a manometric dysfunction that may contraindicate surgery. The best option is **C**.

REFERENCES

1. Mathias SD, Castell DO, Elkin EP, Matosian ML. Health-related quality of life of patients with acute erosive reflux esophagitis. Dig Dis Sci 1996;41:2123–2129.

2. Nebel OT, Fornes MF, Castell DO: Symptomatic gastroesophageal reflux: Incidence and precipitation factors. Am J Dig Dis 1976;21:953–956.
3. Spechler SJ: Complications of gastroesophageal reflux disease. In: The Esophagus, Second Edition. Castell DO, editor. Little, Brown, and Company. Boston, MA. pp. 533–546.
4. Cohen S, Harris LD: Does hiatus hernia affect competence of the gastroesophageal sphincter? N Engl J Med 1971; 289:1053.
5. Kahrilas P, Dodds W, Hogan W, Kern M, Arndorfer R, Reece A: Esophageal peristaltic dysfunction in peptic esophagitis. Gastroenterology 1986; 91:897–904.
6. Dodds WJ, et al: Mechanisms of gastroesophageal reflux in patients with reflux esophagitis. N Engl J Med 1982; 307:1547–1552.
7. Kahrilas PJ, Dodds WJ, Hogan WJ. Effect of peristaltic dysfunction on esophageal volume clearance. Gastroenterology 1986; 94:73–80.
8. Leite LP, Johnston BT, Barrett J, Castell JA, Castell DO: Ineffective esophageal motility (IEM): The primary finding in patients with non-specific esophageal motility disorder. Dig Dis Sci 1997; 42:1853–1858.
9. Richter JE, Wu WC, Johns DN, Blackwell JN, Nelson JL, Castell JA, Castell DO: Esophageal manometry in 95 healthy adult volunteers: variability of pressures with age and frequency of "abnormal" contractions. Dig Dis Sci 1987; 32:583–592.
10. Barish CF, Wu WC, Castell DO: Respiratory complications of gastroesophageal reflux. Arch Intern Med 1985;145:1882–1888.
11. Tuchman DN, Boyle JT, Pack AI, et al. Comparison of airway responses following tracheal or esophageal acidification in the cat. Gastroenterology 1984;87:872–81.
12. Dobhan R, Castell DO. Normal and abnormal proximal esophageal acid exposure: Results of ambulatory pressure monitoring. Dig Dis Sci 1993; 38:225–232.
13. El-Serag HB, Sonnenberg A: Comorbid occurrence of laryngeal or pulmonary disease with esophagitis in United States military veterans. Gastroenterology 1997; 113:755–60.
14. Fouad Y, Khoury R, Hatlebakk J, Katz PO, Castell DO: Ineffective esophageal motility (IEM) is more prevalent in reflux patients with respiratory symptoms. Gastroenterology 1998; 114:A123.
15. Schnatz PF, Castell JA, Castell DO: Pulmonary symptoms associated with gastroesophageal reflux: Use of ambulatory pH monitoring to diagnose and to direct therapy. Am J Gastro 1996; 91:1715–18.
16. Harding SM, Richter JE, et al. Asthma and gastroesophageal reflux: Acid suppression therapy improves asthma outcome. Am J Med 1996;100:395–405.
17. Vignieri S, Termini R, Leandro G, et al: A comparison of five maintenance therapies for reflux esophagitis. N Engl J Med 1995; 333:1106–1110.
18. Spechler SJ, the Department of Veterans Affairs Gastroesophageal Reflux Study Group. Comparison of medical and surgical therapy for complicated gastroesophageal reflux disease in veterans. N Engl J Med 1992; 326:786
19. Lundell L, Dalenback J, Hatlebakk J, et al. Omeprazole or antireflux surgery in the long-term management of gastroesophageal reflux disease: Results of a multicentre, randomised clinical trial. Gastroenterology 1998; 114:A207.
20. Klinkenberg-Knol E, et al: Long-term treatment with omeprazole for refractory reflux esophagitis. Ann Intern Med 1994; 121:161.
21. Klinkenberg-Knol E. Eleven years' experience of continuous maintenance treatment with omeprazole in GERD patients. Gastroenterology 1998; 114:A180.

13

DISORDERS OF GASTRIC MOTILITY AND EMPTYING

Konrad Schulze-Delrieu*

University of Iowa
College of Medicine
Iowa City, Iowa

INTRODUCTION

Disorders that affect gastric motility and emptying pose a considerable diagnostic and therapeutic challenge; one has to suspect and demonstrate a disorder, determine its cause and assess its severity, and devise a rational and safe treatment plan. Tests of gastric emptying and motility play a minor role in this process. None provide a comprehensive assessment of the role of the stomach in mechanical digestion.

CASE STUDY

A 55-year-old woman was admitted to the hospital because of abdominal pain and inability to eat for three days. She had been well and ate a regular diet until three days prior to admission. She felt full as if she would burst and had little relief from enemas. Fifteen years earlier she had undergone a partial gastrectomy and vagotomy for bleeding duodenal ulcers. She had undergone regular esophagoscopies until 2 years prior to admission for Barrett's esophagus. She remained overweight on a regular diet.

On physical examination she was a grossly obese woman in moderate distress. Fullness and tenderness were present in the left upper quadrant. (Labs: WBC = 14,000, glucose = 160 mg, otherwise normal.)

1. Which of the following tests would you order?

   A. EGD with biopsy.
   B. Abdominal CAT scan.

---

* Phone 319-351-0187, Fax: 353-6399, email konrad-schulze@int-med.uiowa.edu

Gastrointestinal Motility, edited by Rao *et al.*
Kluwer Academic / Plenum Publishers, New York, 1999.

C. Retroperitoneal ultrasound.
D. Plain abdominal film.

2. Abdominal films showed a dense accumulation of fluid and air in the stomach, proximal to the anastomotic staples which appeared to have widely separated.

   Which of the following measures would you now consider?
   A. Milk of molasses enemas.
   B. Percutaneous ultrasound guided drainage of abscess.
   C. 5 g cellulase in 200 ml saline.
   D. Parenteral erythromycin.
   E. None of the above.

3. Cellulase diluted in saline was administered through a nasogastric tube. Thick fluid was aspirated through the tube, and gelatinous material was retched up by the patient. By the third day, the abdominal mass and tenderness largely resolved and the patient felt ready to eat. An esophagogastroduodenoscopy showed a small and empty gastric remnant and widely patent anastomosis.

   Which of the following tests is unnecessary in the evaluation of this problem?
   A. Esophageal manometry.
   B. Electrolytes.
   C. Thyroid functions.
   D. Glycosylated hemoglobin.

4. Diets that should not be recommended for this patient include the following:

   A. Low fat diet.
   B. Diet high in insoluble fiber.
   C. Mechanically soft diet.
   D. High in refined carbohydrates.

She was discharged with instructions regarding a low residue diet.

PATHOPHYSIOLOGY

The mechanical functions of the stomach involve the storage of food, its breakdown and dispersion and eventually its emptying into the duodenum at a rate commensurate with intestinal processes. Disorders of gastric motility lead to changes in the storage, digestion, and emptying of food (1–4). For didactic purposes (Table 1), these will be discussed individually even though in most clinical conditions they coexist, overlap and integrate with each other.

Disorders of Storage

Normally, the stomach has the ability to accommodate its contents. When food arrives in the stomach, its proximal segment enlarges; consequently, large amounts of food can be stored without an appreciable increase in the intragastric pressure. When food leaves the stomach, the proximal stomach shrinks in size (2,6).

The normal function of the proximal stomach to store food may be impaired because of diseases that alter the compliance of its muscle or damage the nerves which control the adaptation of muscle tension. Stadaas and Aune (6) have showed an example of this in

Table 1. Processes interfering with motility

• Resection of pylorus or antrum (less digestion)
• Resection of body (less storage)
• Interference with muscle function (inflammation, infiltration)
• Denervation (vagotomy, autonomic neuropathy of diabetes)

their studies of intragastric pressure-volume relationships, before and after vagotomy. During filling of the normal stomach, the basal tension rises little over a wide range of volume. Also, upon emptying the stomach, intragastric pressure is similar for each given volume as on filling (Table 2). In contrast, after vagotomy the accommodation and compliance are lost in the stomach: baseline pressure rises with each filling step; conversely, there is no proper pressure recovery when the stomach empties, and at corresponding volumes the gastric pressure is much lower on emptying than on filling. Distention of the normal stomach triggers phasic contractions with an increasing amplitude. This is mediated by the corporo-antral reflex which is diminished or absent after vagotomy (7). These constellation of findings explain how impaired gastric tone can lead to gastric stasis (because of delayed tone recovery and because of diminished grinding activity) at the same time as dumping (because of inadequate fundic relaxation).

Smooth muscle function and compliance can be impaired by inflammation, fibrosis, and atrophy. To what extent gastric smooth muscle develops any such changes in the postoperative state is unclear. The early radiographic literature associated gastritis with an atonic stomach and gastric stasis-that is, a distended, flaccid, and inactive stomach. Bile reflux may be more common after various gastric operations and with pyloric and gastric ulcers (8,11); bile gastritis is widely held to cause a change in gastric motility. Evidence to support this is indirect; for instance, in patients with gastric ulcers antral hypomotility may return to normal after healing of the lesion (9). Also, fundal gastritis slows the rate of gastric emptying in patients with non-obstructing gastric carcinoma (10).

Patients with poor gastric storage capacity may develop gastroesophageal reflux and dumping (rapid gastric emptying). This, in turn, is associated with early satiety, poor food intake, and loss of weight. Severe dumping is associated with diarrhea, malabsorption, and malnutrition. Reflux and dumping are consequences of increased intra-gastric pressure that forces the food out of the stomach early on; after this early rapid emptying, a lack of the later tension recovery by the fundus can lead to gastric stasis and might contribute to formation of gastric bezoars.

Disorders of Mechanical Digestion

Solid gastric contents are normally broken down by contractions of the antrum. As the pylorus closes ahead of the advancing bolus, solids are retropelled in a powerful jet stream. The combined action of the antrum and pylorus has several effects: first, it breaks

Table 2. Pathophysiology of gastroparesis

• Decreased reservoir
• High pressure = dumping
• Profound feedback inhibition of stomach
• Poor tone recovery
• No sustained back pressure = stasis
• Weak peristalsis = poor digestion

Table 3. Effects of vagotomy on gastric motility

• Change in gastric baseline tone, tone adaptations
• Poor relaxation, poor tone recovery (dumping, stasis)
• Absent extrinsic (intrinsic) reflexes
• Poor peristaltic function (poor grinding)

down large particles into small ones; second, it mixes and suspends these small particles with gastric secretions to form a slurry called gastric chyme ("trituration"); and lastly, it separates or sieves the gastric contents so that only chyme escapes while solids of more than 1- to 2-mm diameter are retained (1, 15). This digestive motor activity is replaced by the fasting motor activity once all or most contents have been emptied from the stomach (2, 16). Phase III of the interdigestive motor activity is thought to clear indigestible solids from the stomach.

Poor mechanical digestion of food occurs with loss of function of the antrum and pylorus from resection, from impaired nervous control, or from diseases altering the muscular functions. For instance, antral denervation leads to infrequent and disorderly rhythmic contractions and absence of phase III of the interdigestive complex (2, 16). This leads to inadequate breakdown and delayed emptying of food by the stomach (Table 3). The inadequate mechanical activity is likely to be compounded by insufficient secretions to generate gastric chyme. As a result, contents remain in the stomach for excessive periods; this may lead to chronic distention, and the contents may putrefy or conglomerate into gelatinous or concrete bezoars.

Antral resection or pyloroplasty may result in the delivery of comparatively large particles to the intestine. Because of their smaller surface-area-to-volume ratio, large particles are less suitable for enzymatic digestion than small particles. Nutrients may reach the ileum before being absorbed and hence trigger the powerful inhibition of gastric emptying known as the "ileal brake". The dumping of large particles also can lead to malabsorption and malnutrition (1, 2, 4, 5).

Dumping

Dumping occurs in the setting of vagotomy, gastric resection, operative pyloroplasty or peptic deformity of the pylorus (1, 2, 3, 8, 14, 18). Early dumping is known as a syndrome brought about by rapid filling of the intestine with hypertonic solutions. Diarrhea and pain may be complicated by vascular volume depletion; cardiovascular symptoms may be due to intestinal distention, mesenteric vasodilation, and release of neuropeptides. Late dumping syndrome is known as alimentary hypoglycemia, which typically occurs a couple of hours after the meal. Dumping is a particularly serious clinical problem in patients with altered metabolic homeostasis, and it is one factor in brittle diabetes.

Gravity and meal viscosity have greater effects on gastro-intestinal transit after gastric resections. Water-soluble fiber exerts significant effects on luminal viscosity and on intestinal transit. Management of dumping includes postprandial recumbency, reduction of osmotic load and judicious attention to the viscosity of the meal (18, 19).

Gastric Distention, Residue and Bezoar Formation

Acute gastric dilatation denotes massive distention by fluid and gas which occurs in a matter of hours or days. The stomach may extend into the pelvis and is in danger of per-

Table 4. Symptomatology

Early satiety, fullness
Weight loss, soft diets
Regurgitation, heartburn, nausea
Vomiting of stale food, halitosis
Maldigestion, steatorrhea, malnutrition
Hypoglycemia, circulatory symptoms, poor bioavailability

foration. Acute gastric dilatation as part of gastrointestinal ileus has become rare since the introduction of postoperative gastric decompression. Chronic gastric distention occurs with mechanical obstruction of the gastric outlet or as part of the gastroparesis from damage to the gastric innervation (20). Vomiting, electrolyte disturbances, and bezoar formation are common. Common symptoms are listed in Table 4.

With poor digestion of solids and poor clearance of indigestible material from the stomach, residue may conglomerate and form concretions or bezoars. Bezoars are usually formed from collections of plant matter. Hypochlorhydria, diminished antral motility, and incomplete mastication are the main predisposing factors. Bezoars occur mainly after vagotomy and in (diabetic) autonomic neuropathy. Clinical problems with bezoars relate to the presence of a large gastric residue; this interferes with the capacity for food intake and may contribute to mucosal inflammation in the stomach. Adsorption of nutrients and drugs to the bezoar are additional concerns. Passage of bezoars into the small bowel can lead to bolus impaction and intestinal obstruction.

Gastric distention is suspected if the physical examination in a fasted patient reveals left upper quadrant tympany or a succussion splash. Abdominal films will reveal the distended stomach and a large air-fluid interface. Solid food residue or bezoar may be recognized by numerous air bubbles trapped in the left upper quadrant, and on occasion needs to be differentiated from an abscess.

Stale, partially digested food or bezoar found in an unobstructed stomach on a radiographic or endoscopic study establishes the diagnosis of gastroparesis. Treatment of bezoars includes gastric suction and lavage, endoscopic fragmentation, or chemical dissolution. Chemical dissolution can generally be achieved by the use of cellulase. Avoidance of pulpy fruits and vegetables with good mastication of the remaining diet is recommended for those at risk.

CLINICAL EVALUATION

One should identify clinical conditions that are likely to interfere with gastric motility (3–5, 21–24) and those that can be treated by improving gastric emptying. It is important to rule out mechanical causes of obstruction such as peptic ulcer disease, which can interfere with gastric emptying because of either (a) pylorospasm or (b) rigidity and deformity of the gastric outlet (8–19); alternatively, neoplasms, particularly if arising in the stomach or pancreas, may obstruct the gastric outlet. In infants, a common cause of gastric outlet obstruction is a condition known as idiopathic pyloric stenosis; it consists of hypertrophy of the musculature of the gastroduodenal junction, which can be eliminated by pyloromyotomy. In the absence of mechanical obstruction, vagotomy, gastric resections, and autonomic (particularly diabetic) neuropathy are the most common causes of altered gastric emptying and motility. A suggested plan of assessment is shown in Table 5.

Table 5. Assessment

Plain films: gastric dilatation, residue
Contrast study: operative or peptic changes
Endoscopy: mucosal inflammation
? Emptying scans, manometry, barostat

Less clear is to what degree non-ulcer dyspepsia in otherwise healthy individuals relates to discrete abnormalities of gastric motility and emptying. Up to 60% of patients with complaints of postprandial distress and early satiety have been claimed to have an antral motility problem. Many of these patients also have esophageal reflux, gastritis, or duodenogastric reflux (3–5).

The severity of gastric malfunction should be assessed by the patient's symptoms tests, (Table 6) and its consequences: has it forced a reduction in food intake or a change to a more easily digested diet? Is it accompanied by loss of gastric tone as manifested by gastric distention, or is it accompanied by loss of gastric digestive function as manifested by the formation of bezoars? Are there signs of maldigestion, malabsorption, and malnutrition such as diarrhea, steatorrhea, anemia, neuropathy, and weight loss? Or is there abdominal distention, tympany or a succussion splash? Early satiety may lead to weight loss because the patient may not be able to take in sufficient quantities. Conversely, patients with dumping cannot absorb the food properly due to rapid transit of food. They also avoid eating because they learn that eating precipitates symptoms.

Peptic Ulcer Disease and Gastritis

Mechanical obstruction and mucosal inflammation must not be overlooked as potentially treatable factors in abnormal gastric motility (Table 7). If drugs cause extensive chemical gastritis, they might have to be stopped. Helicobacter gastritis, if severe, may need to be treated.

Ulcer craters and ulcer deformities can interfere with gastric emptying and motility. Acute ulcers cause mechanical obstruction of the gastric outlet through edema and spasm. This typically resolves promptly with nasogastric suction. Peptic strictures may cause early satiety, regurgitation, vomiting and weight loss. Most peptic strictures occur at the junction of the duodenal bulb and the postbulbar duodenum. Occasionally, a prepyloric stricture or antral web leads to mechanical outlet obstruction.

Peptic lesions of the pylorus lead commonly to a keyhole deformity; widening and shortening of the gastric outlet contribute to its incompetence, leading to rapid flow across it and marked duodenogastric reflux (8).

Duodenal ulcer patients have a prolonged secretory phase in response to meals. This is associated with abnormally rapid emptying of the liquid gastric contents (8,13,14). In Zollinger-Ellison syndrome, the acid load to the duodenum is large even during fasting.

Table 6. Classification

• Acute = reversible or temporary
• Chronic = irreversible vs poorly treated
• Minor: dyspepsia, heartburn, fullness, indigestion
• Moderate: early satiety, vomiting, halitosis, weight loss or metabolic derangements, gastric residue
• Decompensated: nutritional problems despite optimal medical R_x

Table 7. Differential diagnosis

Acute ulcer: obstruction from spasm, edema
Ulcer deformity: stasis
Stricture At Bulb> Duodenum; Antral Web
Stiff, hypertrophic antrum
Keyhole deformity of pylorus = dumping

Ingestion of a meal temporarily neutralizes the gastric secretions, and the rates of emptying of liquid and solid normalize. In the late postprandial period when most nutrients have left the stomach, there is a return to the high fractional emptying of a water load.

In gastric ulcer disease (12,13), particularly with ulcers located above the incisura, there is a decreased secretory response to meals with a relatively increased concentration of bile acids. Gastric emptying of the liquid phase of the meal is normal, but the emptying of solids is slow (13). This suggests decreased antral motor activity.

Studies on atrophic gastritis, achlorhydria, and pernicious anemia have shown differing effects of these diseases on gastric motility (17–20). Atrophic gastritis and pernicious anemia both exhibit slow emptying of solids. In achlorhydric patients, liquids are found to empty more rapidly than normal. Finally, in atrophic gastritis without pernicious anemia there is gastric stasis for both solids and liquids. In pernicious anemia it is known that the gastric muscle layers are thinner; the precise mechanism for gastric motility disorders in this group of diseases is again unknown.

Autonomic Neuropathy, Gastroparesis, and Gastric Atony

Gastroparesis is a syndrome characterized by gastric stasis leading to dilatation of the stomach in the absence of mechanical obstruction; inflammation should also be absent except for "stasis gastritis". Gastroparesis is associated with diabetes, postoperative changes, and post-viral illnesses, or it may be idiopathic (35,22,26). The manifestations of gastroparesis appear to be the same whatever its cause, in particular, diabetic gastroparesis produces virtually the same symptoms and signs as found in postvagotomy gastric atony. It is therefore thought to represent a type of "autovagotomy" from autonomic neuropathy. Interruption of vagal pathways is expected to alter (a) sensory signals from the stomach to the central nervous system and (b) excitatory (cholinergic) and inhibitory (nonadrenergic) motor output from the central nervous system to the stomach. Sensory loss may interfere with inability to perceive the symptoms and might explain why many patients with gastroparesis, even when complicated by bezoar formation and esophagitis, offer so few complaints. The excitatory cholinergic input from the vagus nerves maintains baseline gastric tone; in its absence, patients develop abnormal gastric electrical patterns including a lack of migrating motor complexes involving the stomach.

Gastric atony may occur acutely as part of diabetic ketoacidosis, and it usually resolves with resolution of the hyperglycemia and of the hyperosmolar and acidotic environment. In most cases where chronic diabetic gastroparesis manifests itself, metabolic control has been poor for some time and there are other concomitant diabetic problems such as retinopathy, nephropathy, and peripheral neuropathy. Gastric malfunction is often reversible with metabolic control. This has to be rigorously pursued, even if it requires temporary use of a jejunal feeding tube.

Following vagotomy, particularly total abdominal vagotomy, about 5% of patients develop marked gastric atony (2). The paresis may be partly due to the absence of migrating motor complexes. Patients with gastroenteric anastomoses may demonstrate prolonged gastric stasis postoperatively. The difference in this situation is that the abnormality lies not with the stomach but with the efferent limb of the anastomosis. This occurs mainly in patients with Roux-en-Y-type anastomoses with disruption of the migrating motor complexes in the efferent limb (17).

Smooth Muscle Disorders

Little is known about mechanical dysfunction of the stomach from primary muscular diseases; gastric function is frequently assumed to be not affected by diseases which are known to interfere with esophageal, intestinal, and colonic motility. Wherever attempts have been made to systematically record gastric motility, the studies have revealed abnormalities. Examples of muscle disorders include myotonic dystrophy, progressive systemic sclerosis, polymyositis, systemic lupus erythematosus (SLE) and amyloidosis (28–31).

Postviral Gastroparesis

Postviral gastroparesis normally follows a short viral illness in which a gastroenteritis is described. Cytomegalovirus and influenza A viruses have been associated with this disease, though the titers for these viruses have not been high enough to indicate active infection. The Norwalk and Hawaii agents have been shown experimentally to cause temporary gastroparesis in healthy volunteers. In all cases, the viral illness precedes the onset of gastric stasis. There is no change in gastric secretion reported (25,26). While there are no visible gastric lesions, there may be histological changes in the small intestine. Other viral infections that have been associated with gastrointestinal stasis include the herpes varicella-zoster and the Epstein-Barr virus in infectious mononucleosis. Gastroparesis may come on suddenly without binding a specific viral pathogen. Presumably there are other neuropathy viruses which are yet to be identified. Usually the symptoms resolve spontaneously, but this may take up to 3–4 years; sometimes the impaired function may be permanent.

THERAPY

Treatment of Underlying Disorder

Control of an underlying clinical condition is likely to alleviate the gastric motility disorder (Table 8). This is true for diabetic gastroparesis, and for additional metabolic disorders including hypothyroidism, and electrolyte abnormalities.

Nutrition

Dietary manipulation is the key for the management of patients with abnormal gastric motility and emptying. In patients with early satiety and weight loss, dumping may be improved by the introduction of frequent small meals, and from lying down on their left side to slow the early phase of gastric emptying. The osmotic load on the small bowel should be reduced by avoiding refined carbohydrates and beverages during meals. Elimination of dumping helps to avoid the many problems of maldigestion and metabolic control caused by rapid emptying of hypertonic fluid into this proximal duodenum (Table 9).

The addition of soluble fiber to a meal may help to render the emptying from the stomach and the absorption from the intestine less erratic. Increased viscosity of the meal

Table 8. Elimination of aggravating factors

Treat mechanical obstruction, peptic disease
Treat gastritis from corrosive drugs, helicobacter if severe
If possible stop anticholinergics, antihistamines, smooth muscle relaxants
Restore gastric tone by suction if stomach is dilated
Dissolve bezoars with cellulase
Metabolic control : K+, thyroid, Hba_{1c}
Treat constipation
Treat bacterial overgrowth, celiac sprue, or other coexisting conditions

slows the emptying of gastric contents in the normal and in the resected stomach. Soluble fiber might decrease dumping while dragging along solid residue. Meals should have the consistency of mushy pulp or a thin pudding; usage of fiber filler probably works by reducing the liquidity and the osmolarity of a meal while still producing the normal mechanical stimulation of the stomach and intestines (18,19). Pre-feeding with small amounts of oleic acid homogenized in formula dramatically reduces the symptoms of dumping through the ileal brake and release of PYY.

In gastroparesis, the ability of the stomach to digest tough and fibrous meals is impaired. (This problem may be compounded if the patient can not masticate properly.) A mechanically soft diet as obtained by straining and pureeing is indicated. Difficult-to-digest food need to be eliminated; this includes stringy vegetables (particularly when raw), pulpy fruits, and tough meats.

In patients with malabsorption and malnutrition not responding to dietary and pharmacological measures, long-term nutritional support will be required. Enteral supplements, tube feedings, or total parenteral nutrition may be required until weight has been regained and the patient is no longer in a catabolic state.

Pharmacological Agents

Medications that interfere with gastric motility should be replaced by some suitable alternatives wherever possible. This alone may resolve the symptoms and improve emptying. Adrenergic agonists, especially the beta agonists such as isoproterenol and salbutamol, are known to cause tachygastria and tend to delay gastric emptying. Clonidine, an $alpha_2$ agonist used in the treatment of hypertension, has the side effects of nausea and vomiting which occasionally leads to pseudo-obstruction.

Apomorphine, bromocriptine, and L-dopa are potent emetic agents and at subemetic doses retard gastric emptying. L-dopa used to treat Parkinson's disease has a profound effect on delaying gastric emptying which can be effectively alleviated by metoclopramide (32).

Anti-cholinergic drugs such as atropine inhibit gut contractility and delay gastric emptying. Other agents such as tricyclic antidepressants and antiparkinsonian agents that have anticholinergic properties also inhibit gastrointestinal motility. Opioid drugs have emetic actions, inhibit gastric antral motility, and induce bursts of intestinal activity that

Table 9. Dietary modifications

Prevent dumping by postprandial recumbency, frequent small meals with complex carbohydrates and soluble fiber
Reduce digestion and residue by using mechanically soft food (puréed, strained), avoid pulpy fruits, grains, raw vegetables

resemble phase Ill of the MMC. Chronic abdominal pain, vomiting, weight loss, and features of chronic intestinal pseudo-obstruction characterize the narcotic bowel syndrome, which is caused by prolonged use of narcotic analgesics, predominantly opiates. This responds to opiate withdrawal, and the symptoms of withdrawal are relieved by clonidine.

Effective gastrokinetic drugs include several cholinergic agonists (bethanechol, metoclopramide, and cisapride) and motilin-receptor agonists (e.g., erythromycin derivatives). Metoclopramide stimulates gastric contractions, accelerates gastric emptying, and is an antiemetic (32). Cisapride belongs to the same chemical class as metoclopramide, and it seems to act mostly by enhancing the release of acetylcholine (33).

Erythromycin, a macrolide antibiotic, acts like a motilin agonist and dramatically accelerates gastric emptying in diabetic gastroparesis. (35, 36). When used intra-venously, one study showed almost normal gastric emptying for both solid and liquids in 12 patients with diabetic gastroparesis. Oral erythromycin in the same study showed a significant increase in gastric emptying. The value of any of these drugs in the long-term management of gastric motility disorders remains to be established.

ANSWER KEY AND EXPLANATIONS

1. The first and foremost test would be to perform a plain abdomen x-ray. As it proved in this patient, it can be very useful. The film may show evidence for upper gut obstruction, distended stomach, or as we found here separation of the staples. An EGD may be helpful to identify anastomotic stricture or cancer but only after the plain films. EGD is unhelpful and potentially dangerous in a patient with distended stomach full of food debris. An ultrasound may not be useful. The best option is **D**.
2. The correct answer would be to dissolve the bezoar with cellulase. The best option is **C.**
3. Electrolytes, thyroid function and glycosylated hemoglobin are useful for identifying or correcting an underlying metabolic problem. Esophageal manometry would be unhelpful. The best option is **A**.
4. A low fat diet may help to speed transit through the upper gut. Insoluble fiber on the other hand may promote bezoar formation and should be avoided in this patient. A mechanically soft diet can be very useful in the early stages of feeding in this patient. A diet high in refined carbohydrate may produce dumping in a patient with previous gastrectomy and should be avoided. The best option is **B.**

REFERENCES

1. Meyer JH. Motility of the stomach and gastroduodenal junction. In: Johnson LH, ed. Physiology of the gastrointestinal tract, yet 1, 2nd ed. New York: Raven Press, 1987;613 629.
2. Kelly KA. Effect of gastric surgery on gastric motility and emptying. In: Akkermans LMA, Johnson AG, Read NW, eds. Gastric and gastroduodenal motility. New York: Praeger, 1984:241–262.
3. Poitras P, Picard M, Dery R,Giguere A,Picard D,Morais J,Plourde V,Boivin M. Evaluation of gastric emptying function in clinical practice. *Digestive Diseases & Sciences*. 1997; 42(11):2183–9.
4. Read NW, Houghton LA. Physiology of gastric emptying and pathophysiology of gastroparesis. *Gastroenterology Clinics North Am* 1989; 18(2);359–373.
5. Quigley EM. Gastric and small intestinal motility in health and disease. *Gastroenterology Clinics of North America*. 1996; 25(1):113–45.
6. Stadaas J, Aune S. Intragastric pressure/volume relationship before and after vagotomy. *Acta Chir Scand* 1970;136:611 615.

7. Grundy D, Hutson D, Scratcherd T. A permissive role of the vagus nerve in the genesis of corporto-antral reflexes in the anaesthetized ferret. *J Physiol* 1986;381:377- 384.
8. Lu CH, and Schulze-Delrieu K. Pyloric deformation from peptic disease. Evidence for incompetence rather than obstruction. *Dig Dis Sci* 1990;35:1459–1467.
9. Bromster D. Gastric emptying rate in gastric and duodenal ulceration: a clinical study including the changes of the gastric contents after a liquid test meal. *Scand J Gastroenterol* 1969;4:193–201.
10. Tatsuta M, Oishi H, Okuda S. Gastric emptying in patients with fundal gastritis and gastric cancer. *Gut* 1990;31:767 769.
11. Davies WT, Kirkpatrick JR, Owen GM, Shields R. Gastric emptying in atrophic gastritis and carcinoma of the stomach. *Scand J Gastroenterol* 1971;6:297–301.
12. Halvorsen L, Dotevall G, Walan A. Gastric emptying in patients with achlorhydria or hyposecretion of hydrochloric acid. *Scand J Gastroenterol* 1973;8:395–399.
13. Miller LJ, Malagelada JR, Longstreth GF, Go VL. Dysfunctions of the stomach with gastric ulceration. *Dig Dis Sci* 1980; 11: 857-864.
14. Dubois A, Van Erdewegh P, Gardner JD. Gastric emptying and secretion in Zollinger Ellison syndrome. *J Clin Invest* 1977;59:255–263.
15. Meyer JH, Ohaski H, Jehn D, Thompson JB. Size of liver particles emptied from the human stomach. *Gastroenterology* 1981;80:1489–1496.
16. Sarna SK, Daniel EE. Vagal control of gastric electrical control activity and motility. *Gastroenterology* 1975;68:301–308.
17. Mathias JR, Fernandez A, Sninsky CA, Clench MH, Davis RH. Nausea, vomiting and abdominal pain after Roux-en-Y anastomosis: motility of the jejunal limb. *Gastroenterology* 1985;88:101–107.
18. Ehrlein HJ, Wulschke S, Thoma G, Schumpelick V. Gastrointestinal motility and gastric emptying after Billroth II gastrectomy in dogs. *Dig Dis Sci* 1989;34(g):1199–1210.
19. Reppas C, Meyer JH, Siros PJ, Dressman JB. Effect of hydroxypropylmethylcellulose on gastrointestinal transit and luminal viscosity in dogs. *Gastroenterologv* 1991; 100: 217–1223.
20. Hom S, Saff MG, Kelly KA, Hench V. Postoperative gastric atony after vagotomy for obstructing peptic ulcer. *Am J Surg* 1989; 1 57:282–286.
21. Goldberg KB, Lasichak A, Rock CL. Dietary adequacy in patients with diabetic gastroparesis. *Journal of the American Dietetic Association.* 1997;97(4):420–2.
22. Thumshirn M. Bruninga K. Camilleri M. Simplifying the evaluation of postprandial antral motor function in patients with suspected gastroparesis. *Am J Gastroenter.* 1997;92(9):1496–500.
23. Malagelada JR Distrutti E. Management of gastrointestinal motility disorders. A practical guide and appropriate ancillary measures. *Review. Drugs* 1996;52:494–506.
24. Soergel KH. Greenberger NJ. Nausea and vomiting in the diabetic patient, *Hospital Practice* (Office Edition). 1998;33(2):43–5, 49–50, 56–61; discussion 61–2.
25. Oh JJ, Kim CH. Gastroparesis after a presumed viral illness: clinical and laboratory features and natural history. *Mayo Clin Proc* 1990;65:636–642.
26. Meeroff JC, Schreiber DS, Trier JS, Blacklow NR. Abnormal gastric motor function in viral gastroenteritis. *Ann Intern Med* 1980;92:370–373.
27. Nowak TV, Anuras S, Brown BP, Lonasescu V, Green JB. Small intestinal motility in myotonic dystrophy patients. *Gastroenterology* 1984;86:808–813.
28. Maddem GJ, Horowitz M, Jamieson GG, Chatterton BE, Collins PJ, Roberts-Thompson P. Abnormalities of esophageal and gastric emptying in progressive systemic sclerosis. *Gastroenterology* 1984;87:922–926.
29. Raffin SB. Bezoars. In: Sleisenger MH, Fordtran JS, eds. *Gastrointestinal Disease*, Vol 1, Philadelphia:WB Saunders, 1989;741–744.
30. Horowitz M, McNeill JD, Maddem GJ, Collins PJ, Shearman DJC. Abnormalities of gastric and esophageal emptying in polymyositis and dermatomyositis. *Gastroenterology* 1986;90:434.
31. Ikeda SI, Makishita H, Oguchi K, Yanagisawa N, Nagata T. Gastrointestinal amyloid deposition in familial amyloid polyneuropathy. *Neurology* (NY) 1982;32:1364–1368.
32. Schulze-Delrieu K. Metoclopramide. In: Koch-Weser J, ed. Drug Therapy. *N Engl J Med* 1981; 305(1):28–33.
33. Brown CK, Khanderia U. Use of metoclopramide, domperidone and cisapfide in the management of diabetic gastroparesis. *Clin Pharm* 1990;9:357–365.
34. Janssens J, Peeters TL, Vantrappen G, et al. Improvement of gastric emptying in diabetic gastroparesis by erythromycin. *N Engl J Med* 1990;322(15):1028 103 1.
35. Mozwecz H, Pavel D, Pitrak D, Orellana P, Schlesinger PK, Layden TJ. Erythromycin steorate as prokinetic agent in postvagotomy gastroparesis. *Dig Dis Sci* 1990;35(7):902–905.

14

DISORDERS OF SMALL INTESTINAL MOTILITY

Robert W. Summers*

Department of Internal Medicine
4545 JCP
James A. Clifton Center for Digestive Diseases
University of Iowa College of Medicine
200 Hawkins Drive
Iowa City, Iowa 52242

INTRODUCTION

Many believe that the small intestine is not commonly affected by disorders of motor function. However, disordered small intestinal motility may contribute to multiple symptoms and clinical problems. For example, dysmotility either directly or indirectly can lead to the development of small bowel bacterial overgrowth.

Intestinal motor dysfunction is commonly due to a systemic disorder involving not only the small bowel, but also other hollow organs of the digestive tract. In these circumstances, symptoms caused by gastric emptying disorders or colonic dysfunction may predominate and the physician may not consider investigating a small bowel motor disorder. Although less common, the small bowel may be the only digestive organ involved. The clinician frequently is asked to differentiate mechanical obstruction from pseudo-obstruction. This is often a very difficult distinction and actually, both problems may coexist. In *mechanical obstruction,* an anatomic blockage of the bowel may be partial or complete, preventing or retarding the passage of gas, liquid or solid material. Total obstruction always causes symptoms which are aggravated by eating. Partial small bowel obstruction frequently causes bacterial overgrowth, but impaired transit does not always produce symptoms. Physical findings include abdominal distention, tympany, and abnormal bowel sounds. Radiologic findings include dilation of intestinal loops, air fluid levels or demonstration of a narrowing or blockade of the bowel through the use of radiocontrast media.

Intestinal dysmotility may be defined as abnormal contractile patterns or pressure wave patterns that may or may not be associated with disordered transit or cause symptoms. It may be classified as a *primary motor disorder*, if only the digestive tract is in-

* Phone: 319/356-2130, Fax: 319/353-6399, E-mail: robert-summers@uiowa.edu

Gastrointestinal Motility, edited by Rao *et al.*
Kluwer Academic / Plenum Publishers, New York, 1999.

volved, or a *secondary motor disorder* if it is a gastrointestinal manifestation of a systemic disorder. *Clinically, Intestinal pseudoobstruction* may appear indistinguishable from mechanical obstruction, but no anatomic or mechanical abnormalities can be found by imaging studies or at the time of laparotomy.

CASE STUDY

A 22-year-old woman who was well until two years ago presents with intermittent nausea and vomiting. She would feel full about an hour after an ordinary sized meal and then progressively become uncomfortable with nausea over the next four to five hours. These symptoms would continue unabated until she finally vomited a large amount of food and fluid. This would relieve her symptoms until she ate again. In order to avoid these symptoms she started eating smaller amounts of food. Consequently she lost about 20 pounds in weight. When her abdomen became distended, she would experience mild diffuse abdominal pain, but the pain was not the most prominent feature. Although she was bothered by mild constipation, she did not consider this to be a problem because she had this most of her life. Review of systems was unremarkable except for numerous urinary tract infections over the past ten years. These had been treated promptly, but had never been investigated. Other than this, she had no other serious illnesses or hospitalizations. In the family history there were no serious illnesses but her mother and sister were experiencing similar symptoms. The patient was employed as a registered nurse and did not smoke or drink alcohol.

Physical examination revealed a thin otherwise healthy-appearing young woman. Vital signs were normal. The chest and heart exam were normal. The abdominal contour was rounded without scars and was tympanitic in the epigastrium. There was a succussion splash in the left upper quadrant. The bowel sounds were present but hypoactive. The liver and spleen were not enlarged, but there was fullness in the hypogastrium consistent with an enlarged bladder. No tenderness was appreciated. The remainder of the examination including the neurologic examination was within normal limits.

1. In the further evaluation of this patient, which diagnostic test would you first perform?

   A. Gastric emptying study
   B. Upper endoscopy
   C. Barium meal and follow through
   D. Antroduodenojejunal manometry
   E. CAT scan of the upper abdomen.

2. An upper endoscopy was normal. Barium follow through study showed some impaired transit through the small bowel with pooling of barium and minimally distended loops but no definite obstruction The most helpful study in establishing an underlying etiology would be:

   A. Mucosal biopsy of the distal duodenum.
   B. Creatinine phosphokinase.
   C. Deep rectal biopsy.
   D. Full thickness small bowel biopsy.
   E. Anti-nuclear antibodies.

3. The most likely cause of this lady's clinical problem is:
   A. Familial visceral neuropathy.
   B. Viral-induced neuropathy.
   C. Familial visceral myopathy.
   D. Viral-induced myopathy.
   E. Amyloidosis.

HISTORY AND PHYSICAL EXAMINATION

Complete, partial, or pseudo-obstruction may cause symptoms of abdominal fullness, a sense of being bloated, borborygmi, abdominal pain, usually of a cramping nature, and nausea and vomiting (Table 1). The pain or upper abdominal discomfort often is periumbilical or sometimes epigastric and exacerbated by eating, especially solids. Avoiding eating or vomiting may reduce the symptoms of impaired transit. Other symptoms may originate from the small bowel, which are not related to slowing of transit. It should be emphasized that accelerated gastric emptying or "dumping" may produce identical symptoms to those caused by slowed small bowel transit. Retrograde propulsion of the proximal bowel may transport bile and digestive enzymes into the stomach causing symptoms of gastritis. Accelerated transit may move nutrients into the colon where they are digested by colonic bacteria to form short chain fatty acids, osmotically active substances and gas. This will cause diarrhea, gaseous distension, excessive flatus and even steatorrhea. Because symptoms are not specific, measurement of motor function and transit are essential if rational therapy is to be applied.

Impaired transit of chyme may cause "backward" failure resulting in burning epigastric pain, early satiety, prolonged fullness after meals, nausea and vomiting from regurgitation of intestinal contents into the stomach. In addition, heartburn and regurgitation of acidified food into the esophagus and hypopharynx may occur in the presence of impaired intestinal motility. These same symptoms may also represent involvement of multiple organs including the stomach and esophagus. Symptoms alone are unreliable indicators and independent esophageal manometry, antroduodenal manometry and scintigraphic studies of gastric emptying should be done to determine whether esophageal and gastric motor function are normal. If the colon is involved by the same disease process, symptoms may include constipation or diarrhea, difficulty expelling gas or incontinence. As the disease process becomes more advanced, malnutrition and weight loss become increasingly more apparent. Again, it is important to assess both motor function and transit or emptying in order to apply therapy appropriately.

Family History

Probably one of the most important aspects of the history is the family history. Familial visceral myopathies may be either autosomal recessive or autosomal dominant, but all of the reported cases of familial visceral neuropathies are autosomal dominant. The severity of symptoms varies considerably in individuals with proven involvement so that careful questioning is necessary in order to suspect genetically determined diseases. It is likely that these disorders are more common than they are recognized. They may not cause symptoms until later in life and may be entirely asymptomatic. The propositus is often unaware of unexplained gastrointestinal symptoms (Table 2) in even close relatives and direct interviews of the family members may be necessary to find the genetic influence.

Table 1. Symptoms related to the small bowel

Upper-mid abdominal pain or discomfort
Nausea, vomiting
Cramping, bloating, distension
Burning distress, dyspepsia
Diarrhea, steatorrhea (bacterial overgrowth)

Special Features of the Physical Examination

There may be features of the physical examination, which may be helpful in determining the presence or underlying etiology of a motor disorder (Table 3).

DOCUMENTATION OF DYSMOTILITY

Radiography

Plain Radiographs. The importance of abdominal radiographs without contrast cannot be overstated. Plain abdominal X-rays can suggest that a motility problem exists and should be one of the first studies ordered in evaluation of a patient with suspected dysmotility. If the patient is complaining of bloating and distension *after a meal,* upright and supine abdominal films can demonstrate normal or abnormal distribution and patterns of air or gas. Organs or loops of bowel become distended from both swallowed air or gas produced by bacterial digestion of ingested nutrients. Plain films suggest the organs that are involved, the severity of involvement and may yield information about the underlying cause.

Contrast Radiographs. Barium contrast studies remain the procedure of choice to define the anatomy of structural or functional abnormalities of the small bowel. Careful examination showing persistent narrowing or blockade of the intestinal lumen is necessary. If a stenosis is suspected clinically and not apparent on a study that relies on the natural passage of orally ingested barium, then a dilute barium suspension can be injected into the duodenum via a nasogastric tube. This type of study is called a small bowel enema or enteroclysis and it is more sensitive in demonstrating areas of stenosis than the routine small bowel series. The diameter of the lumen proximal to the involved area is often dilated in both mechanical obstruction and in impaired motor function. With impairment of motility, the passage of barium may be delayed although the interpretation of delayed passage is subjective and qualitative and not strictly quantitative.

Accelerated transit may be suspected by rapid transit of barium, however, caution must be exerted regarding interpretation because barium is not a nutrient material and the normal feedback mechanisms are not active in contrast to both neural and humoral effects that occur after a meal. In addition, there is a wide range of normal transit.

Table 2. Symptoms related to other digestive organs

Esophagus	Dysphagia for solids and liquids, heartburn, eructation
Stomach	Early or prolonged satiety, nausea, vomiting, regurgitation fullness, discomfort, distension
Colon	Constipation / episodic diarrhea, difficulty expelling gas or stool
Colon/bladder	Urgency / incontinence
Any or all	Anorexia, weight loss, malnutrition

Table 3. Special feature of the physical examination

Skin	Pigmented nevi are sometimes associated with the visceral myopathies Rashes may indicate collagen vascular diseases
Eyes disorders	Mydriasis, ptosis, ophthalmoplegia suggest neuropathies or myopathies
Abdomen	Distended abdomen may indicate dilatation of the stomach, intestine or colon Organomegaly or masses may indicate infiltrative disorders (e.g. amyloid) or neoplasms causing mechanical obstruction. Pay special attention to scars and hernias, which may contribute to obstruction
Muscles	Muscular dystrophies, myotonic dystrophies and myositis
Neurologic	Neurofibromatosis, autonomic neuropathies, paraneoplastic syndromes, Parkinson's disease, multiple sclerosis

Nuclear Markers of Transit

The study of transit is too often neglected when evaluating symptoms that suggest dysmotility. Nuclear methods are the gold standards to measure transit and they are quantitative. Unfortunately, they are not always available, they are relatively costly, radiation exposure is inherent in their use and prohibits repetitive measurements, and the methods are incompletely standardized. Nevertheless, if an institution has the technical capability, the experience and expertise, and the patient can afford the study, nuclear tests of transit can be invaluable in understanding the pathophysiology and effects of dysmotility.

Hydrogen Breath Tests

Breath tests are more readily available, less costly to perform and can easily and safely be performed on multiple occasions. They depend on the conversion of a non-absorbed sugar such as lactulose to hydrogen when the carbohydrate reaches the cecum. There are a number of pitfalls in their use, however. In patients with impaired motor function, the small bowel is often overgrown with anaerobic and gram negative bacteria. These organisms metabolize the ingested sugar and produce an early rise in breath hydrogen. This falsely suggests rapid transit, but overgrowth may be suspected by doing another breath test with another sugar such as glucose. The interpretation also assumes that gastric emptying of the test substance is normal when it actually may be delayed or accelerated. Finally, prior antibiotic therapy may alter colonic flora and the patient should have been off antibiotics for a week or more before a breath study is attempted.

Table 4. Visceral myopathies

Primary visceral myopathies
Familial and non-familial
Connective tissue diseases
Progressive systemic sclerosis/CREST
Dermatomyositis
Lupus erythematosus/MCTD
Muscular dystrophies
Duchenne's dystrophy
Myotonic dystrophy/other types
Infiltrative diseases
Amyloidosis
Sarcoidosis
Diffuse lymphoid infiltration

Table 5. Visceral neuropathies

Primary visceral neuropathy
Familial and non-familial
Neurological diseases
Diabetic neuroenteropathy/polyneuropathy
Parkinson's disease
Primary autonomic neuropathy/Shy-Drager
Multiple sclerosis
Post-viral neuropathy (CMV, EBV, Herpes?, other)
Ganglioneuromatosis/neurofibromatosis
Spinal muscular atrophy
Chagas' disease
Paraneoplastic syndrome

Small Bowel Manometry

(See Chapter 6 on antroduodenojejunal manometry)

CAUSES OF DYSMOTILITY

Once dysmotility has been established, a quest should be initiated to find an underlying etiology. The search is especially important if effective or curative therapy is available. In many cases, the primary systemic disease is obvious as in the case of diabetes mellitus. In others, it is occult or unexpected as in some paraneoplastic syndromes, subtle familial disorders or unknown as in certain post-viral states. A wide spectrum of causes has been reported and that classification is listed in Tables 4 and 5. Some can be readily diagnosed, but some will be idiopathic and no underlying cause can be found. It is not possible to discuss the vast number of causes, but an approach to the search for etiologies is found in Tables 6 and 7. The most effective therapy for dysmotility states is to treat the underlying disorder. Unfortunately, treatments of many causes are unavailable or only palliative and the impairment of motility increases as the primary disease advances.

TREATMENT

A graded approach to therapy is preferable in the management of these patients.

Table 6. Other causes of motor dysfunction

Drug-related	Endocrine/ metabolic	Unclassified
Opiates	Uremia	Post-operative
Tricyclic antidepressants	Porphyria	Organ transplant
Anticholinergics	Hypokalemia	Radiation injury
Alpha-2 agonists	Hyperglycemia	Celiac sprue
Cytotoxic agents	Hypothyroidism	Small bowel diverticula
Antihypertensives	Pheochromocytoma	
Anti-Parkinson drugs		
Phenothiazines		
Laxatives (?)		
Vinca alkaloids		

Table 7. Search for an underlying cause

Review medications, radiation, and travel history
In depth exploration of the family history
Laboratory tests for systemic disease
Immunologic profile
ESR, ANA, SCL-70, RAF, SSA, SSB, others
Myopathy profile
CPK, aldolase, EMG, muscle biopsy
Metabolic/endocrine profile
FBS, T3, T4, TSH, K+, Ca++, BUN
Infection serologies
CMV, T.cruzi, Ebstein-Barr, others

Stage I — "Modified Life Style"

In this early stage, symptoms are mild and intermittent and the nutritional state of the patient is compensated. Disease management consists mainly of modification in life style (Table 8).

Stage II— "Formulas and Pharmacology"

In this stage, symptoms are moderately severe and the patient begins to experience episodes of nutritional decompensation manifest by weight loss because of inadequate intake. Ordinary food is not tolerated in sufficient amounts to maintain nutrition. Therapy requires significant physician input and intervention and the services of a dietitian are essential in proper management. Small bowel bacterial overgrowth is very common at this stage and must be recognized and treated. Antibiotics that are directed against gram negative and anaerobic organisms can reduce gas and distension as well as diarrhea and maldigestion. If the underlying dysmotility cannot be improved significantly with prokinetic drugs and other measures, the problem recurs and may require repetitive therapy. It may be useful to observe the effects of prokinetic drugs on the manometry to determine whether the medication is having an effect (Table 9).

Stage III — "Intestinal Failure"

In this stage, gut motility is severely compromised and symptoms are far-advanced. The patient's small bowel is decompensated so that little or no nutrition can be administered by mouth. Enteral feedings should be tried and continuous infusions may permit postponement of parenteral feeding. The costs and risks of central venous feeding are sufficient to try to delay initiation of this mode of therapy as long as possible (Table 10).

Table 8. Stage I — "modified life style"

Mechanical soft diet, avoid "roughage", that is, raw fruits and vegetables, skins, seeds, nuts hulls, tough meats, fatty meals
Avoid or discontinue motor inhibiting drugs Anticholinergics, tricyclic antidepressants, antihistamines clonidine, beta-adrenergic antagonists, calcium channel blockers narcotics, dextropropoxyphene
Strict control of blood glucose in diabetes mellitus, intermittent use of prokinetic drugs

Table 9. Stage II — "formulas and pharmacology"

Formula diet by mouth, nasogastric tube, gastrostomy or jejunostomy
Intermittent use of antibiotics for small bowel bacterial overgrowth directed against anaerobes and gram negative organisms
Long-term reliance on prokinetic drugs–cisapride, metoclopramide, macrolide antibiotics (e.g. erythromycin or other analogs) octreotide, CCK antagonists (loxiglumide), opioid kappa-receptor agonists

Surgical Intervention

In this stage, the intestine is totally decompensated and may not even be able to handle endogenous secretions or gas. The abdomen becomes distended and the patient may experience significant discomfort or frequent vomiting in spite of maximal pharmacologic therapy. A venting procedure may be necessary to relieve these symptoms. If only a segment of bowel is involved, it may be possible to resect the segment with a resultant improvement in function, which is sufficient to allow resumption of enteral feeding. In some patients, especially children, liver failure occurs secondary to parenteral nutrition and small bowel-liver transplantation becomes the only alternative (Table 11).

Management of Accelerated Transit

It is often assumed that abnormal motility causes impaired transit. However, as emphasized above, many of the symptoms of impaired motility are indistinguishable from those of rapid or accelerated transit. Most clinicians are familiar with the so-called dumping syndrome after vagotomy and a drainage procedure. The post-prandial symptoms include abdominal discomfort, distension, a sense of fullness or excessive gastrointestinal gas and diarrhea. These symptoms may occur in diabetics with neuropathy and it is noteworthy that emptying of solids may be retarded at the same time that emptying of liquids is accelerated. Emptying studies must include both the study of solid and liquid emptying. If liquid emptying is accelerated, it may be very helpful to institute therapy to slow the emptying or transit of liquids through implementation of the "ileal brake" and the "jejunal brake". Administration of a small volume (3–5 ml) of oleic acid homogenized with 30–40 ml of formula may provide dramatic relief of dumping symptoms.

CONCLUSIONS

The small intestine is only recently being recognized as a potential source of symptoms because of either altered intestinal motor activity or transit. The duodenum and jejunum can now be readily studied using transit studies or manometry. It may be surprising to some that both accelerated and retarded transit may be responsible for the development of the same symptoms. Hence, transit must be studied before institution of therapy. Many

Table 10. Stage III — "intestinal failure"

Episodic or permanent bowel rest - nothing is allowed by mouth
Nasogastric suction is often helpful or required to relieve distension
Parenteral nutrition is required temporarily or permanently

Table 11. Stage IV — surgical intervention

Venting gastrostomy, jejunostomy or multiple ostomies
By-pass or resection of most severely affected segments
Small intestinal transplantation

disorders of small bowel motility are usually part of a systemic disorder that involves multiple digestive organs. Symptoms which arise from these other organs may be the focus of attention, yet when investigating these symptoms the small bowel should not be overlooked; often treating dysmotility may help both problems.

ANSWER KEY AND EXPLANATIONS

1. The first step in the evaluation of this patient with signs and symptoms suggestive of partial intestinal obstruction would be a plain film of the abdomen. But this choice was not given. Among the choices given the best answer would be an upper endoscopy to identify any mucosal disease that may be responsible for her symptoms and to obtain mucosal biopsies. A CAT scan may reveal a gross obstructive pathology or reveal other lesions such as lymphoma but this appears less likely. Manometry may be useful to define the underlying mechanism for the stasis but only after excluding an obstruction. After excluding an upper GI source a barium study can be quite useful. Some may prefer to order a barium study in the first instance and this is acceptable provided there is no bezoar. The best option is **B**.
2. Once it is clear that the underlying problem is a muscle or nerve dysfunction of the upper gut, the best method to define the cause is a full thickness biopsy of the small bowel performed with laparoscopy. The other options may be helpful to identify celiac disease or amyloidosis or autoimmune process but the definitive answer would be a biopsy with special immunohistochemical stains to identify any viral, metabolic, degenerative or inflammatory nerve or muscle disease. The best option is **D**.
3. Although a family history may help, most of these disorders are autosomal recessive and a family history is not always present. Full thickness biopsy revealed extensive degeneration of the muscles indicating visceral myopathy. The best option is **C**.

REFERENCES

1. Wilmer A, Van Cutsem E, Andrioli A, Tack J, Coremans G, Janssens J: Ambulatory gastrojejunal manometry in severe motility-like dyspepsia: lack of correlation between dysmotility, symptoms, and gastric emptying. Gut 1998;42(2):235–242.
2. Jebbink HJ, Van Berge-Henegouwen GP, Bruijs PP, Akkermans LM, Smout AJ: Gastric myoelectrical activity and gastrointestinal motility in patients with functional dyspepsia. Eur J Clin Inv 1995;25:429–437.
3. Jebbink HJ, vanBerge-Henegouwen GP, Akkermans LM, Smout AJ: Small intestinal motor abnormalities in patients with functional dyspepsia demonstrated by ambulatory manometry. Gut 1996;38:694–700.
4. Kellow JE, Gill RC, Wingate DL: Prolonged ambulant recordings of small bowel motility demonstrate abnormalities in the irritable bowel syndrome. Gastroenterology 1990;98:1208–1218.
5. Schmidt T, Hackelsberger N, Widmer R, Meisel C, Pfeiffer A, Kaess H: Ambulatory 24-hour jejunal motility in diarrhea-predominant irritable bowel syndrome. Scand J Gastroenterol 1996;31(6):581–589.
6. Scolapio JS, Camilleri M, Fleming CR: Gastrointestinal motility considerations in patients with short-bowel syndrome (Review). Dig Dis 1997;15:253–262.

7. Annese V, Bassotti G, Napolitano G, Usai P, Andriulli A, Vantrappen G: Gastrointestinal motility disorders in patients with inactive Crohn's disease. Scand J Gastroenterol 1997; 32:1107–1117.
8. Layer P, von der Ohe MR, Holst JJ, Jansen JB, Grandt D, Holtmann G, Goebell H: Altered postprandial motility in chronic pancreatitis: role of malabsorption. Gastroenterology 1997;112:1624–1634.
9. Stotzer PO, Bjornsson ES, Abrahamsson H: Interdigestive and postprandial motility in small-intestinal bacterial overgrowth. Scand J Gastroenterol 1996;31:875–880.
10. Hackelsberger N, Schmidt T, Renner R, Widmer R, Pfeiffer A, Kaess H: Ambulatory long-term jejunal manometry in diabetic patients with cardiac autonomic neuropathy. Neurogastroenterology & Motility1997;9:77–83.
11. Russo A, Fraser R, Horowitz M: The effect of acute hyperglycaemia on small intestinal motility in normal subjects. Diabetologia 1996;39:984–989.
12. Evans PR, Bak YT, Kellow JE: Effects of oral cisapride on small bowel motility in irritable bowel syndrome. Alimen Pharmacol & Therap 1997;11:837–844.
13. Bassotti G, Germani U, Calcara C, Spinozzi F, Roselli P, Morelli A: Effects of octreotide on manometric variables in patients with neuropathic abnormalities of small bowel. Dig Dis & Sci 1997;42:1634–1639.
14. Schonfeld J, Evans DF, Wingate DL: Effect of viscous fiber (guar) on postprandial motor activity in human small bowel. Dig Dis & Sci 1997;42:1613–1617.

15

POST SURGICAL GASTRIC DYSMOTILITY SYNDROMES

John H. Pemberton*

Mayo Clinic
Rochester, Minnesota

INTRODUCTION

After gastric surgery, there are three major problems; slow gastric emptying (atony), rapid gastric emptying (dumping), and bile reflux. In the normal stomach, gastric motor activity and emptying are regulated such that the stomach mixes food and propels this into the duodenum at an optimal rate for digestion and absorption. Disruption of this function by surgery may lead to a variety of disorders. Operations such as vagotomy, gastric resection or gastroenterostomy can cause dumping, small gastric remnant syndrome, postvagotomy diarrhea, chronic gastric atony, reflux gastritis and Roux stasis syndrome. Each of these problems will be discussed.

CASE STUDY

A 45-year-old school teacher presents with a two year history of recurrent nausea, bilious vomiting, early satiety, post-prandial fullness, abdominal pain and a 25 pound weight loss. During the previous 6 months, she had been hospitalized seven times with recurrent fever and septicemia. Because of her inability to keep food down, she was placed on total parenteral nutrition. Subsequently, she had two line infections but the current I.V. line has been functioning satisfactorily for 2 months. She also had a surgical jejunostomy.

Five years ago she had a cholecystectomy. Two years ago she underwent vagotomy, partial gastrectomy and gastrojejunostomy for recurrent bleeding and severe peptic ulcer disease. One month ago she had an endoscopy which showed mild gastritis, bile in the stomach, patent anastomoses, without ulceration. She is an ex-smoker. Because of aforementioned problems, she has not been able to teach for one year.

On exam, she was thin built, weighed 92 lbs, height 5'6", BP 116/98; PR 90 min. In the abdomen, scars of previous surgery were present and a jejunostomy tube was seen.

---

* Phone: 507-284-2359, Fax: 507-284-1794, E-mail: pemberton.john@mayo.edu

Gastrointestinal Motility, edited by Rao *et al.*
Kluwer Academic / Plenum Publishers, New York, 1999.

There was a distinct succussion splash. Bowel sounds were sluggish. There was no hepatosplenomegaly. Stool was negative for occult blood.

1. Which of the following conditions is most likely responsible for her current symptoms:

   A. Blind loop syndrome.
   B. Small bowel bacterial overgrowth.
   C. Bile reflux gastritis with poor gastric emptying.
   D. Dumping syndrome.
   E. Anastomotic ulceration.

2. Of the following, which test would you order that in your opinion may guide further management?

   A. Upper endoscopy.
   B. Upper GI barium series.
   C. Supine and erect abdomen films
   D. Gastric emptying study.
   E. Glucose – H_2 breath test.

3. A glucose – H_2 breath test showed an early and a sustained rise in breath hydrogen values suggesting bacterial overgrowth. An aspirate from the proximal bowel loop grew >10^5 colonies of Torulopsis glabrata. A solid and liquid phase gastric emptying study was performed. The half time for liquid emptying was 10 min and the half time for solid-emptying was 208 minutes (NR 90–120 min.).

   What is your next step in the management of this patient?

   A. Stop oral feeds and continue TPN only.
   B. Add cisapride and continue oral feeds.
   C. Treat the bowel infection with oral fluconazole.
   D. Treat with high dose oral nystatin suspension.
   E. Encourage feeding with low refined carbohydrates and avoid fluids with meals.

4. After antifungal therapy, the patient gained ten pounds in weight. However, her postprandial symptoms persisted. She continued to bring up bilious material, and had post prandial pain and fullness. What would you now advise this patient?

   A. Stop TPN and oral feeds and start jejunostomy feeding.
   B. Total gastrectomy and Roux-en-Y gastroenterostomy.
   C. Lifelong TPN.
   D. Reassure that these symptoms will resolve and continue on oral fluids and cisapride.
   E. Convert to Billroth I gastrectomy.

EFFECTS OF SURGERY ON GASTRIC MOTOR FUNCTION

Operations that are commonly used to treat patients with peptic ulcer disease (who fail medical management) include vagotomy, gastric resection, "drainage procedure", and a combination of these approaches. Vagotomy (proximal gastric, total gastric, and truncal) may cause the following problems. Proximal gastrovagotomy (PGV) destroys gastric receptive relaxation and impairs storage of food in the proximal stomach. Consequently, gastric filling can induce large intra-gastric pressure waves that lead to rapid gastric emptying of liquids. But this usually does not result in diarrhea or dumping. Importantly, antral peristaltic contractions and gastric emptying are not altered by PGV. The effects of total gastric vagotomy and truncal vagotomy on the proximal stomach are similar to that of proximal gastric vagotomy. However, unlike PGV, both total gastric vagotomy and truncal vagotomy weaken antral peristalsis and delay the gastric emptying of solids. This slow emptying nearly always occurs and requires another operation (a drainage procedure) such as pyloroplasty or gastroenterostomy to facilitate emptying.

Distal Gastric Resection

Distal gastric resection does not significantly affect the motor physiology of the proximal stomach, although gastric emptying of solids may be more rapid. Additionally, duodeno gastric reflux can occur. On the other hand, *Proximal* gastrectomy changes the ability of the stomach to receive and store food. There is a larger postprandial pressure gradient between the stomach and the duodenum that can lead to rapid gastric emptying of liquids.[1]

Drainage Operations

Pyloroplasty has little effect on the function of the proximal stomach. However, disruption of the physical barrier between the stomach and duodenum can result in rapid gastric emptying of solids and increased duodeno gastric reflux. Gastrojejunostomy results in the opposite problem; the net rate of gastric emptying of solids and liquids is slow.[2] Both proximal and distal gastric motor physiology are impaired after vagotomy combined with either a gastroduodenostomy, gastrojejunostomy, or pyloroplasty. After vagotomy and gastroduodenostomy or gastrojejunostomy, gastric emptying of both liquids and solids is rapid, whereas after vagotomy and Roux gastrectomy, gastric emptying of solids and liquids is slower.[3,4,5]

THE POST GASTRECTOMY SYNDROMES

Gastric resection with either vagotomy or a drainage procedure may cause symptoms that are secondary to disrupted gastric motor physiology. These are commonly grouped as the post gastrectomy syndromes and has been reported to occur in 5–50% of patients.[6,7] Table 1 summaries the pathophysiology and treatment options for patients with post gastrectomy syndromes.

Dumping Syndrome

There are two types; early and late dumping. Both forms occur as a consequence of ingesting hyperosmolar, high carbohydrate meals. Early dumping begins within 10–30 minutes after eating while late dumping manifests two to four hours later. Both forms produce vasomotor symptoms such as diaphoresis, weakness, dizziness, flushing, and palpitations. In addition, early dumping causes fullness, crampy abdominal pain, nausea, vomiting, and explosive diarrhea.[8]

The etiology of this problem is rapid emptying of hyperosmolar carbohydrates into the small intestine due to a loss of gastric reservoir function. Vagotomy reduces the reservoir capacity of the stomach and the "drainage procedure" causes a loss of control of gastric emptying. These changes facilitate dumping or the rapid transit of large amounts of carbohydrate-rich material into the small intestine that can lead to excessive fluid shifts and produce symptoms noted above. In addition, high carbohydrate food stimulates the re-

Table 1. Pathophysiology and treatment of postgastrectomy syndromes

Problem	Pathophysiology/etiology	Treatment (medical, surgical)
Dumping	Rapid emptying of hyperosmolar content	Frequent small meals
	↓	Increased refined carbohydrates
	Increases enteroglucagon	Hypothetical: Duodenal passing
	↓	
	Increases insulin	
	↓	
	Lowers blood sugar = "late" dumping	
Small gastric remnant	Extensive gastric resection	Dietary therapy
Postvagotomy diarrhea	Incontinence	Diet
	Truncal vagotomy 20%	Reversed jejunal segment
	Selective vagotomy 5%	
	Highly selective vagotomy 4%	
Gastric atony	Gastric resection or postoperative ileus	Time
	↓	Diet
	Decreases gastric wall tone	Near total gastrectomy
	↓	
	Delayed emptying	
Alkaline reflux gastritis	Gastrojejunostomy destroys or bypasses the pylorus	Bilroth II convert to Bilroth I
		Bilroth I convert to Bilroth II or to Roux-en-Y
Roux stasis syndrome	Distal gastric resection and vagotomy	Hypothetical - "uncut Roux" [will prevent] Roux stasis
	↓	Shorten a very long roux limb
	stasis	Near total gastrectomy

lease of Enteroglucagon from the small bowel causes insulin hypersecretion. The hyperinsulinemia causes hypoglycemia which in turn causes symptoms of late dumping. Epinephrine is also released by the hypovolemia of both early and late dumping which further contributes to these symptoms. After a gastrojejunostomy, the incidence of dumping exceeds 50%.[8] In contrast, after Billroth I gastroduodenostomy and pyloroplasty the duodenal brake mechanism is retained and therefore there is a lower incidence of dumping. Roux-en-Y gastrojejunostomy rarely causes dumping.

Small Gastric Remnant Syndrome

When more than 80% of the stomach is resected, symptoms such as early satiety, pain, vomiting, severe weight loss, anemia, vitamin B12 and iron deficiency can occur.

Postvagotomy Diarrhea

Postvagotomy diarrhea is not uncommon after vagotomy or gastric resection. Truncal vagotomy is responsible for the highest incidence followed by selective and proximal vagotomy (20%, 5%, and 4% respectively). The diarrhea is usually episodic and lessens after the first year.

Gastric Atony

Decreased gastric motor tone causes delayed gastric emptying. Patients complain of pain, nausea, vomiting, and epigastric fullness. Patients cannot tolerate solids, but usually can tolerate liquids. Both postoperative ileus and gastric surgery can cause gastric atony. After gastric surgical procedures, there can be a loss of gastric vagal innervation. Vagotomy results in the loss of muscle contractions leading to disrupted antral mixing. Acute postoperative gastric atony results in the activation of neural reflexes that lead to the release of norepinephrine from the gastric wall. Norepinephrine acts directly on the gastrointestinal smooth muscle, inhibits gastric action potentials, and thus, contractions.[12] The postoperative form of gastric atony lasts three to five days and is in distinct contrast to chronic gastric atony that is seen after gastric operations.

Alkaline Reflux Gastritis

Five to fifteen percent of patients undergoing gastric surgery tend to develop bile or alkaline reflux gastritis.[10] Their symptoms include burning epigastric pain, nausea, bilious vomiting, weight loss, and anemia. This problem occurs when the pylorus is resected, bypassed or destroyed. This form of postgastrectomy syndrome requires surgical treatment more often than any other problem.[11] Bilroth II gastrojejunostomy has the highest incidence of reflux gastritis followed by loop gastrojejunostomy, gastroduodenostomy, and pyloroplasty in that order.

Roux Stasis Syndrome

Roux-en-Y gastrojejunostomy or distal gastric resection with vagotomy may cause a delay in the emptying of solids and liquids from the stomach in about 10 to 50% of patients.[12,13] Surgical construction of the Roux limb requires transection of the jejunum. This process disturbs the normal pattern of aborally propagated potentials.[14] Consequently, ectopic pacemakers develop in the Roux limb which can induce pacesetter potentials that migrate orally.[15] Thus, there is slow transit through the Roux limb, resulting in symptoms. Recently, an operation called "uncut Roux" is being investigated as a management option for patients with stasis related to Roux-en-Y gastroenterostomy.[16]

CONCLUSIONS

Operations which impair proximal gastric motility can cause rapid gastric emptying of liquids and result in both dumping and diarrhea. On the other hand operations that impair distal gastric contractions, such as truncal vagotomy can slow gastric emptying and cause chronic gastric atony. Knowledge regarding the physiology of the stomach, both in health and after surgery should enable the gastroenterologist or the surgeon to better manage patients with the post gastrectomy syndromes. It is worth reiterating that many of these patients have profound motor dysfunction and are best assessed and treated by specialists with major interest in gastric physiology and surgery.

ANSWER KEY AND EXPLANATIONS

1. A blind loop syndrome with bacterial overgrowth is possible and could account for some of her symptoms. Similarly, early satiety and post prandial bloating suggests dumping syndrome. The recurrent nausea and vomiting may suggest gastric efferent loop obstruction or anastomotic ulceration. But she had a normal EGD recently and hence this is unlikely. The symptoms of gastric stasis, the bilious vomiting and the succussion splash would suggest bile reflux gastritis with poor gastric emptying. The best option is **C.**

2. An upper endoscopy is superfluous as she had a normal exam recently. On the other hand, an upper GI barium series may be useful to delineate the anatomy and demonstrate any luminal obstruction, ulceration or stasis. But, barium studies are often unhelpful for assessing motor function. A supine and erect abdominal film may show the presence of a bezoar, small intestinal obstruction, or fluid levels. A breath test may help to evaluate the presence of bacterial overgrowth. The overwhelming issue is whether the gastric remnant is functioning satisfactorily or in fact is significantly contributing to her current illness. This can be best assessed by a solid and liquid phase gastric emptying study. The best option is **D.**
3. The tests reveal fungal overgrowth in the small bowel with impaired gastric emptying. Although stopping feeding would help her symptoms. Actually, in the long term they would not help. Cisapride is unlikely to be helpful in this situation. Fluconazole is often implicated in the overgrowth of torulopsis, a candida-like fungus and hence this is not the preferred antifungal agent. Oral feeding is unlikely to help. The best option is **D.**
4. In the presence of refractory symptoms and documented gastric stasis, the best option would be to remove the gastric remnant and perform a Roux-gastroenterostomy. This procedure was performed with substantial improvement. The patient was able to resume oral feeding, and return to work. The best option is **B.**

REFERENCES

1. Cullen JJ, Kelly KA: Gastric motor physiology and pathophysiology. Surg Clinics of North America 1993;73:6:1145–1160.
2. Dozois RR, Kelly KA, Code CF: Effect of distal antrectomy on gastric emptying of liquids and solids. Gastroenterol 1971;61:675–681.
3. Fich A, Neri M, Camilleri M, et al: Stasis syndromes following gastric surgery: Clinical and motility features of 60 symptomatic patients. J Clin Gastroenterol 1990;12:505–512.
4. Hinder RA, Kelly KA: Canine gastric emptying of solids and liquids. Am J Physiol 1977;233:E335-E340.
5. Hinder RA, Kelly KA: Human gastric pacesetter potential: Site of origin, spread and response to gastric transection and proximal gastric vagotomy. Am J Surg 1977;133:23–33.
6. Hocking MP, Vogel SB, Falasca CA, et al: Delayed gastric emptying of liquids and solids following Roux-en-Y biliary diversion. Ann Surg 1981;194:494–501.
7. Karlstrom L, Kelly KA: Ectopic jejunal pacemakers and gastric emptying after Roux gastrectoy: Effect of intestinal pacing. Surgery 1989;106:867–871.
8. Kelly KA: The motility of the stomach and gastroduodenal junction. *In* Johnson LR, Christensen J, Grossman MI, et al (eds): Physiology of the Gastrointestinal Tract, vol 1. New York, Raven Press, 1981, pp 383–410.
9. Wilbur BG, Kelly KA: Effect of proximal gastric, complete gastric, and truncal vagotomy on canine gastric electric activity, motility and emptying. Ann Surg 1973;178:295–303.
10. Wilbur BG, Kelly KA, Code CF: Effect of gastric fundectomy on canine gastric electrical and motor activity. Am J Physiol 1974;226:1445–1449.
11. Ritchie WP Jr: Alkaline reflux gastritis: An objective assessment of its diagnosis and treatment. Ann Surg 1980;192:288–298.
12. Britten JP, Johnston D, Ward DC, et al: Gastric emptying and clinical outcome after Roux-en-Y diversion. Br J Surg 1987;74:900–904.
13. Herrington JL, Scott HW, Sawyers JE: Experience with vagotomy-antrectomy and Roux-en-Y gastrojejunostomy in surgical treatment of duodenal, gastric, and stomal ulcers. Ann Surg 1984;199:590–597.
14. Richter HM III, Kelly KA: Effect of transection and pacing on human jejunal pacesetter potentials. Gastroenterology 1986;91:1380–1385.
15. Karlstrom L, Soper NJ, Kelly KA, et al: Ectopic jejunal pacemakers and enterogastric reflux after Roux gastrectomy: Effect of intestinal pacing. Surgery 1989;106:486–495.
16. Stoddard CJ, Duthie HL: Effect of vagotomy on the response of gastric myoelectrical activity to glucagon and food. Scand J Gastroenterol 1976; 11 (suppl 42):77–83.

16

DYSPEPSIA AND UPPER GASTROINTESTINAL MOTILITY IN CHILDREN

Carlo Di Lorenzo*

University of Pittsburgh
Children's Hospital of Pittsburgh
Pittsburgh, Pennsylvania

INTRODUCTION

Symptoms suggestive of intestinal dysmotility in children are always a source of great anxiety for both the patient and the parents. Chronic vomiting or persistent diarrhea may have serious consequences on growth organ. Infants double their body weight during the first five months of life and triple it by the end of one year and even a marginal loss of calories can cause failure to thrive. Having normal bowel movements is seen as an essential sign of health in children of all ages. Any deviation from what is felt to be normal triggers a call to the pediatrician. Abdominal pain in a child is particularly worrisome for the parents who worry that a serious disease is causing the symptom. In reality, most children presenting with either constipation, diarrhea, abdominal pain, or vomiting do not suffer from any serious organic disease and have what is considered a "functional" disorder. For long time, it was felt that motility abnormalities were the primary cause of many pediatric functional bowel disorders. Disturbances in gastrointestinal motility have been described in children with recurrent abdominal pain, gastroesophageal reflux, toddler's diarrhea, nonulcer dyspepsia, intestinal pseudo-obstruction, achalasia, Hirschsprung's disease, and functional constipation and rumination syndrome (1, 2). These entities encompass more than 50% of the children presenting to pediatric gastroenterologists. In some conditions such as achalasia, Hirschsprung's disease and pseudo-obstruction the motility abnormality is almost entirely responsible for the symptoms, in others it constitutes only one of the many pathogenic factors. It is now possible to study gastrointestinal motility in children using most of the same diagnostic techniques used by adult gastroenterologists. Antroduodenal manometry, electrogastrography, nuclear medicine investigations, the electronic

* Tel: (412) 692-5180, Fax: (412) 692-7355, e-mail: dilorec@chplink.chp.edu

Gastrointestinal Motility, edited by Rao *et al.*
Kluwer Academic / Plenum Publishers, New York, 1999.

barostat, and transit studies using the ^{13}C octanoic acid breath test have now been used in children and infants. The paucity of normal controls and the patient's lack of cooperation during invasive and often tedious tests continue to create problems in the interpretation of study results in the pediatric age. Differentiation of artifacts from motor events and the understanding of the effect of stress and discomfort on the test require considerable expertise which cannot be developed in most pediatric gastroenterology training programs. The correct interpretation of manometry studies in children requires also an understanding of the differences between gastrointestinal motility in children and adults and experience in dealing with children and their diseases. It is probably advisable that some of the most sophisticated and innovative tests to study gastrointestinal motility in children continue to be performed in few selected tertiary pediatric referral centers.

After discussing a case, I will briefly review some of diagnostic techniques available for the study of gastrointestinal motility in children, with special attention to the factors that make the tests peculiar to the pediatric age. I will then review some of the childhood conditions associated with gastrointestinal dysmotility.

CASE STUDY

A 14-year-old adolescent girl presents with a one year history of postprandial non bilious emesis. She describes sudden, effortless regurgitation few minutes after each meal. There is no abdominal pain or headache, but minimal nausea. She has lost 16 lb. in weight. The episodes of vomiting had started suddenly one year prior to the visit and have persisted unchanged in frequency or intensity. Past medical history reveals that she had scoliosis, but was otherwise healthy. Review of system was unremarkable. Treatment with cisapride, domperidone, metronidazole, ondansentron, sucralfate, and omeprazole did not help. Screening laboratory tests including a complete blood count, chemistry panel, stool evaluation for ova and parasites, and urinalysis were normal. Physical examination reveals an apprehensive 14 year old girl, well nourished and not acutely distressed. Abdominal and neurological exam were normal.

1. What is the next step in the evaluation of this patient?

   A. Upper gastrointestinal series
   B. Esophagogastroduodenoscopy
   C. Radionuclide gastric emptying study
   D. Ultrasound of the abdomen

All these tests were performed and were normal, except for the gastric emptying study, which was difficult to interpret because of postprandial vomiting. The vomiting and the weight loss continued and the patient was hospitalized. A nasojejunal catheter was placed for nighttime infusion of a high caloric density formula.

2. What would be your next step?

   A. Psychiatric evaluation
   B. MRI head
   C. Hepatobiliary scintigraphy
   D. CT of the abdomen
   E. 24 hour intraesophageal pH monitoring
   F. Antroduodenal manometry

The MRI revealed no evidence of intracranial pathology. The psychiatric evaluation was not in favor of an eating disorder. An antroduodenal manometry was performed (Figure1). There was a normal fasting pattern and an adequate motor response to the meal. There were many episodes of regurgitation associated with simultaneous increases of pressure in all the recording sites.

3. What diagnosis does this study suggest?

   A. Intestinal pseudo-obstruction
   B. Mechanical obstruction
   C. Munchausen's syndrome by proxi
   D. Rumination syndrome
   E. Scleroderma

THE STUDY OF GASTROINTESTINAL MOTILITY IN CHILDREN

The evaluation of a child with a suspected motility disorder often begins with a radiologic study. The information derived from observing the transit of barium should not be relied upon to determine if a child has a gastrointestinal motor disturbance. Radiologic contrast studies should be used only to clarify the gastrointestinal tract anatomy. Among the common anatomical abnormalities associated with pediatric motor disorders are malrotation and the superior mesenteric syndrome. Malrotation is found in one third of children with congenital intestinal pseudo-obstruction. A compression of the superior mesenteric artery on the third portion of the duodenum is often found in children with rapid weight loss. The superior mesenteric artery syndrome may be both a consequence of an acquired motility disorder and contribute to aggravate the symptoms.

Gastric motility may be indirectly evaluated by using radionuclide techniques. Gastric emptying is often measured in infants in conjunction with the performance of gastroesophageal scintigraphy for the evaluation of reflux. Standardization of this test is particularly difficult in children. Milk or formula labeled with ^{99m}Tc is used in infants. Ideally, the composition and volume of milk should vary according to body size. In practice, many children are on modified feeding regimens and may be able to drink only a selected type of formula or smaller volumes than required by the test. Forceful restraint is often needed to keep the infant motionless under the gamma camera for an extended period of time. Since healthy children are not studied, the exact normal values for gastric emptying are difficult to establish in pediatrics. There is no experience in pediatrics with the use of scintigraphy to evaluate small bowel transit.

Electrogastrography (EGG) has been used in preterm and term infants and in older children (3). Abnormalities of gastric electric rhythm have been described in children with nonulcer dyspepsia and intestinal pseudo-obstruction (4, 5). Improvement of symptoms has been associated with restoration of normal gastric rhythm (6). The use of EGG as a reliable diagnostic technique in children still requires standardization of methods such as where to place electrodes, the duration of test, and how to analyze and interpret the study.

Manometry represents the most direct approach to study the strength, duration, frequency, and propagation of gastrointestinal contractions. It is a technique, which has received much interest during recent years in pediatrics, and it might provide information which can clarify the diagnosis, direct treatment and have prognostic value (Table 1). There are unique challenges associated with the use of manometry in children. The size of patients and the lack of cooperation complicate the placement of manometry catheters and

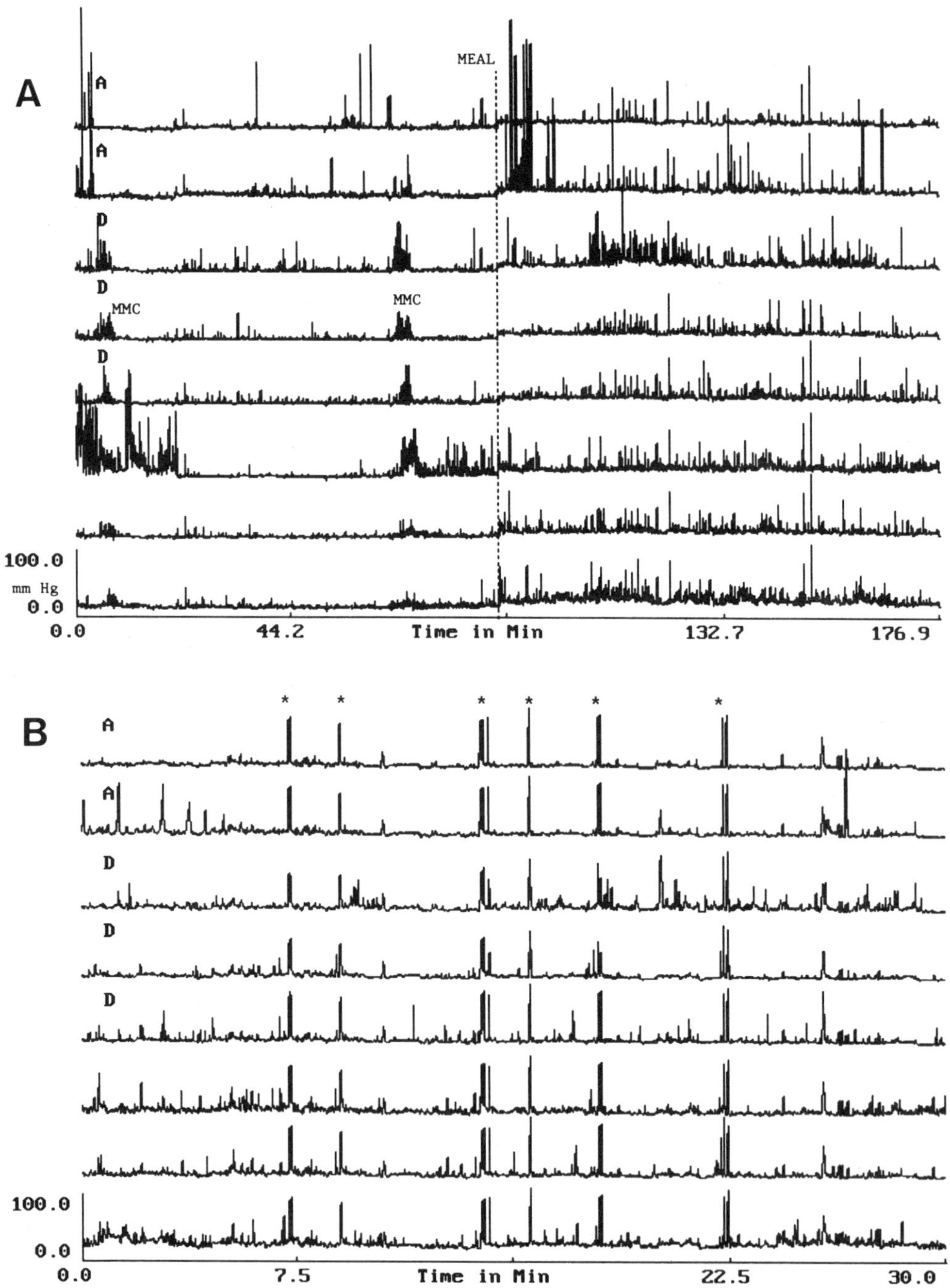

Figure 1. Antroduodenal manometry from the patient described in the case study (refer to question 3). A: Most of the study is shown. It is possible to notice normal MMCs occurring during fasting and a normal gastric and intestinal response following the ingestion of a meal. B: A segment of the postprandial study in the same subject. There are synchronous phasic pressure waves in all recording channels (*) associated with episodes of regurgitation. A= antrum; D=Duodenum.

Table 1. Information provided by gastrointestinal manometry in children

Diagnosis
Normal (consider emotional and behavioral disorders, Munchausen's by proxi syndrome) (27)
Obstruction (repetitive postprandial clusters of small bowel contractions) (28)
Rumination (simultaneous postprandial pressure increases in stomach and small bowel, normal fasting motility) (2)
Neuropathy vs. myopathy (normal amplitude uncoordinated contractions vs low amplitude coordinated contractions) (29)
Assessment of the extent of the disease (isolated gastric vs. isolated small bowel disease vs. both)
Prognosis
MMCs (presence of MMCs associated with the ability to tolerate enteral feedings) (23)
Motility index (higher motility index associated with better prognosis) (30)
Treatment
Effect of drugs (31)
Success of jejunal feedings (23)
Before small bowel transplantation for CIP (confirm the diagnosis and assess need for isolated small bowel vs. multiviscera transplant)

Information provided by gastrointestinal manometry in children.

the completion of studies. Although catheters with implanted transducers that are small enough to be used in toddlers are now available, most data in the pediatric literature have been acquired using water-perfused catheters. One should remember that the standard rate of water infusion with adult manometry studies might place infants at risk of overhydration and electrolyte imbalance. In younger children or during prolonged studies we often perfuse the catheters with dilute oral rehydrating solutions and administer electrolyte-containing fluids intravenously. Fluoroscopy or endoscopy are used to facilitate catheter placement in the sedated child. The study should be initiated only when the child has completely recovered from the sedation and from the stress of waking up with a naso-duodenal tube. In most instances, it is advisable to perform the study the day after catheter placement. The duration of test is based on the patient's age. Motor migrating complexes (MMCs) have a shorter cycle in children than in adults and four hours of fasting recording is usual sufficient to observe at least one MMC in the younger patient. The study is then completed by a two-hour postprandial recording. Drugs, such as erythromycin and the long acting somatostatin analogue, octreotide, are often administered during a motility study to test their potential as therapeutic agents. The long duration of the manometry test provides also a unique opportunity to evaluate the parent-child interaction under stressful circumstances. How the child and the parent(s) react to stress and to symptoms may play an important role in many childhood functional bowel diseases. A trained observer should be present during every pediatric study to ensure the child's safety, to correlate symptoms with discrete manometric events, and to observe the child and the parent's behavior.

Irritable Bowel Syndrome, Recurrent Abdominal Pain and Nonulcer Dyspepsia

Irritable bowel syndrome (IBS) and functional or recurrent abdominal pain of childhood are two very common and often overlapping entities in children. Recurrent abdominal pain (RAP) is traditionally defined as the presence of at least three episodes of abdominal pain interfering with daily activities occurring every month over a period of at least 3 months (7). The pain is located in the periumbilical region without radiation to the chest or to the back and there is no correlation with meals, exercise or position. It has been estimated that approximately 13% of middle school students and 17% of high school students suffer from weekly episodes of abdominal pain (8). A diagnosis of IBS, based on the criteria used in adults, may be made in 6% of middle school and 14% of high school students, a prevalence similar to that found in the adult population. In the same study, it was noted that approximately 8% of all students had visited a physician because of abdominal pain during the previous year (8). Recurrent abdominal pain may also be associated with nausea, bloating, and early satiety (9), overlapping with the symptoms which are characteristic of nonulcer dyspepsia (NUD). For many years, children with NUD have been considered to suffer from a severe form of RAP. In childhood NUD, eating may cause symptoms severe enough to require special means of alimentation (10), making it arduous to differentiate between severe NUD from mild form of intestinal pseudo-obstruction. It has been suggested that a disturbance in the regulation of gastrointestinal motility in response to a heterogeneous group of physical or psychosocial stimuli may be the major contributing factor in the development of functional abdominal pain (11). Several investigators have described heterogeneous abnormalities in antroduodenal motility in children with functional bowel diseases (12–14). However, it remains unclear whether the motor disorders have any etiologic role in causing symptoms or are merely epiphenomena found in an organ, the intestine, capable of only a limited repertoire of responses to different stimuli. There is a poor temporal correlation between motility abnormalities and symptoms. Some children have symptoms when manometry is normal. Other children have abnormal contractions but no symptoms. Only in a minority of children with nonulcer dyspepsia one may detect a significant delay in gastric emptying. It is currently believed that manometric studies rarely provide direct insight into the pathogenesis of visceral pain. Children with IBS, RAP or NUD who for the most part have only minor or no abnormalities of gastrointestinal motility demonstrate an abnormal perception of noxious and non-noxious visceral stimuli. The use of the electronic barostat has demonstrated that the same stimulus causing minimal discomfort in normal individuals can be perceived as extremely painful by children with functional bowel diseases (15). The same children have a normal or even decreased threshold for somatic pain.

Toddler's Diarrhea

This is an entity described as a persistent, unexplained diarrhea typically occurring in children between 1 and 4 years of age. Stools are loose, occur up to 6–8 times each day and often contain undigested material, without blood or pus. Children with this condition are healthy, active and thriving and the diarrhea rarely lasts after the age of 5 years. High fat diet may be quite helpful to decrease frequency of bowel movements. It has been suggested that children with toddler's diarrhea have an inadequate postprandial motor response with MMCs which continue to cycle even after feeding, causing a more rapid small bowel transit time (16). The intestinal content dumped in the colon under these circumstances may be larger than usual and contain partially digested food. Undigested nutrients in the colon are fermented and produce gaseous distension, cramps and abdominal discomfort. Sudden distension of the colon is one of the factors which can trigger high amplitude propagated colonic contractions which can rapidly carry the feces to the rectum and contribute to the evacuation of frequent, liquid stools.

Chronic Intestinal Pseudo-Obstruction

The term chronic intestinal pseudo-obstruction (CIP) describes a heterogeneous group of gastrointestinal nerve and muscle disorders. Childhood forms differ from adult forms of

CIP in many aspects. The majority of children with this condition suffer from congenital and primary CIP. At times, childhood CIP can even be diagnosed *in utero*, when ultrasonography demonstrates polyhydramnios and megacystis (17). Secondary forms, such as those found in adults with diabetes, scleroderma, paraneoplastic syndromes, and other extrinsic neuropathies are uncommon in children. Familial forms are rare and most cases of childhood CIP can be considered sporadic. Pseudo-obstruction carries a much higher morbidity in children than in adults. One third of children with CIP die in the first year of life (18). Several factors interact to cause such high mortality. Urinary bladder neuromuscular disease is present in one third of children with congenital CIP and is particularly prevalent in children with the myopathic form of CIP. These children are at high risk for recurrent urinary tract infections and kidney damage. Bacterial overgrowth is common in children with dilated bowel loops and poor intestinal motility, and may be responsible for translocation of intestinal flora leading to sepsis. Children with venting gastrostomies and ileostomies frequently suffer from fluid and electrolytes imbalance. Because of the very high growth rate in the newborn period, parenteral nutritional support is often necessary to provide appropriate caloric intake. Total parenteral nutrition (TPN) carries a significant health risk and is responsible for most morbidity and mortality in children with CIP. Infants who require TPN from birth are at high risk for cholestasis and progressive liver damage. Repeated episodes of sepsis and thrombotic complications may cause progressive difficulty in maintaining vascular access sites in children who require many years of TPN.

There is no single test diagnostic of CIP. A high index of suspicion and the demonstration that there is no mechanical obstruction are key to an early diagnosis. Often, children undergo many exploratory laparotomies before a diagnosis of CIP is made. Esophageal manometry is abnormal only in a minority of children with CIP and should not be relied upon to make a diagnosis of CIP. Antroduodenal manometry is always abnormal when the disease involves the upper gastrointestinal tract (19) (Figure 2). When the manometry study is normal, one should suspect factitious disorders or behavioral problems in the child or the family (17). In the absence of a dilated bowel, manometry allows differentiation between a neuropathy and a myopathy. Coordinated, persistently low amplitude contractions suggest myopathy, whereas normal amplitude, uncoordinated contractions and abnormal patterns of contractions have been associated with neuropathy. The presence of propagating clustered contractions and non propagated, prolonged contractions during the postprandial period suggests partial bowel obstruction and should prompt a renewed search for a mechanical obstruction. In children, there are motility characteristics correlating with the ability to tolerate enteral feeding and a more favorable response to cisapride (20) (Table 1).

Treatment of childhood CIP is challenging and at times frustrating. Careful evaluation of the motor function of the entire intestine should be performed before proceeding with any surgical intervention. Gastrostomies are often placed to decrease the frequency of vomiting, or to allow gastric decompression after an antireflux surgery had made vomiting impossible (21). Decompression may improve gastrointestinal motility as the dilated bowel resumes a normal diameter. Jejunostomies are used to bypass the stomach and provide slow continuous infusion of nutrients (22). Parenteral support is reserved for children who fail trials of gastrostomy and jejunostomy feedings. Prokinetic drugs are occasionally effective in children with CIP without irreversibly dilated bowel. Cisapride improves small bowel motility in some children with CIP and may help transitioning from parenteral nutrition to enteral feedings (20, 23). Other children may benefit from the use of erythromycin and/or octreotide (24). Less favorable, in children with CIP, has been the experience with metoclopramide, domperidone, leuprolide, and misoprostol (25). Isolated small bowel and multiviscera transplantations have been used to treat children with CIP complicated by life

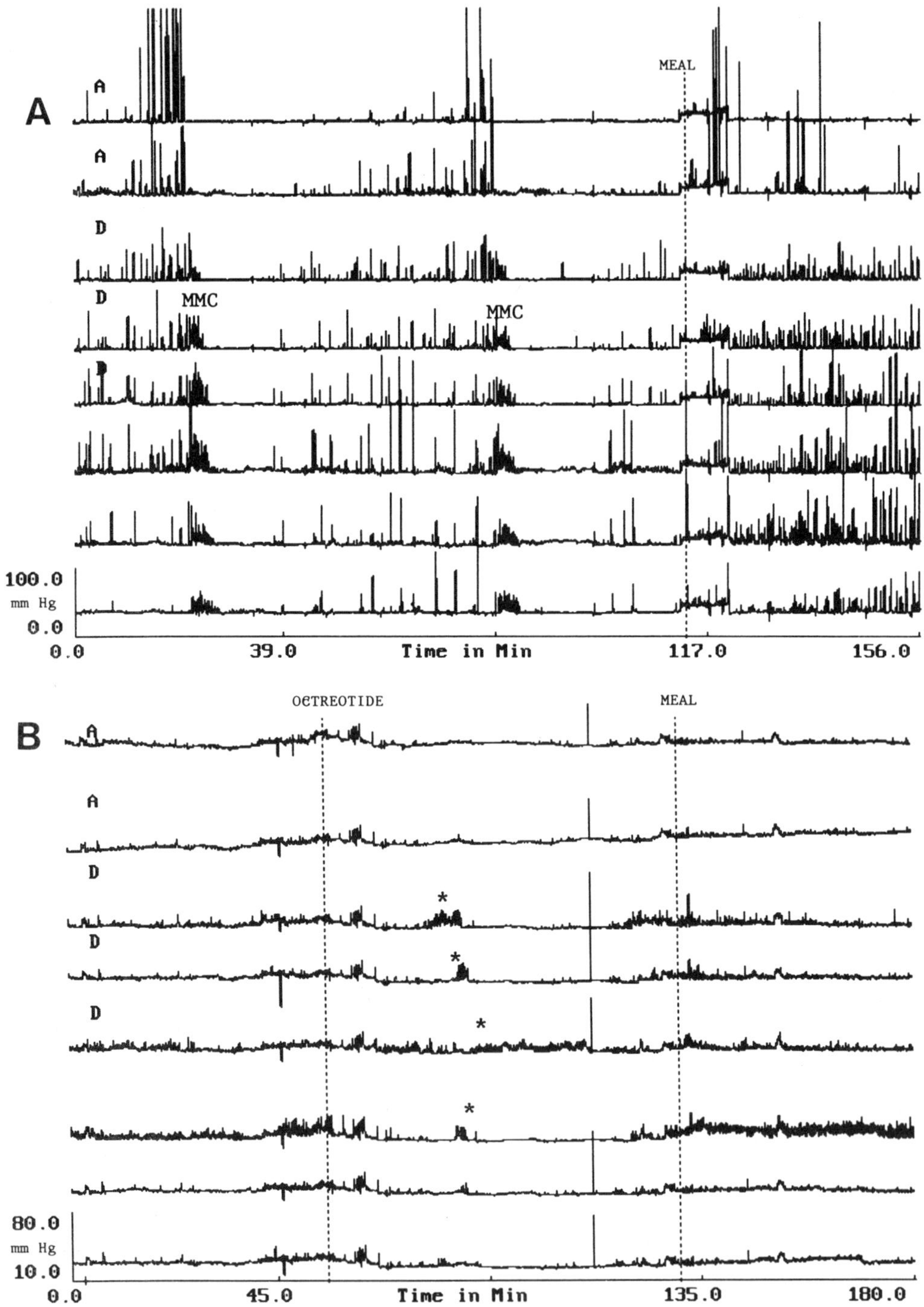

Figure 2. A. Antroduodenal manometry from a patient without CIP. The three phases of the MMCs can be easily recognized. There is a normal motor response to the ingestion of a meal. B. Antroduodenal manometry from a 9 year old girl with a myopathic CIP. There is very minimal motor activity during the baseline study and after ingestion of a meal. Injection of 2 mcg octreotide subcutaneously induces a low amplitude cluster of contractions (*). A= antrum; D=Duodenum.

threatening conditions such as liver failure, recurrent sepsis, and lack of venous access for nutritional support (26). Long term outcome of small bowel transplantation performed in children with CIP is similar to that of bowel transplantation performed for other indications.

CONCLUSIONS

Motility disorders are common in children. Some of the pediatric conditions associated with gastrointestinal motor dysfunction are similar to those found in adults; others are peculiar to the pediatric age. Gastrointestinal manometry studies are feasible in children and currently should not be considered "experimental". In experienced hands, they provide a wealth of information which directly aid in the evaluation and treatment of a child with a suspected motility disorder.

ANSWER KEY EXPLANATIONS

1. This young girl has lost significant amount of weight, is symptomatic and has not responded to a variety of therapeutic agents. At this stage it is imperative to perform appropriate tests to exclude upper gut obstruction and/or a motility disorder. All of the proposed tests are reasonable but the best option would be an upper endoscopy followed by a gastric emptying study. The best option is **B**.
2. Because initial evaluations were unrevealing further work up is required. At this stage one should look for local GI as well as extra GI causes that may lead to protracted vomiting and weight loss. Hence it is appropriate to evaluate the CNS and an MRI scan would be an appropriate step. The best option is **B**.
3. Based on the history and results of the manometry study and the other tests a diagnosis of rumination syndrome was made. We recommended treatment with amytriptiline and a trial of biofeedback and relaxation techniques with a behavior psychologist. She improved remarkably during the next 8 weeks and she is now completely asymptomatic. Rumination is a syndrome characterized by repetitive postprandial regurgitation of small amounts of gastric content. The food is then either reswallowed or expelled. This syndrome has been extensively described in neurologically handicapped infants and in children with abnormal mother-child interactions but only recently has been reported in adolescents and adults (2). Antroduodenal manometry demonstrates a characteristic pattern with episodes of vomiting associated with synchronous pressure increases termed "R waves" in all recording sites. In most other disorders associated with vomiting, the emesis is usually associated with high pressure antral contractions. Reassurance and behavioral therapy are currently the most effective interventions in patients with this condition. The best option is **D.**

REFERENCES

1. Di Lorenzo C, Hyman PE. Pediatric application of pH monitoring. In Richter J ed. Ambulatory esophageal pH monitoring: Practical approach and clinical applications. 2nd edition. William and Wilkins, Baltimore, MD, 1997: 195–205.
2. Malcom A, Thumshirn MB, Camilleri M, Williams DE. Rumination syndrome. Mayo Clin Proc 1997; 72: 646–52.

3. Koch KL, Tran TN, Stern RM, et al.. Gastric myoelectrical activity in premature and term infants. J Gastrointest Mot 1993: 5, 41–7.
4. Cucchiara S, Riezzo G, Minella R, et al.. Electrogastrography in nonulcer dyspepsia. Arch Dis Child 1992: 67, 613–7.
5. Di Lorenzo C, Reddy SN, Flores AF, Hyman PE. Is electrogastrography a substitute for manometric studies in children with functional gastrointestinal disorders?. Dig Dis Sci 1997; 42: 2310–2316.
6. Cucchiara S, Minella R, Riezzo G, et al.. Reversal of gastric electrical dysrhythmias by cisapride in children with nonulcer dyspepsia. Dig Dis Sci 1992: 37, 1136–40.
7. Apley J, Naish N. Recurrent abdominal pain: a field survey of 1000 school children. Arch Dis Child 1958: 33, 165–70.
8. Hyams JS, Burke G, Davis PM, Rzepski B, Andrulonis PA. Abdominal pain and irritable bowel syndrome in adolescents: a community based study. J Pediatr 1996; 129: 220–6.
9. Stone RT, Barbero GT. Recurrent abdominal pain in childhood. Pediatrics 1970: 45, 732—8.
10. Di Lorenzo C, Hyman PE, Flores AF, et al.. Antroduodenal manometry in children and adults with severe nonulcer dyspepsia. Scand J Gastroenterol 1994: 29, 799–806.
11. Boyle JT. Chronic abdominal pain. In: Pediatric gastrointestinal disease. Ed: Walker WA, Durie PR, Hamilton JR, Walker—Smith JA, Watkins JB; BC Decker Inc, Philadelphia, PA, 1994, 45–54.
12. Pineiro-Carrero VM, Andres JM, Davis RH, et al. Abnormal gastroduodenal motility in children and adolescents with recurrent functional abdominal pain. J Pediatr 1988: 113, 820–5.
13. Hyman PE, Napolitano J, Diego A, et al.. Antroduodenal manometry in the evaluation of chronic functional gastrointestinal symptoms. Pediatrics 1990: 86, 39–44.
14. Cucchiara S, Bortolotti M, Colombo C, et al.. Abnormalities of gastrointestinal motility in children with nonulcer dyspepsia and in children with gastroesophageal reflux disease. Dig Dis Sci 1991: 36, 1066–73.
15. Di Lorenzo C, Sigurdsson L, Griffiths J, Scharff L, Wald A. Site specific hyperalgesia in children with functional bowel diseases. Am J Gastroenterol 1998; 93: 1668.
16. Fenton TR, Harries JT, Milla PJ. Disordered small bowel motility: a rational basis for toddler' diarrhea. Gut 1983; 24: 897–903.
17. Hyman PE. Chronic intestinal pseudo-obstruction. In: Hyman PE, Di Lorenzo C eds. Pediatric gastrointestinal motility disorders. Academy Professional Information Services, Inc., New York, NY, 1994, 115–28.
18. Vargas JH, Sachs P, Ament ME. Chronic intestinal pseudo-obstruction syndrome in pediatrics: results of a national survey by members of the North American Society of Gastroenterology and Nutrition. J Pediatr Gastroenterol Nutr 1989: 7, 323–32.
19. Hyman PE, McDiarmid SV, Napolitano J, et al.. Antroduodenal motility in children with chronic intestinal pseudo-obstruction. J Pediatr 1988: 112, 899–905.
20. Hyman PE, Di Lorenzo C, McAdams L, et al.. Predicting the clinical response to cisapride in children with chronic intestinal pseudo-obstruction. Am J Gastroenterol 1993: 88, 832–6.
21. Pitt HA, Mann LL, Berquist WE, et al. Chronic intestinal pseudo-obstruction: Management with total parenteral nutrition and a venting enterostomy. Arch Surg 1985; 120, 614–8.
22. Di Lorenzo C, Reddy SN, Villanueva-Meyer J, et al.. Cisapride in children with chronic intestinal pseudo-obstruction. Gastroenterology 1991: 101, 1564–70.
23. Di Lorenzo C, Flores AF, Buie T, et al. Intestinal motility and jejunal feedings in children with chronic intestinal pseudo-obstruction. Gastroenterology 1995: 108, 1379–85.
24. Di Lorenzo C, Flores AF, Tomomasa T, et al. Effect of erythromycin on antroduodenal motility in children with chronic functional gastrointestinal symptoms. Dig Dis Sci 1994: 39, 1399–405.
25. Starzl TE, Todo S, Tzakis A, et al. The transplantation of gastrointestinal organs. Gastroenterology 1993: 104, 673–9.
26. Rudolph CD, Hyman PE, Altschuler SM, et al. Diagnosis and treatment of chronic intestinal pseudo-obstruction in children. J Pediatr Gastroenterol Nutr 1997; 24: 100–1.
27. Hyman PE, Di Lorenzo C, Beck D, Hamilton AB, Zeltzer LK. Discriminating chronic intestinal pseudo-obstruction from Munchausen's syndrome by proxy in toddlers with intestinal failure. Gastroenterology 1995; 108: A619.
28. Camilleri M. Jejunal manometry in distal subacute mechanical obstruction: significance of prolonged simultaneous contractions. Gut 1989; 30: 468–75.
29. Cucchiara S, Annese V, Minella R, et al.. Antroduodenojejunal manometry in the diagnosis of chronic intestinal pseudo-obstruction in children. J Pediatr Gastroenterol Nutr 1988; 7: 319–22.
30. Fell JM, Smith VV, Milla PJ. Infantile chronic intestinal pseudo-obstruction: the role of small intestinal manometry as a diagnostic tool and prognostic indicator. Gut 1996; 39 306–311.
31. Verne GN, Eaker EY, Hardy E, Sninsky CA. Effect of octreotide and erythromycin on idiopathic and scleroderma-associated intestinal pseudoobstruction. Dig Dis Sci 1995; 40: 1892–901.

17

PROKINETICS

Joseph A. Murray*

Mayo Medical School
Rochester, Minnesota 55905

INTRODUCTION

The gastrointestinal tract is an active conduit, not a passive tube. Movements of the muscle which lines this tube are not only responsible for the mechanical breakdown of food particles and the mixing of chyme with digestive enzymes but also for the transport of material. In addition, motility provides an important housekeeping function that occurs predominately at night and during fasting. Fasting motor activity clears the stomach and small intestine of undigested large particles, and prevents bacterial overgrowth. The colon, of course, has the important function of water absorption, storage, and elimination of stool. These complex motor events are controlled by an interplay of intrinsic and extrinsic nerves and hormones that act upon different hollow organs, each with its own distinct structure and function.[1] Neural input can be divided into excitatory and inhibitory innervation. Although, one may presume that excitation would lead to propulsion and inhibition to stasis, this is not always true. There is a complex interaction between these two mechanisms that allows integrated function of the GI tract. Furthermore, the motor functions do not occur independent of the secretory and absorptive functions of the GI tract. For example, rapid transit can lessen the time for effective absorption to occur, thereby causing diarrhea. Thus, optimal gut function depends on a well-coordinated gastrointestinal motor activity.

PROKINESIS

This chapter will address only those agents that may enhance the normal motor contractions of the gastrointestinal system. It will not address those agents that inhibit motor activity such as loperamide or the smooth muscle relaxants such as clinidinium or atropine like agents.

---

* 507-284-2467(Tel), 507-284-0538(Fax), Email <murray.joseph@mayo.edu

Gastrointestinal Motility, edited by Rao *et al.*
Kluwer Academic / Plenum Publishers, New York, 1999.

Prokinesis can be defined as the promotion of beneficial motor activity of the gastrointestinal system. Implicit is the enhancement of propulsive activity though this is not necessarily the result. The aim is to enhance or restore normal motor function, not simply increase contraction (squeezing) and forward motion. It should be organ specific, i.e., targeting the organ that is dysfunctional, be that the esophagus, the lower esophageal sphincter, stomach, small intestine, or colon. For example, a patient may not benefit with enhanced gastric emptying if he suffers with disabling diarrhea.

LIMITATIONS OF PROKINETIC THERAPY

Our therapeutic armentarium is quite limited with no more than 5–6 agents currently in use as prokinetics, very few of which are FDA approved. There are many agents with prokinetic effects in development, but as we have come to realize that there can be a lag period between the development and marketing of new agents especially in areas that are not subject to fast track approval process that we have seen with AIDS and cancer agents.

There are many problems with prokinetic therapy. Of the many candidate substances that affect motor function that have been identified in lab research, the vast majority don't make it to the bedside. This can be due to species differences in effects, unacceptable side effects either on other parts of the GI tract or on other organ systems. Another limitation to the development of effective prokinetic therapies is the lack of real understanding of the complex pathophysiologic processes that cause a motility problem. As has been pointed out elsewhere in this book, derangements are quite complex and extend far beyond mere motor function but encompasses much more complex brain-gut interaction particularly those affecting sensation. The other important concept is that squeezing does not equal propulsion. Just because your gut will squeeze doesn't mean it will propel the bolus caudally. Often an organ will at one time retain the bolus for mixing, trituration, or absorption and the same organ will subsequently need to propel the bolus. Another important concept is the inherent plasticity of the enteric nervous system (ENS). The ENS is both plastic in its function and in its structure and can produce tolerance to drugs rather quickly. The ENS is a very complex system with multiple neurotransmitters, central connections, as well as interactions with cytokines and other non-neurologic cells within the gut. Altering a single component of the system—or using a drug that changes one mechanism may have a short-lived effect as tolerance may occur. Other limitations are non-GI side effects and drug-drug interactions. The drug interactions are discussed below.

CLASSES OF PROKINETIC MEDICATIONS

There are several classes of drugs that have been used as prokinetic agents. (Table 1). Most of these were not designed as prokinetics and few are approved by the FDA specifically for prokinesis.

Specific Drugs

I will only discuss those agents that are currently available in the USA.

The original or prototypic agents are the cholinomimetic agents: –, bethanechol and neostigmine. These could be termed "squeezers". These act directly on cholinergic receptors and induce contraction. These agents do not produce coordinated contractions. They may not

Table 1. List of prokinetic agents

Cholinomimetics
Bethanechol*
Neostigmine
Cholinergics
Cisapride*
Metoclopramide*
Mosapride
Dopamine antagonists
Domperidone
Metoclopramide
Macrolides (motilin agonists)
Erythromycin*
EM-523 (synthetic macrolide)
Azithromycin* (not practical due to long t ½)
Prostaglandin analogues
Miscellaneous
Somatostatin analogues (octreotide)*

* Available for use in the USA though not necessarily approved as motility agents.

enhance propulsive force and are largely reserved for gastroparesis and are rarely used. The most commonly used agents are primarily cholinergic: –metoclopramide or cisapride. These agents increase the release or the effect of acetylcholine in the GI tract. Motilin agonists are one of the more exciting groups of agents. The prototype motilin agonist is erythromycin (the antibiotic). Derivatives of erythromycin and other motilin agonists currently are in phase 2 trials. Opioid receptor agonists have some potential prokinetic effect, though the major interest in these compounds has been in the area of visceral pain. Another area of drug development is 5 HT_4 agonists – cisapride is one of these compounds and others are in development.

Another class of drugs is the dopamine receptor antagonists—such as metoclopramide or domperidone. Domperidone has been available in Europe and elsewhere for a number of years. Its major use is in disorders of gastric emptying and non-ulcer dyspepsia. Its advantages are the improvement in gastric emptying together with a reduction in the feeling of post-prandial dyspepsia. CCK antagonists, such as loxiglumide and other peptides, in particular octreotide have a limited role in the treatment of motility disorders. There are other miscellaneous agents that are useful, such as misoprostol for severe constipation[2] and unconfirmed reports of leuprolide for delayed gastric emptying in women.[3]

Clinically Available Agents

While one may have some optimism about what will be available tomorrow, we have to make do with those agents that are available today. The agents that are discussed here, with a couple of exceptions have not been approved as prokinetics but rather are approved for specific disease states or symptoms. Many of these agents are available both in oral and IV form. Some oral formulations are liquid which may be preferable in patients with markedly delayed gastric emptying. Somatostatin is available only as an injection, which limits its use.

Cisapride. This is a substituted benzamide that was developed from metoclopramide. It does not cross the blood brain barrier and hence is free from the central effects

Table 2. Use of Cisapride is contraindicated with the following drugs

P450 3A4 inhibitors
Macrolides*
Erythromycin
Azithromycin
Clarithromycin
Troleandomycin
Azole antifungals*
Ketoconazole
Fluconazole (diflucan)
Itraconazole (sporanox)
Protease inhibitors
Norvir
Serzone (nefazodone)
Drugs that prolong the QT interval
Antiarrythmics
class 1A (quinidine and procainamide)
class III (sotalol)
tricyclic antidepressants (amytriptyline)
tetracyclics (maprotiline)
Antipyschotics (includes phenothiazines)
Astemizole (Hismanal)
Sparfloxacin
Terodiline
General anesthetics

*The azoles and the macrolides may also prolong the QT interval.

so common in chronic therapy with metoclopramide. It exerts it effects by enhancing the release of acetycholine from enteric neurons. It probably does this by activating $5HT_4$ receptors on the neurons.[4] There may also be direct effects on smooth muscle as its effects are reduced but not blocked by atropine.[5] It improves esophageal clearance, increases LES tone, accelerates gastric emptying, and enhances coordinated antro-duodenal contractions. It also speeds up gall bladder emptying and small intestinal and colonic transit[6].

Cisapride has been tested in many clinical situations, including gastro-esophageal reflux disease, delayed gastric emptying, pseudobstruction with intact vagal function and chronic constipation due to colonic inertia or impaired rectal sensation.[7,8,9,10] Ironically, in the United States, the only approved indication cisapride is for nocturnal heartburn.

Cardiac Arrhythmias and Cisapride. There have been a number of case reports describing cardiac arrhythmias in patients receiving cisapride.[11] The cardiac arrhythmias include torsades de Pointes, ventricular tachycardia and fibrillation.[12,13] These are associated with a prolonged QT interval. Cisapride prolongs cardiac repolarization by inhibiting the delayed rectifier current (I_{Kr}).[14] This effect is a concentration-related phenomenon. The reports of arrhythmias are rare but some of these reports have had a fatal outcome. This risk seems to be extremely small in otherwise healthy patients who are on therapeutic doses of cisapride and not receiving a drug that also prolongs the QT interval, has inherent arrhythmogenic potential or inhibits the metabolism of cisapride by the cytochrome P450 3A4 system (Table 2).

Some of the cases reported involved either higher doses of cisapride or were described in children.[15] The effect can be rapid. Cisapride levels can rise many fold after the

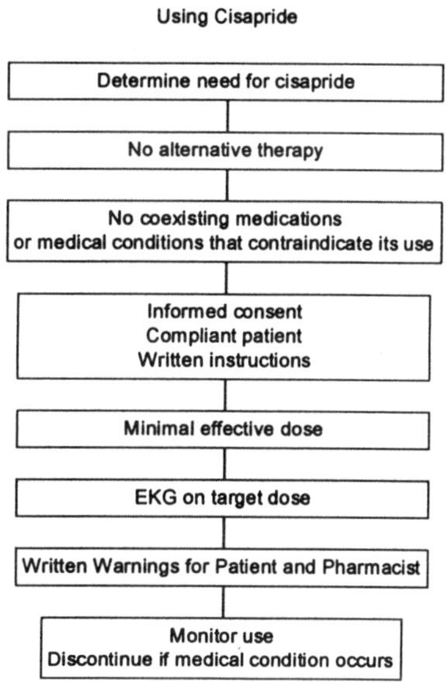

Figure 1. Suggested schema for initating chronic therapy with cisapride.

administration of a single dose of erythromycin. While the cardiac arrhythmias are rare, the manufacturer has expanded the warnings to reduce this possibility. Many patients are on cisapride and the vast majority will likely never suffer any ill effects from this medication.[16] However, as suggested by the manufacturers, one should exercise caution when prescribing cisapride for a patient who is on other medications. A suggested scheme for initiating cisapride therapy is shown in Figure 1.

Other side effects are similarly rare. Headache occurs in a small proportion of patients usually half to one hour after taking the drug. The headache may be severe enough to cause discontinuation of the medication. Diarrhea can occur at higher dosage. The manufacturer has suggested that cisapride not be used as a first line agent for reflux. Ordinarily an acid reducing agent should be tried first.

Metoclopramide. This is the original piperidine. It was the standard prokinetic prior to the introduction of cisapride in the US and domperidone elsewhere. It is a central and peripheral dopamine antagonist. It may also act via a cholinergic mechanism. It has been shown to speed gastric emptying and increase LES pressure.[16] It also enhances small bowel peristalsis. Its major utility is in delayed gastric emptying. It can be used acutely; either IV or SC, especially in-patients with intractable vomiting associated with exacerbation of their gastroparesis. It does have a central anti-emetic effect that cisapride does not have and may help the morning nausea in diabetics. It has been used for reflux but these not justified, as there are many potent acid-suppressing agents that are more potent. Both cisapride and metoclopramide have been used in patients with anorexia nervosa, who frequently have delayed gastric emptying. The major limitations of metoclopramide are the frequent neurological side effects.[16]

Acute dystonic reactions are rare but can start within 48 hours of starting metoclopramide. This can be very painful. Patients complain of tightening in their throat, choking, and trimus. More serious and especially in the older patients are tardive dyskinesias. They occur more often with chronic therapy than with short-term therapy. Because their onset is gradual, many patients may not notice this by themselves. Because this is only slowly reversible, one should be concerned about chronic therapy with metoclopramide. Fatigue and depression are more common and debilitating side effects.

Erythromycin. The motility effects of erythromycin were first observed as a common side effect of oral and IV administration, which often produced nausea, vomiting and diarrhea. It was subsequently discovered that the erythromycin releases motilin in the intestine resulting in increased contractions.[18] At lower doses (0.6mg/ Kg), erythromycin triggers phase III migrating motor complexes and enhances the fasting pattern of gastric and duodenal contractions.[19] At higher doses (3–5 mg/kg) that approximate an antibiotic dose, more continuous contractions are produced.[21] This may be caused by a cholinergic effect of the erythromycin producing contraction.[20] A small dose of erythromycin can trigger a house keeping or a phase 3 "migrating motor complex" that extends from the stomach into the small intestine. It is important to start patients with a very small dose (50 mg) as an oral suspension, about a half-hour before eating and sometimes just at night in patients with recurrent bacterial overgrowth due to small bowel stasis. If therapy is begun with a higher dose, 250 mg or more, then the patient may experience side effects without the therapeutic benefits. The dose can be titrated gradually by using the pediatric suspension, until the patient either benefits or experiences side effects.

Erythromycin has been shown to increase the rate of gastric emptying in gastroparesis by stimulating the abrupt emptying of solids. This property can be used to help move a bezoar from the stomach after mechanical or enzymatic digestion. Lower doses seem to enhance small bowel motility and may increase small intestinal transit in patients with chronic pseudo-obstruction. There has been a few anecdotal reports of benefit in acute colonic pseudo-obstruction, though many studies show little effect on the colon.[22] Other macrolides also have motilin agonist effects but most of these have too long a ½ life to be useful triggers[23]. Cost is also an issue with the newer agents. A synthetic macrolide derivative is in clinical trials and is a more potent prokinetic than erythromycin.[24]

Other Agents. Newer classes of agents include peripheral Kappa opioid agonists. Of this class, *Fedotozine* (a partial agonist) has been most widely studied in animals and humans. While effects of these agents appear to be on the afferent (sensory) innnervation, it may reverse morphine-induced ileus. It may reduce pain and motor responses to stress. Its main potential is in functional GI pain syndromes.[25]

Serotonin (5-HT) plays many complex roles in the GI tract. It is released from enterochromaffin cells, enteric nerves, in response to many diverse stimuli. The effects of serotonin vary depending on which receptor is activated. In the GI tract, 5-HT_3 and 5-HT_4 receptors predominantly mediate serotonin effects on neural and muscular function. Antagonists of 5-HT_3 and 5-HT_4 reduce GI motility and may find use in functional diarrhea. Cisapride is believed to exert some of its effects via 5-HT_4 agonism. A partial 5-HT_4 agonist, has prokinetic effects in both the small and large intestine improving symptoms of functional constipation.[26]

Octreotide. Somatostatin is active in many parts of the GI tract and has potent effects on blood flow, secretion and motility.[27] Octreotide is a long acting somatostatin analogue that can be given by injection only. While it has been used for carcinoid tumors and acute variceal bleeding, there have been anecdotal case studies that have described its use in post vagotomy stasis and scleroderma bowel. It induces migrating motor complexes when used in small doses. Reports of both relief of symptoms, decreased bacterial overgrowth and even reduced need for parenteral nutrition has been observed.[28] Higher doses can however induce gastric intestinal or gallbladder stasis and steatorrhea. Sandostatin also has effects on sensory afferent receptors reducing colonic sensation in irritable bowel syndrome.

Cholecystokinin (CCK) antagonists such as loxiglumide may reduce colonic motor responses in irritable bowel syndrome and may reduce meal-induced symptoms in these patients. CCK antagonists are not prokinetics but may find use in functional GI conditions.

The gonadotrophin-releasing hormone analogue, *Leuprolide* (Lupron) has been used in functional gastrointestinal disorders in women. It blocks the release of progesterone in the latter half of the menstrual cycle. A single center study showed benefit in women who had symptoms of delayed gastric emptying that were associated with the menstrual cycle[29]. It has been used primarily in menstruating women. The expense, the inconvenience of injection, and the estrogen-like effects greatly limit its potential for wider use in patients with functional bowel disease.

Administration of Prokinetics

The potency of a prokinetic varies from drug to drug and from organ to organ. In the stomach, the most potent drug is probably erythromycin. Cisapride and metoclopramide have equivalent effects on the stomach. Metoclopramide has an advantage when treating coexisting nausea. In general, IV preparations are more potent than oral preparation. This may be because of the delayed absorption from impaired motility. It is critical to administer the prokinetic at an optimal time of the day, in order to achieve the desired effect. In patients with small bowel bacterial overgrowth (due to small bowel stasis), what is needed is housekeeping contractions at night. This can be achieved by low dose erythromycin or octreotide at night. If the aim is to accelerate gastric emptying then the prokinetic should be given before meals. Generally prokinetics that enhance gastric emptying are given half hour or one hour before eating.

A rule of thumb is to ask the patients to identify when they feel their stomach churning after taking the medication. Approximately ten minutes later would be the right time to eat. If they take this medication a half hour before eating, but do not receive any benefit, and they feel full, maybe they should take it sooner – perhaps an hour beforehand. The patient may be a better judge of the optimal timing for medication.

Combination Therapy

There is a trend to utilize combinations of medications to maximize the prokinetic effect without side effects. One often has to end up using an antiemetic; hopefully an antiemetic that does not have anti-cholinergic effects as it may reduce the cholinergic effects of the prokinetic. Many common combinations are no longer recommended due to the potential for drug interactions. An example would have been to add erythromycin to cisapride in diabetic gastroparesis. Erythromycin can be added to metoclopramide, which might be a useful strategy in hospitalized patients with intractable gastroparesis.

Currently there are no published trials that have addressed combination medications.

Drug Tolerance

One of the significant problems with any prokinetic is the problem of drug tolerance. Apparent drug tolerance could be due to a progression of the underlying process, worsening glycemic control in diabetics with gastroparesis. Down regulation of receptors may be a common mechanism for the loss of effect of erythromycin.[30] Drug tolerance may be age or condition specific. How can drug tolerance be avoided? Several strategies including intermittent dosing, once nightly dosing or drug holidays may be tried, if the patients can tolerate it. Pharmacological therapy should be combined with other treatment modalities such as dietary manipulation, and nutritional support.

CONCLUSIONS

Prokinetic drugs are at present the mainstay of treatment for gastroparesis, small intestinal dysmotility and pseudo-obstruction. Although useful, today's armamentarium is limited. Currently, several exciting studies are in progress with a wide array of drugs. Soon, some of these drugs will reach the clinical arena and this should provide better treatment for patients with GI dysmotility.

REFERENCES

1. Ouyang A. Cohen S. Effects of hormones on gastrointestinal motility. Medical Clinics of North America. 1981; 65(6):1111–27.
2. Soffer EE, Metcalf A, Launspach J: Misoprostol is effective treatment for patients with severe chronic constipation. Dig Dis Sci 1994;39:929–933.
3. Mathias JR, Clench MH, Roberts PH, et al: Effect of leuprolide acetate in patients with functional bowel disease: Long term follow-up after double-blind, placebo controlled study. Dig Dis Sci 1994;39:1163–1170.
4. McCallum RW, Prakash C, Campoli-Richards DM, et al: Cisapride: A preliminary review of its pharmacodynamic and pharmacokinetic properties, and therapeutic use as a prokinetic agent in gastrointestinal motility disorders. Drugs 1988;36:652–681.
5. Wiseman LR, Fauld D: Cisapride: An updated review of its pharmacology and therapeutic efficacy as a prokinetic agent in gastrointestinal motility disorders. Drugs 1994;47:116–152.
6. Marzio L, DiFelice F, Laico MG, et al: Gallbladder hypokinesia and normal emptying of liquids in patients with dyspeptic symptoms: A double-blind placebo-controlled clinical trial with cisapride. Dig Dis Sci 1992;37:262–267.
7. Toussaint J, Gossuin A, Deruyterre M, et al: Healing and prevention of relapse of reflux oesophagitis by cisapride. Gut 1991;32:1280–1285.
8. Badiali D, Corazziari E, Habib FI, et al: A double-blind controlled trial on the effect of cisapride in the treatment of constipation in paraplegic patients. J Gastrointest Motil 1991;3:263–267.
9. Camilleri M, Malagelada J, Abell TL, et al: Effect of six weeks of treatment with cisapride in gastroparesis and intestinal pseudoobstruction. Gastroenterology 1989;96:704–712.
10. Staiano A, Cucchiara S, Andreotti MR, et al: Effect of cisapride on chronic idiopathic constipation in children. Dig Dis Sci 1991;36:733–736.
11. Wysowski DK. Bacsanyi J. Cisapride and fatal arrhythmia [letter]. [Letter] N Eng J Med 1996;335(4):290–1.
12. Olsson S, Edwards IR. Tachycardia during cisapride treatment. BMJ 1992;305:748–9
13. Ahmad SR, Wolfe SM. Cisapride and torsades de pointes. Lancet 1995; 345:508
14. Drolet B, Khalifa M, Daleau P, Hamelin BA, Turgeon J. Circulation, 1998;97:204–10.
15. Bran S, Murray WA, Hirsch IB, Palmer JP. Long QT syndrome during high-dose cisapride. Arch Intern Med 1995;155:765–8
16. Schulze-Delrieu K. Metoclopramide. In: Koch-Weser J, ed. Drug Therapy. *N Engl J Med* 1981; 305(l):28–33.
17. Wagner E, Tooley PJ. Pearce GL. Wilton LV. Mann RD. A comparison of two cohort studies evaluating the safety of cisapride: Prescription-Event Monitoring and a large phase IV study. Eur J Clin Pharmacol 1997;52(2):87–94.

18. Itoh Z, Nakaya M, Suzuki T, et al: Erythromycin mimics exogenous motilin in gastrointestinal contractile activity in the dog. Am J Physiol (Gastrointest Liver Physiol) 1984;247:G688-G694.
19. Janssens J, Peeters T, Vantrappen G, et al: Improvement of gastric emptying in diabetic gastroparesis by erythromycin. N Engl J Med 1990;322:1028–1031.
20. Zara GP, Thompson HH, Pilot MA, et al: Effects of erythromycin on gastrointestinal tract motility. J Antimicrob Chemother 1985;16:175–179.
21. Chaussade S, Michopoulos S, Sogni P, et al: Motilin agonist erythromycin increases human lower esophageal sphincter pressure by stimulation of cholinergic nerves. Dig Dis Sci 1994;39:381–384.
22. Jameson JS, Rogers J, Misiewicz JJ, et al: Oral or intravenous erythromycin has no effect on human distal colonic motility. Aliment Pharmacol Ther 1992;6:589–595.
23. Peeters TL: Erythromycin and other macrolides as prokinetic agents. Gastroenterology 1993;105:1886–1899.
24. Depoortere I, Peeters TL, Vantrappen G: The erythromycin derivative EM-523 is a potent motilin agonist in man and rabbit. Peptides 1990;11:515–519.
25. Dapoigny M. Abitbol JL. Fraitag B. Efficacy of peripheral kappa agonist fedotozine versus placebo in treatment of irritable bowel syndrome. A multicenter dose-response study Dig Dis & Sci. 1995;40(10):2244–9.
26. Sanger GJ. 5-Hydroxytryptamine and functional bowel disorders. Neurogastroenterology & Motility. 1996;8(4):319–31.
27. Owyang C: Octreotide in gastrointestinal motility disorders. Gut 1994;3(suppl):S11-S14.
28. Soudah HC, Hasler WL, Owyang C: Effect of octreotide on intestinal motility and bacterial overgrowth in scleroderma. N Engl J Med 1991;325:1461–1467.
29. Mathias JR, Clench MH, Reeves-Darby VG, et al: Effect of leuprolide acetate in patients with moderate to severe functional bowel disease: Double-blind, placebo controlled study. Dig Dis Sci 1994;39:1155–1162.
30. Tack J, Janssens J, Vantrappen G, et al: Effect of erythromycin on gastric motility in controls and in diabetic gastroparesis. Gastroenterology 1992;103:72–79.

18

GALLBLADDER DYSMOTILITY

Glen A. Lehman*

Division of Gastroenterology/Hepatology
Department of Medicine
Indiana University Medical Center

GALLBLADDER

Normal Function of the Gallbladder

The role of the gallbladder is to store and concentrate bile during the interdigestive intervals and to deliver the concentrated bile to the duodenum during feeding. The neurohumoral regulation of these events is only partially understood. The stimulation of the duodenal mucosa by intraluminal long chain fats and aminoacids releases cholecystokinin. This peptide is the major factor which causes gallbladder contraction. Although cholinergic and sympathetic innervation may play a role, truncal vagotomy and celiac ganglionectomy do not appear to alter gallbladder function. Numerous conditions and drugs can adversely affect gallbladder motility and this list includes pregnancy, obesity, prolonged fasting, estrogen administration, octreotide, anticholinergics, bile supersaturated with cholesterol, presence of gall stones, octreotide, and spinal injuries (1).

Table 1 lists the types of common gallbladder motility disorders. Hypermotility disorders are rare and commonly manifest as asymmetrical gallbladder contraction that is best seen on oral cholecystogram. Since oral cholecystograms are rarely done these days this diagnosis is rarely made. Nuclear scintigraphy studies with 99Tc labeled iminodiacetic acid derivatives (e.g. HIDA) have reported that the normal gallbladder ejection is 35–70% after cholecystokin stimulation. Whether a nuclear scintigraphy study that shows greater than 70% ejection fraction represents the same entity is unknown.

Cystic duct obstruction may cause faulty gallbladder contraction and is very understandable if a cystic duct stone is present. Most such situations are not credited as being motility disorders, but rather stone disease. Yap, et al (2) showed that low gallbladder ejection fraction may occur from cystic duct narrowing as evidenced by narrow caliber

* Telephone: 317-274-4821, Facsimile: 317-278-0164

Gastrointestinal Motility, edited by Rao *et al.*
Kluwer Academic / Plenum Publishers, New York, 1999.

Table 1. Types of gallbladder dysmotility

Hypermotility	Hypomotility
Segmental	Smooth muscle
Diffuse	Neurohumoral
Obstructive	Inflammatory
Cystic duct	
Sphincter of Oddi	

duct to probe passage. (See below.) Whether sphincter of Oddi dysfunction can be such that the gallbladder cannot adequately evacuate is uncertain. Ruffolo, et al. (3) studied a group of predominantly right upper quadrant abdominal pain patients with nuclear scintigraphy and sphincter of Oddi manometry via ERCP. Approximately half of the patients had each abnormality, although only one fourth of the patients had both abnormalities.

The most recognized form of gallbladder dysmotility is the sluggish or hypomotile gallbladder. It is generally thought that this entity most commonly occurs from inflammatory changes in the wall of the gallbladder with or without cholesterolosis. While there are several techniques for studying gallbladder motility (Table 2) , nuclear scintigraphy with CCK stimulated gallbladder ejection fraction has become the most widely accepted test in the medical community. The technique has been standardized and after IV administration of one of the IDA agents, and the gallbladder fills passively. The synthetic octapeptide of CCK is given intravenously as a bolus and the evacuation of the radionuclide from the gallbladder is monitored for the subsequent 20 minutes. In addition to examining the ejection fraction whether the patient's symptoms can be provoked is also noted.

Normal control patients eject 35–70% of the radionuclide. Several studies show that patients with low ejection fraction and typical right upper quadrant biliary colic have a 70–90% probability of receiving symptom relief with cholecystectomy. In a prospective randomized trial of patients with upper abdominal pain and low gallbladder ejection fraction, Yap et al. (2) performed cholecystectomy on one half of the patients (Table 3) and observed the other half of patients. Observation and nonspecific medical therapy gave no relief while cholecystectomy rendered 10/11 patients asymptomatic. Cystic duct narrowing or chronic cholecystitis was found in resected specimens.

Especially with the common use of laparoscopic cholecystectomy, most clinicians are now comfortable with requesting or performing cholecystectomy in a patient with low ejection fraction and a clinical pain syndrome that is acceptable for biliary colic. Patients with more atypical pain are less likely to have pain relief. The other most commonly used test for subtle gallbladder disease is analysis of duodenal, common duct or gallbladder bile aspiration and microscopic analysis for crystals. These may be cholesterol or calcium bilirubinate crystals. Such crystals are thought to reflect subtle calculous disease and have only modest correlation with gallbladder motility problems.

Table 2. Tests of gallbladder motility and ejection fraction

Nuclear scintigraphy
Ultrasonography
Oral cholecystogram
Duodenal drainage
Magnetic Resonance Imaging
(Plus symptoms during provocation with CCK)

Table 3. Cholecystectomy in symptomatic patients with acalculous gall bladder disease as evidenced by abnormal gall bladder ejection fraction

103 patients with upper abdominal pain. All had normal oral cholecystogram and transcutaneous ultrasound of gallbladder.

↓

21 patients had abnormal HIDA gallbladder ejection fraction of <35% (Mean 18%)

↓

Randomized Controlled Study

↙ ↘

11 Cholecystectomy*		10 Nonspecific medical therapy
↓	Pain Relief at	↓
10/11	34 months follow-up	1/10

Ref 2: *Abnormal gallbladder histology 11/11; Narrow cystic duct 10/11.

The relationship of gallbladder hypomotility to stone disease is well known. Hypomotility may lead to stasis which in turn may lead to supersaturation of bile, crystallization and stone formation. With cholecystectomy, after stone formation, the pathologist reports the presence of stones and inflammation. The contribution of gallbladder dysmotility as a primary or a contributing factor is often ignored. However, today, in the absence of frank stone disease, such motility disorders are being increasingly recognized (4).

SPHINCTER OF ODDI DYSFUNCTION

Sphincter of Oddi dysfunction (SOD) refers to benign, noncalculous obstruction to flow of bile or pancreatic juice through the pancreaticobiliary junction, i.e., the sphincter of Oddi (SO). SOD may manifest clinically by right upper quadrant pain, pancreatico biliary pain, pancreatitis, or deranged liver function tests. Often, there is a significant overlap of symptoms between patients with gallbladder dysmotility, SOD and sphincter of Oddi stenosis. SO dyskinesia refers to a motor abnormality of the sphincter of Oddi which causes a hypertonic sphincter. In contrast, SO stenosis refers to a structural alteration of the sphincter, probably from an inflammatory process with subsequent fibrosis. Because it is often impossible to distinguish patients with SO dyskinesia from those with SO stenosis, the term SOD has been used to incorporate both groups of patients. In an attempt to deal with this overlap in etiology, and also to determine the appropriate utilization of SO manometry (SOM), a clinical classification system has been developed for patients with suspected SOD (5) (Hogan-Geenen SOD classification system) based on clinical history, laboratory results, and ERCP findings. A variety of less accurate terms are listed in the medical literature to describe this entity such as papillary stenosis, ampullary stenosis, biliary dyskinesia, and post-cholecystectomy syndrome. The latter term is somewhat of a misnomer, as sphincter of Oddi dysfunction may clearly occur with an intact gallbladder.

A detailed description of SOD can be found in Chapter 19. In brief, our knowledge of Sphincter of Oddi dysfunction and manometric techniques to assist in this diagnosis are evolving. Successful endoscopic SOM requires good general ERCP skills and careful attention to the main details listed above. If SOD is suspected in a Type III or mild to moderate pain level Type II patient, medical therapy should generally be tried. If medical therapy fails or is bypassed, ERCP and manometric evaluation are recommended. The role of less-invasive studies remains uncertain due to undefined sensitivity and specificity. Sphincter ablation is generally warranted in symptomatic Type I patients and in Type II or III patients with abnormal manometry. The symptom relief rate varies from 55–95%, depending on the clinical presentation and selection of patient for therapy. Initial non-responders require thorough evaluation of the pancreatic sphincter and the pancreatic parenchyma. SOD patients have relatively high complication rates after invasive studies or therapy. It is important to review the risk:benefit ratio of this procedure with each patient.

REFERENCES

1. Bilhartz LE, Horton JD. Gallstone disease and its complications. In: Feldman M, Scharschmidt BF, Sleisenger MH, eds. Sleisenger & Fordtran's Gastrointestinal and Liver Disease. Philadelphia: WB Saunders 1998:948–972.
2. Yap L, Wycherley AG, Morphett AD, Toouli J. Acalculous biliary pain: Cholecystectomy alleviates symptoms in patients with abnormal cholescintigraphy. Gastroenterology 1991; 101:786–793.
3. Ruffolo TA, Sherman S, Lehman GA, Hawes RH. Gallbladder ejection fraction and its relationship to sphincter of Oddi dysfunction. Dig Dis Sci 1994;39:289–292.
4. Bilhartz LE. Acute acalculous cholecystitis, adenomyomatosis, cholesterolosis, and polyps of the gallbladder. In: Feldman M, Scharschmidt BF, Sleisenger MH., eds. Sleisenger & Fordtran's Gastrointestinal and Liver Disease. Philadelphia: WB Saunders 1998: 993–1005.
5. Hogan W, Sherman S, Pasricha P, Carr-Locke DL: Position paper on sphincter of Oddi manometry. Gastrointest Endosc 1997, 45:342–348.

19

SPHINCTER OF ODDI AND PANCREATIC SPHINCTER DYSMOTILITY

Frederick C. Johlin*

University of Iowa College of Medicine
Iowa City, Iowa 52242

INTRODUCTION

How a sphincter that controls the flow of bile and pancreatic juice may cause pain and suffering is based on an understanding of fluid hydraulics. The basic concept that allows one not only to investigate, but to understand the mechanisms and the etiologies that revolve around sphincter of Oddi dysfunction (SOD) is the concept that fluids are non-compressible. While fluid can exert pressure, it is not compressible. Therefore, when one exposes fluid that is contained in the bile duct or the pancreatic duct to an obstructive force, the consequence is that the fluid will exert its pressure on the tissues within and around those ductal systems. The advent of manometric systems using micro-bore tubing in concert with very low compliance, high fidelity transducers, has allowed one the opportunity to investigate the level and character of the obstruction in these systems. Most investigators believe that when a sphincter of Oddi malfunctions, the type of malfunction can be divided into either a fixed high-pressure zone, which we will call an area of fibrosis, or an area of *Papillary Stenosis*. The alternative dysfunction of the sphincter Oddi is that of a motor dysfunction or a *Dyskinesia*. Therefore, in this text, we will be discussing diseases which evolve around fixed areas of fibrosis in sphincter of Oddi vs. a sphincter of Oddi that has an abnormal pattern of muscle function. Some patients may have a component of both diseases. This chapter will cover in some detail how and when to evaluate these patients. The main tool discussed in this chapter will be how and when to use manometry in the diagnosis and direction of treatment in patients with SOD.

* Phone: 319/356-4030, Fax: 319/353-6399

Gastrointestinal Motility, edited by Rao *et al.*
Kluwer Academic / Plenum Publishers, New York, 1999.

CASE STUDY

A 45-year-old female presents with a 2-year history of recurrent knife-like R.U.Q. pain radiating to the back. In 1990, she had a cholecystectomy for " gall bladder inflammation". Past history consists of several ER visits with nonspecific abdominal pain. Labs: SGOT = mild elevation once. Amylase – mild elevation once. Normal CAT scan. Trial of IBS therapy was unhelpful.

1. What is your next diagnostic step?

   A. Trial of antidepressants
   B. Gastroduodenal motility studies
   C. Nuclear scintigraphy –HIDA scan
   D. ERCP

2. HIDA scan showed slight delay in the duodenal arrival of radionuclide. (Patient took time-released oxycodone 8 hours before study.) Patient continues to complain of severe pain. What is your next step in the evaluation of this patient?

   A. Trial of calcium channel blocker
   B. Magnetic resonance cholangio pancreatography [MRCP]
   C. ERCP
   D. ERCP with manometry
   E. Empiric biliary sphincter balloon dilation

3. ERCP showed that there was no stenosis, duct obstruction or calculus but the CBD was dilated up to 10 mm consistent with post-cholecystectomy. Patient had a normal ventral pancreas. Biliary manometry showed a basal sphincter pressure of 85 mmHg with phasic wave amplitude of 290 mmHg. What is your next step?

   A. Biliary sphincterotomy
   B. Minor papilla cannulation
   C. Pancreatic duct manometry
   D. Balloon dilation of sphincter
   E. Sphincteroplasty

DEFINITION

Sphincter of Oddi Dysfunction (SOD) is a drainage disorder of the biliary tree or the pancreatic ductal system that results in pain. During periods of pain the patient may have clinical consequences of obstruction including: abnormal clinical chemistries and/or ductal dilation and/or clinical signs of obstruction including pancreatitis and jaundice.

TERMINOLOGY

Sphincter of Oddi Dysfunction (SOD), Papillary Stenosis, Biliary Dyskinesia, stenosing papillitis, biliary dyssynergy and post-cholecystectomy syndrome are terms used to describe patients with this phenomena.

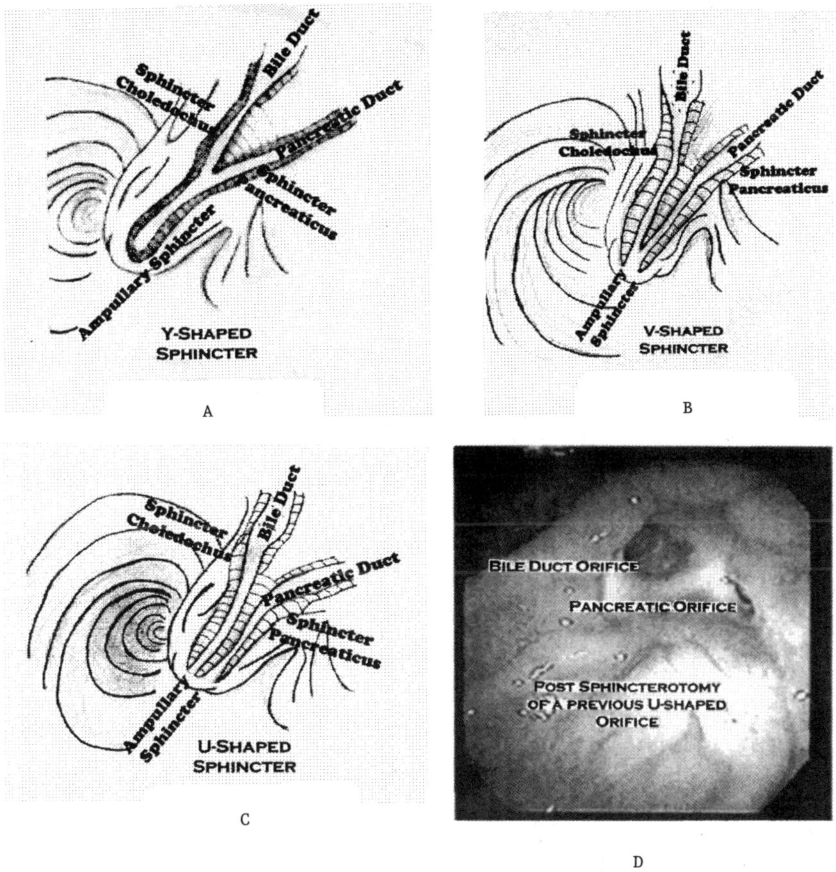

Figure 1. Y-shaped sphincter. The most common variant of the sphincter of Oddi. The key feature is the length of the common channel into which both the bile and pancreatic juice drain. V-shaped sphincter. The least common variant of the sphincter of Oddi. The bile duct orifice is under a hooded lip which could make it difficult to access the bile duct, but the pancreatic orifice is straight and easily accessible. U-shaped sphincter. The third variant of the sphincter of Oddi where it appears like a double banded orifice opening into the duodenum. A generous sphincterotomy of a previously defined U-orifice. Two large orifices can now be seen.

HISTORICAL BACKGROUND

In 1887, Ruggero Oddi, a medical student, first dissected the sphincter of Oddi (SO) and published his findings (1). In 1937, Boyden extended that work by describing the different segments within this anatomic entity (Figure 1). During the following decades, when surgeons examined this sphincter, they gauged its elasticity by passing small tapered dilators through the SO. Resistance to passage of these dilators was considered diagnostic of a fibrotic sphincter of Oddi. Subsequently, with the introduction of the endoscopic retrograde examination of the biliary and pancreatic ducts, one learned to examine the sphincter less invasively than performing exploratory surgery. This endoscopic technique would later allow the development of biliary and pancreatic manometry and a better delineation of the sphincter morphology (Figure 1).

SOD SYNDROME

This syndrome for decades was considered synonymous with post-cholecystectomy syndrome (PCS). Post cholecystectomy syndrome is a collection of symptoms that occurs in 10 to 20 percent of patients after they have had their gallbladder removed. Unfortunately, PCS also includes symptoms of diarrhea, crampy abdominal distress and persistence of the patient's pain that led to their gallbladder removal in the first place. Sphincter of Oddi dysfunction may be the cause of pain in these patients. There are however, a myriad of other causes to explain the lack of improvement or the worsening of symptoms in post cholecystectomy patients, including the possibility that the cholecystectomy itself might have been an error. Potential problems include ulcer disease, antroduodenal dysmotility, retained bile duct stones or malabsorptive diseases. The most common new symptom after a cholecystectomy that brings the patient to medical attention is the development of worsening diarrhea and worsening fatty meal intolerance. The diarrhea of post-cholecystectomy syndrome usually responds to cholestyramine at bedtime and a reduced fat diet.

PHYSIOLOGY

Stenosis

Papillary stenosis is due to fibrosis of the sphincter of Oddi or the orifice of the duct of Santorini that restricts the flow of bile or pancreatic fluid. Histologically, one can demonstrate that papillary stenosis is associated with fibrosis and narrowing of the native sphincter. The biopsy requires tissue from the sphincter, not from the surface epithelium. Therefore one needs to open the sphincter to obtain the biopsies. This is not practical as a diagnostic tool as it requires a sphincterotomy before the diagnosis can be ascertained. Fibrosis can cause a buildup of back pressure, and if the pressure exceeds an absolute threshold, it can cause pain and suffering. This threshold is different for the pancreatic duct and for the bile duct, as these ductal systems have different resting pressures. When an absolute threshold is exceeded, either because the flow has increased through the fixed sized orifice or because the orifice is now partially plugged by some form of debris or cellular material, then the patient will develop symptoms. This material might include small stones, mucus, sloughed epithelium or a combination thereof. When this cast of material becomes lodged in the area of fixed obstruction, then that area will become even more obstructed. It is, presumably, when this occurs that one can detect the symptoms that allows one to classify patients into the different types of SOD. One feature proposed is that all patients must have some form of pain during the obstructive event. While this is an over simplification of the syndrome of SOD, it does allow classification of the vast majority of patients with this phenomena. Thus, pain is an important feature of all forms of SOD.

Dyskinesia

Sphincter of Oddi dyskinesia is a disease in which the coordination of the pumping function of the sphincter of Oddi, or the orifice of the duct of Santorini, is impaired. This impairment can include the lack of relaxation when it is normal for these sphincters to relax, or an event during which the sphincter actually increases its pressure at a time when the pressure is supposed to be decreasing. The other forms of dyskinesia include rapid fre-

Table 1. SOD: purpose of the ERCP

ERCP serves multiple purposes:
Defines the anatomy
Documents the absence of stones
Provides bile duct drainage time
Provides pancreatic drainage times
Defines the duodenal mucosa
Helps detect small tumors
Reproduces the pain

quent pressure waves (Tachyoddia) in the sphincter which are too frequent to allow the delivery of fluid or particulate material, retrograde contractions which occur too frequently or paradoxical responses to gut hormones such as cholecystokinin (CCK) or secretin (Secretin®-Ferring). In contrast, these types of sphincter dysfunction should respond to potent agents that are capable of relaxing smooth muscle, such as nitrates.

Laboratory Testing during Pain

If one assumes that the basic disease process in SOD is one that produces symptoms by increasing back pressure in the affected organ, then one should be able to obtain laboratory parameters to detect damage within that organ. If the organ that is being affected is the biliary tree and thereby the liver, then one should obtain liver chemistries to detect any increased pressure within the smallest of the bile ducts. The liver chemistries that become abnormal early in an obstructive process are the transaminases i.e. aspartate aminotransferase (AST) and alanine aminotransferase (ALT). It is only after prolonged obstruction that the ALP (alkaline phosphatase) and the GGT (Gamma glutamyl transferase) become elevated. If the organ that is being affected is the pancreas, then one should obtain an amylase and lipase. The next most useful test is ERCP. The advantages of ERCP are shown in Table 1. After defining the biliary and pancreatic ducts, the next most useful step is to perform biliary and pancreatic manometry.

CLASSIFICATION

Hogan and Geenen, in 1988, provided the most widely used classification that allows one to diagnose and treat patients with sphincter of Oddi dysfunction (2,3). While some authors dispute the value of this system, it allows a clear pigeonholing of patients that undergo evaluations for this problem (Table 2).

This classification takes into account the presence of pain that is either considered pancreatic or biliary in character, and the development of liver function testing abnormalities or pancreatic enzyme elevation during pain events. This classification also requires the demonstration of the biliary and pancreatic duct diameter by either a direct imaging technique or by ultrasound. These criteria are then used to suggest when a manometry study of the sphincter of concern is appropriate or essential. Using this set of criteria has allowed these authors to demonstrate that predictive factors at the time of manometry will allow one to determine which patients might benefit from a sphincterotomy. The best predictive factor is an elevated basal pressure of > 40 mm Hg. Biliary and pancreatic manometry have therefore evolved as the gold standard for the diagnosis and direction of treatment for patients

Table 2. SOD:the clinical syndrome

SOD group	Characteristic findings in patients with PAIN plus
I	Abnormal labs and Dilated duct and Poor drainage
II	Abnormal labs or Dilated duct or Poor drainage
III	Pain alone

with sphincter of Oddi dysfunction. This technology is not widely available because it is technically difficult, time-consuming and has significant inherent risk.

Risk factors for sphincter of Oddi dysfunction vary dramatically based on the population group being studied. Tables 3, 4, and 5 contain data obtained over a five-year period from patients undergoing sphincter of Oddi evaluation at a University tertiary referral center. This data documents that the vast majority of patients are female and that fibrosis is the most common etiology for obstruction. However, when HIV infected patients are included, the relative proportion of males with sphincter of Oddi dysfunction changes. A likely reason for this is that while the general population does not contain a high percentage of patients with unusual ampullary infections, the specific HIV population includes predominantly those with infection related complications. Post liver transplant patients primarily include individuals in whom there is evidence for acute papillitis. There is also a difference in the type of disease detected in the general population based on whether the patient is male or female. A diagnosis of primary sclerosing cholangitis was a relatively frequently reported cause of sphincter of Oddi obstruction in males, but a very uncommon source in females. Recent data from our center has demonstrated that when screening for sphincter of Oddi dysfunction in patients with idiopathic relapsing pancreatitis, one cannot underestimate the importance of celiac disease.

Table 3. SOD: risk factors in the general population[a]

Etiology of SOD-biliary	Females (N = 200)	Males (N = 19)
Fixed lesion	177 (89%)	5 (26%)
Paradoxical responses to Kinevac	8 (4%)	0
Tachyoddia	6 (6%)	0
Narcotic bowel related	5 (2.5%)	3 (16%)
Primary sclerosing cholangitis associated	3 (1.5%)	6 (32%)
Small carcinoid tumor	1 (0.5%)	3 (16%)
Celiac disease related	0	2 (10%)

[a]University of Iowa dyskinesia data

Table 4. SOD: risk factors in the HIV associated patients[a]

Etiology of SOD	Females (N = 0)	Males (N = 8)
Cryptosporidia	0	4 (50%)
Narcotic bowel	0	2 (25%)
CMV associated*	0	2 (25%)

*Buffey coat culture positive
[a]University of Iowa dyskinesia data

BILIARY AND PANCREATIC MANOMETRY

It is generally accepted that the most important pressure that one can measure when evaluating patients for sphincter of Oddi dysfunction is the basal pressure. An elevated basal sphincter of Oddi pressure is the cardinal feature of patients with sphincter of Oddi stenosis (Table 6). If this sphincter is stenotic, then the sphincter cannot relax when exposed to glucagon, kinevac, nitrates such as amyl nitrate, or secretin (Figure 2). When one measures an elevated basal pressure that disappears with the treatment of amyl nitrate, then one can interpret that elevated basal pressure as sphincter spasm rather than sphincter stenosis. Glucagon will not only lower the basal pressure but will typically abolish the phasic activity of the sphincter. Sphincter of Oddi dyskinesia can include abnormal frequent phasic contractions (tachyoddia), abnormal propagation of phasic contractions, and/or paradoxical responses to analogs of cholecystokinin (CCK) such as kinevac (Table 7).

Tachyoddia

Iachyoddia, as referred to in cardiac terminology, is believed to shorten the diastolic filling time of the sphincter of Oddi so severely that the sphincter is no longer able to propagate any material that is contained within it. The peristaltic pumping action of the sphincter of Oddi is not sufficient to empty the common bile duct of the volume of bile produced each day, therefore it is proposed that the peristaltic action of the sphincter actually removes particulate material from the bile duct. Only a small subset of patients actually have non-narcotic tachyoddia, while a significant percentage of patients who have evidence for tachyoddia based on biliary manometry will have positive urine toxicology screens for nar-

Table 5. SOD: risk factors in organ transplants

Etiology of SOD	Females (N = 8)	Males (N = 5)
Denervation	3 (38%)	3 (60%)
Acute papillitis** (Includes re-perfusion injury)	5 (62%)	2 (40%)

**All developed fibrosis on follow-up
[a]University of Iowa dyskinesia data

Table 6. Sphincter of Oddi manometry reference range

Manometric variable	Normal range range (median)	Abnormal
Common bile duct pressure (mmHg)	3-6 (5)	>12
Basal pressure (mmHg)	8-15 (10)	>40
Phasic amplitude (mmHg)	60-220 (140)	>240
Phasic duration (seconds)	3-6 (5)	>10
Phasic frequency (#/min)	4-6 (5)	>12
Antegrade contraction (%)	50-85 (60)	>50

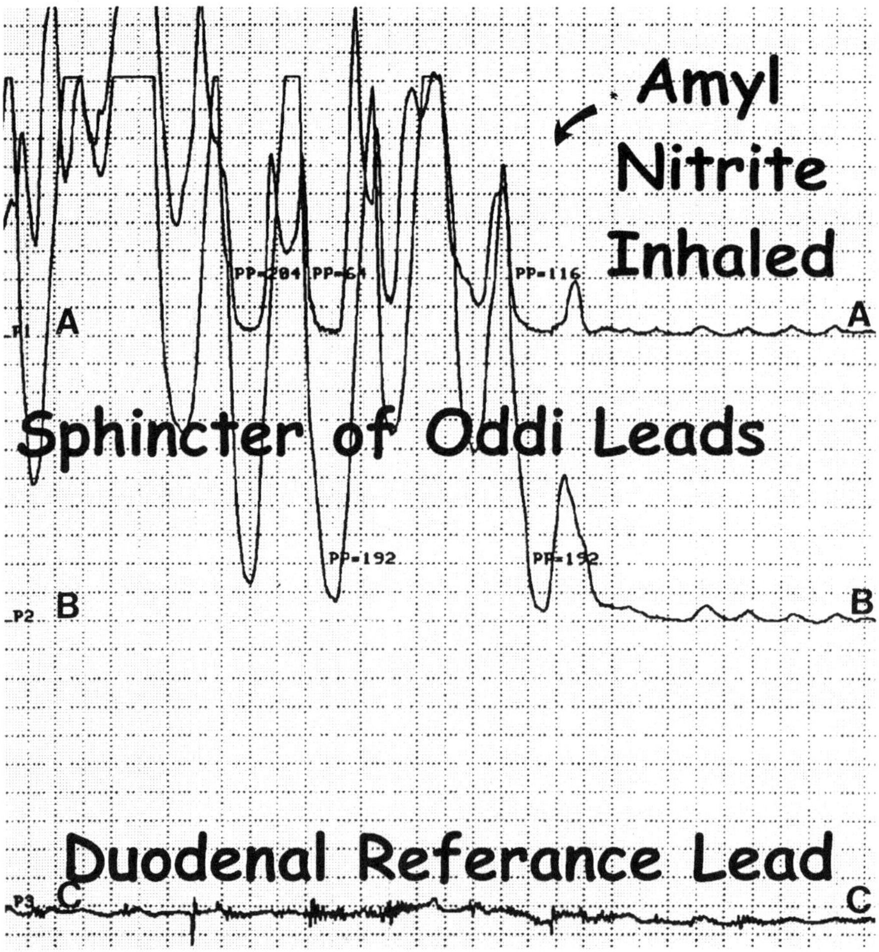

Figure 2. This manometry tracing of the sphincter of Oddi has 3 leads or channels: a-the medial sphincter lead; b-the lateral sphincter lead; c-the duodenal reference lead. Note that after inhalation of 0.2 ml of amyl nitrite, the higher pressure segment in the sphincter of Oddi no longer has either phasic or elevated basal pressure. This elevated pressure segment was therefore not a fixed stenosis.

Table 7. SOD: the manometric lesion

Stenosis
A FIXED lesion
Chronic inflammation and fibrosis
Acute obstruction (liver and kidney transplants)
Dyskinesia
SOD-spasm
Tachyoddia
Narcotic bowel
Paradoxical response to Kinevac (CCK)
Abnormal phasic wave pattern

cotics, which they have failed to report. While it is assumed that the demonstration of tachyoddia at the time of a biliary manometry is a pathologic finding, it is unclear which therapy is useful for these patients. We encourage them to avoid narcotics. It is reported that the phase III inter-digestive duodenal activity front also can physiologically yield tachyoddia. In our experience, patients that exhibit tachyoddia rarely benefit from a sphincterotomy.

Retrograde Contractions

The peristaltic pump of the sphincter of Oddi should propagate fluid and particulate matter into the duodenum. This requires antegrade contractions directed towards the duodenum. If the majority of contractions are retrograde, then the assumption is that the sphincter is dysfunctional. Simultaneous or retrograde contractions may prevent adequate clearance of the biliary and pancreatic contents. This appears to be more of a diagnostic dilemma than a true clinical entity. Hence, based on our current understanding, performing an endoscopic sphincterotomy does not appear to be justified.

Paradoxical Responses to Gut Hormones

The sphincter of Oddi relaxes when exposed to slow infusion of either cholecystokinin analogues such as kinevac or slow infusion of secretin (Secretin®-Ferring). A significant minority of patients with signs and symptoms of SOD will in fact have a paradoxical response to gut hormones and yet the SO may be able to relax when exposed to nitrates. Many of these patients will also have elevated basal pressures. Those in whom the basal pressure is also elevated may respond to an endoscopic sphincterotomy, however some of these patients do not have isolated dysfunction of the sphincter of Oddi, but rather have a global dysfunction of the antro-duodenal motor unit. In these patients, the response to endoscopic sphincterotomy is at times disappointing, and in fact, many have worse symptoms, post-sphincterotomy (4). Evaluation of these patients demonstrates that two-thirds will show high phasic duodenal pressures that can be sustained for several minutes. Based on our experience, we would reserve endoscopic sphincter ablation for those patients who are truly disabled by their sphincter dysfunction and in whom antro-duodenal dysmotility is absent. Fortunately, this comprises of a small subset (less than 10%) of patients.

Non-Invasive Testing

The most widely available non-invasive test for SOD is probably the biliary ultrasound. In patients with biliary symptoms, a dilated bile duct, especially one that is over 12

Table 8. SOD: provocative testing with ultrasound

Fatty meal stimulated ultrasound	
Lipomul, 1.5 m./kg PO	
Scan the CBD and CHD at times 0, 30, 45, and 60 minutes	
Measure the change in the CBD diameter	
Results:	Value
Bile duct diameter increased by ≥ 2 mm	positive
Bile duct diameter increased by < 2 mm	equivalent
Bile duct diameter decreases	negative

RM Darweesh, WJ Dodds, WJ Hogan et al. *Gastro.* 1988, **94**:779-786.

mm in diameter, is strongly suggestive of some pathology in the biliary tree. Provocative ultrasound examinations, including a fatty meal stimulated ultrasound (Table 8), or secretin stimulated ultrasound, provides a low-risk evaluator of the pancreas and biliary tree.

The Hopkins group has suggested that scintigraphic examination based on visual criteria and areas of scintigraphic interest after pre-medication with Kinevac may allow diagnosis of this disorder with 100% sensitivity (5). The criteria for scoring are included in Table 9. However, at our center we have found that the sensitivity of this technique is only 50% for patients with a bile duct diameter of less than 6 mm. Hence, individual centers must evaluate the utility of this diagnostic test. When we correct for the bile duct diameter, based on an ultrasound of the biliary tree, we were able to predict those patients who may have a positive scintigraphic examination, but not those who may have an abnormal manometry. Patients with a bile duct diameter that was greater than 12mm were more likely to have a positive Hopkins protocol scintigram. Therefore, in our experience, scintigraphy has not given us enough information to justify its cost.

MEDICAL THERAPY

Sphincter of Oddi dysfunction presents in patients with pain that is chronic and/or recurrent in character. The sphincter of Oddi has rich innervation and is composed exclusively of smooth muscle. These two characteristics enable us to use medications that may alter the tone or completely relax the sphincter of Oddi. This type of therapy would be effective in patients that have sphincter dysfunction that is primarily a motor disturbance. Therapy designed to help patients with fibrotic or stenotic sphincter of Oddi would be designed to enhance pain tolerance. It is widely accepted that calcium channel blockers and nitrates can relax smooth muscle. In our experience, therapy with tricyclic antidepressants is effective in improving pain in patients with less severe pain syndromes. Antidepressants such as amitriptyline and imipramine also have anti-cholinergic side effects that may also reduce smooth muscle tone. These therapies avoid the risk of endoscopic examination and endoscopic sphincterotomy (Table 10). Side effects of these medications are considerably less than the potential risks of a possible ERCP-induced pancreatitis.

At our center, most patients with SOD that is disabling, receive a six-week trial of medical therapy before endoscopic examination(s). Other medications that can be helpful are the anti-cholinergics and low fat diet. Patients with pancreatic sphincter dysfunction are started on non-enteric, pancreatic enzyme therapy and anti-oxidants such as Vitamin C 250 mg three times per day plus Vitamin E, 400 IU three times per day. However, whether these recommendations are beneficial has not been evaluated in a blinded fashion.

Table 9. Hopkin's scoring of scintigraphy

Time to peak liver activity	
Less than 10 minutes	0
10 minutes or more	1
First intrahepatic bile duct appearance	
Less than 15 minutes	0
15 minutes or more	1
Dialated bile duct	
Not dilated	0
Dilation of the extrahepatic biliary tree	1
Dilation of the entire biliary tree	2
Time to the first bowel visualized	
Less than 15 minutes	0
15–30 minutes	1
More than 30 minutes	2
Common bile duct emptying	
By more than 50%	0
By less than 50%	1
No change	2
Increasing activity	3
Common bile duct to liver ratio	
CBD-60 < liver-60	0
CBD-60 > liver-60 but < liver-15	1
CBD-60 > liver-60 but = liver-15	2
CBD-60 > liver-60 and liver-15	3

S Soste, AN Kalloo, EJ Spiegler et al. *J Nucl Med* 1992, **33**:1216–1222.

ENDOSCOPIC SPHINCTEROTOMY

Surgical or endoscopic sphincterotomy provides relief from pain and suffering in many patients with suspected sphincter of Oddi obstruction, but a careful patient selection is imperative to ensure success. With the advent of endoscopically guided sphincter of Oddi manometry, one may clearly select patients in whom a sphincterotomy can offer the greatest benefit. Current data suggests that morbidity and mortality of endoscopic sphincterotomy are statistically acceptable, when performed by a sufficiently experienced endoscopist. However, endoscopic evaluation and treatment of patients with presumed SOD does have significant risk for post-procedural pancreatitis.

Table 10. SOD: medical therapy

Medical therapy
Change the cholinergic tone (tricyclic antidepressants)
Improve pain tolerance (tricyclic antidepressants)
Smooth muscle relaxation
PRN medications:
Nitrates
Calcium-channel blockers
While less than 25–35% will respond, the risk of this therapy is very, very small and the cost is also small

Table 11. Response to sphincterotomy or sham sphincterotomy

Basal pressure (mm Hg) and treatment	Number in this group	% pain improved at one yr.	% pain improved at four yrs.
> 40 and ERS	11	91	96
> 40 and sham	12	25	40
< 40 and ERS	12	42	41
< 40 and sham	12	33	40

JE Geenen, WJ Hogan, Wj Dodds et al. *NEJM* 1989; 320:82.

Prospective studies have shown that the risk of pancreatitis is less than 12% (6). This risk can be reduced by the concomitant placement of a short pancreatic stent (7). However, prophylactic pancreatic stenting in the main pancreatic duct does not appear to reduce the risk of pancreatitis in other centers (8). Multiple factors may cause pancreatitis in patients undergoing sphincter of Oddi manometry. One key element of a successful endoscopic technique is to prevent over-distension of the pancreatic duct when the perfusion catheter or injection catheter is placed within the duct. The dead space volume for the normal pancreatic duct is less than 2 ml. If the pancreatic duct is perfused with a catheter at a perfusion volume of 0.5 ml per minute, then overfilling of the pancreatic duct can occur within a few minutes. Over-filling of the pancreatic duct can result in damage to the pancreatic acinar cells, and cause pancreatitis. Geenen and colleagues performed a hallmark study, which demonstrated the efficacy of endoscopic sphincterotomy in the treatment of SOD (2). In a sham controlled, randomized, prospective trial, the authors have shown that the response to endoscopic sphincterotomy in patients with abnormal basal pressures (>40 mm Hg) was approximately 90% (Table 11). In patients without an elevated basal pressure, the failure rate was higher than the placebo controls. Also, the reduction in pain was sustained.

Surgical sphincteroplasty has been effective in a large number of historical series. The surgical literature suggests that a transduodenal sphincteroplasty can effectively ablate the entire biliary and pancreatic sphincter. These uncontrolled reports demonstrate a response rate of 40 to 50% (9). Surgical ablation is difficult to justify, given the inherent risks of infection and general anesthesia when compared to the reduced morbidity of an endoscopic technique. In cases where the anatomy is technically unfavorable for the endoscopic approach, then a surgical approach may be more appropriate. Stenosis of the sphincterotomy site can occur with either the endoscopic or the surgical approach. A stenotic sphincterotomy site can be revised either endoscopically or surgically. While it may be argued that balloon dilation of the sphincter is less permanent than an endoscopic sphincterotomy, one report suggests a higher risk of pancreatitis (10). Also recent studies demonstrate that stenting of the bile duct has a higher risk of pancreatitis (11). However, our experience suggests that this risk is no different from that which occurs with gallstone pancreatitis and that the pancreatic orifice has to be protected before the common channel is dilated. Again, the pancreatic mini stent may provide a useful and innovative approach that could prevent pancreatitis.

CONCLUSIONS

The advent of advanced endoscopic techniques has facilitated the investigation of patients with sphincter of Oddi dysfunction. It is essential that patients are selected carefully and the risks of invasive and diagnostic testing are weighed against the benefits of therapy. Some patients benefit from medical therapy alone, while others may require manometry studies or endoscopic interventions. One must remember that no one dies from biliary sphincter dysfunction but there have been several deaths and litigations because of ERCP-related complications.

ANSWER KEY AND EXPLANATIONS

1. It is too premature to start this patient on antidepressants even if it proved ultimately that there was no definable pathology for her symptoms. Gastroduodenal motility studies are appropriate but again in the presence of signs and symptoms suggestive of biliary etiology, ERCP or scintigraphy would be preferable to look at ductal flow and identify if there is any calculous disease. Nuclear scintigraphy is preferable because it is less invasive and if normal would exclude serious biliary dysfunction. The best option is **C.**

2. Given her long history of symptoms and delayed excretion with HIDA scan, it is now appropriate to proceed with invasive testing such as ERCP. In some centers MRCP is available and could be considered but again would not add any further diagnostic information in this patient. One should be prepared to perform a manometry along with ERCP, if the ductal architechture is normal. Empiric balloon dilation is fraught with problems and should be avoided. The best option is **D**.

3. Based on her history and manometric findings, this patient fulfills criteria for SOD dysfunction Type II. She would benefit with biliary sphincterotomy. Some patients may continue to remain symptomatic. In this situation it would be worth pursuing pancreatic manometry and pancreatic sphincterotomy, but pancreatic studies are fraught with a higher incidence of complications. The best option is **A**.

REFERENCES

1. Oddi R. D'une disposition a sphincter speciale de l'ouverture du canal choledoque. Arch Ital de Biol 1887;8:317.
2. Geenen JE, Hogan WJ, Dodds WJ, et al. The efficacy of endoscopic sphincterotomy in patients with sphincter of Oddi dysfunction. N Engl J Med 1989;320:82–87.
3. Hogan WJ, Geenen JE. Biliary dyskinesia. Endoscopy 1988;1:179–83.
4. Soffer EE, Johlin FC. Intestinal dysmotility in patients with sphincter of Oddi dysfunction. A reason for failed response to sphincterotomy. Dig Dis Sci 1993;39:593–595.
5. Sostre S, Kalloo AN, Spiegler EJ, Camargo EE, Wagner HN. A noninvasive test of sphincter of Oddi dysfunction in postcholecystectomy patients: The scintigraphic score. J Nucl Med 1992;33:1216–22.
6. Sherman S, Ruffolo TA, Hawes RH. Complications of endoscopic sphincterotomy. A prospective series with emphasis on the increased risk associated with sphincter of Oddi dysfunction. Am J Gastroenterol 1994;89:303–305.
7. Tarnasky PR, Palesch Y, Cunningham JT, Mauldin PD, Cotton PB. Pancreatic stenting prevents pancreatitis after biliary sphincterotomy in patients with sphincter of Oddi dysfunction. Gastroenterology 1998;115(6):1518–1524.
8. Smithline A, Silverman W, Rogers D. Effect of prophylatic man pancreatic stenting on the incidence of biliary endoscopic sphincterotomy-induced pancreatitis. Gastrointest Endoscop 1993;39:652–657.
9. Nardi GL, Michelassi S, Zannini P. Transduodenal sphincteroplasty. Ann Surg 1983;198:453–461.

10. Geulrud M, Siegel JH. Endoscopic cholangiopancreatoplasty. Hydrostatic balloon dilatation in the bile duct and pancreas. Gastrointest Endosc 1983;29:99–103.
11. Bader M, Geenen JE, Hogan WJ, Dodds WJ, et al. Endoscopic balloon dilatation of the sphincter of Oddi in patients with suspected biliary dyskinesia: results of a prospective randomized trial. Gastrointest Endosc 1986;32:158–162.

20

CONSTIPATION

Satish S.C. Rao*

University of Iowa
College of Medicine
University of Iowa Hospitals and Clinics
Iowa City, Iowa 52242

INTRODUCTION

Constipation affects 5–20% of the population and is more common in females. Each year, it accounts for more than 2.5 million patient visits to physicians [1,2]. In 1994, $840 million was spent on laxatives [3] and in 1985, $354 million was spent on health care services for patients with constipation [4]. Today, the expenditure is likely to be much higher because of escalating health care costs, an increasing array of diagnostic tests and an increasing recognition of this problem. However, in the community, many patients self-treat their "constipation", often successfully, using either home remedies or laxatives. Hence, those who seek medical help merely constitute the tip of the "constipation ice-berg". Most of these patients have a significant problem, which merits further evaluation.

After discussing a clinical problem, I will review the physiology of defecation and the clinical approach to constipation.

CASE STUDY

A 30-year-old female, jail matron presents with a history of constipation since childhood. She describes excessive straining, passage of hard, pellet-like stool, and difficulty with stool expulsion. Without laxatives, her stool frequency varies between one bowel movement a week to once every three weeks. Several years ago, she developed Insulopenic (Type I) diabetes mellitus. Over the previous three years she was hospitalized 15 times with ketoacidosis, five of which were within the previous 6 months. Her current medications include GoLYTELY, ½ gallon every 3rd day, fleets phospho soda, enema on

* Telephone: 319/353-6602, Fax: 319/353-6399, E-Mail: satish-rao@uiowa.edu

Gastrointestinal Motility, edited by Rao *et al.*
Kluwer Academic / Plenum Publishers, New York, 1999.

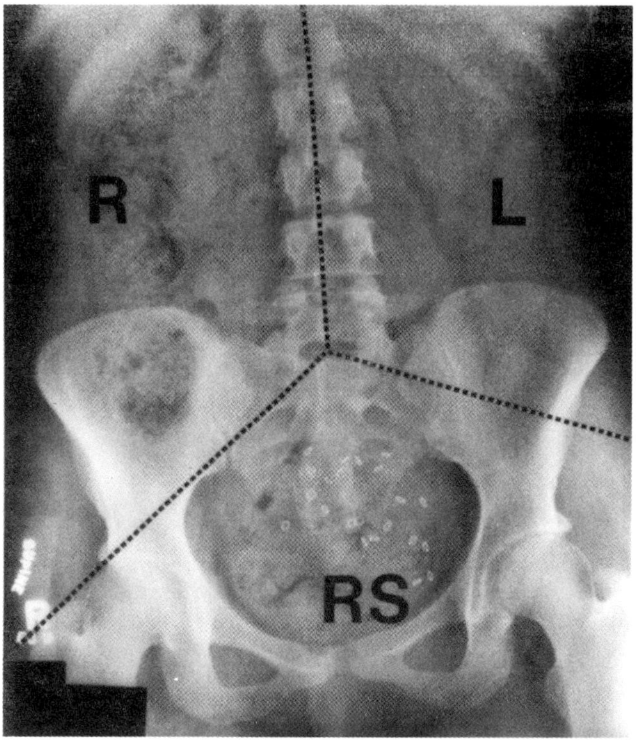

Figure 1. Colon transit study from the patient in the case study (refer to question 2).

alternate days, urecholine and insulin. Her past history was significant for a hysterectomy and a lumbar disc prolapse for which she had back surgery.

1. What is the next step in the evaluation of this patient?

   A. Colonoscopy
   B. Defecography
   C. Digital rectal and perianal neurological examination
   D. Colon transit study
   E. Anorectal manometry

Physical Examination revealed palapable stool in the right colon and mild tenderness in the left upper and lower quadrants. The bowel sounds were normal and there was no succussion splash. Neurological examination showed no peripheral neuropathy but mild retinopathy. The anocutaneous reflex was present in all four quadrants. Digital rectal exam revealed a moderate amount of pellet-like stool that was a guaiac negative. The anal sphincter tone was normal but when asked to bear down there was no anal relaxation or perineal descent. Subsequently, a flexible sigmoidoscopy was performed which was normal. A metabolic profile including thyroid function test was normal. A colon transit study and anorectal manometry were performed.

2. A single capsule containing 24 radiopaque markers was administered on day 1 and a flat plate film of the abdomen was obtained on day 6 [Figure1]. How would you interpret this colon transit study?

   A. Normal colon transit study.
   B. Obstructive or dyssenergic defecation.

C. Whole gut transit time of 62 hours.
D. Slow transit constipation.
E. Diabetic colopathy.

3. This is an excerpt from the patient's anorectal manometry [Figure 2]. The rectal and anal pressure changes when bearing down, and when a rectal balloon was inflated with 60 cc air is shown [Fig. 2]. How would you interpret the manometric changes?

A. Normal bearing down maneuver with normal recto anal inhibitory reflex.
B. Abnormal bearing down maneuver with Hirschsprung's disease.
C. Obstructive defecation with normal rectoanal inhibitory reflex.
D. Obstructive defecation with Hirschsprung's disease.
E. Obstructive defecation with spinal cord injury.

This lady had dyssynergic defecation; a problem that affects 30–50% of patients with chronic constipation. Because of chronic fecal stasis, she developed sub-acute intestinal obstruction which led to recurrent nausea and vomiting. These events precipitated her admissions with diabetic ketoacidosis. She was treated by neuromuscular conditioning with biofeedback techniques and her pelvic floor function improved. After biofeedback therapy, over the next three years she has never been admitted to the hospital and remains well with a regular bowel habit.

PHYSIOLOGY OF DEFECATION

The colon mixes ileal effluent, ferments and salvages unabsorbed carbohydrate, and dessicates intraluminal contents to form stool. These functions are carried out by several mechanisms that include complex movements such as short duration and long duration contractions, segmental, propagated and retrograde pressure waves (5) and high amplitude propagated contractions (5,6,7). How stool is propelled from the proximal colon to the rectosigmoid region is not clearly understood? It is believed that mass movements which occur few times a day may be involved in this process (7,8).

The next stage consists of several stereotyped events that are initiated by a number of reflexes and are possibly controlled by a centre in the brain stem. The basic regulatory mechanism(s) is present in the newborn but the art of controlled defecation develops through training and is controlled by higher cortical centers. Arrival of stool in the rectum causes rectal distension. This induces a desire to defecate and initiates a defecation reflex consisting of rectal contraction and internal anal sphincter (IAS) relaxation. These events push the stool into the proximal anal canal, i.e. the "firing position" for defecation. When the sensitive anoderm perceives stool, a strong urge to defecate is induced that can only be resisted by a vigorous contraction of the external anal sphincter (EAS) and puborectalis. If social and environmental conditions are favorable, the subject sits or squats, contracts the diaphragm, abdominal muscles and rectal muscles (9) and simultaneously relaxes the external anal sphincter and puborectalis. This straightens the anorectal angle, opens the anus and facilitates stool expulsion.

DEFINITION

Constipation is a symptom, not a disease. It is defined as the occurrence of two or more of the following symptoms for at least twelve months, when not using laxatives: a) fewer than three bowel movements per week, b) excessive straining during at least 25% of

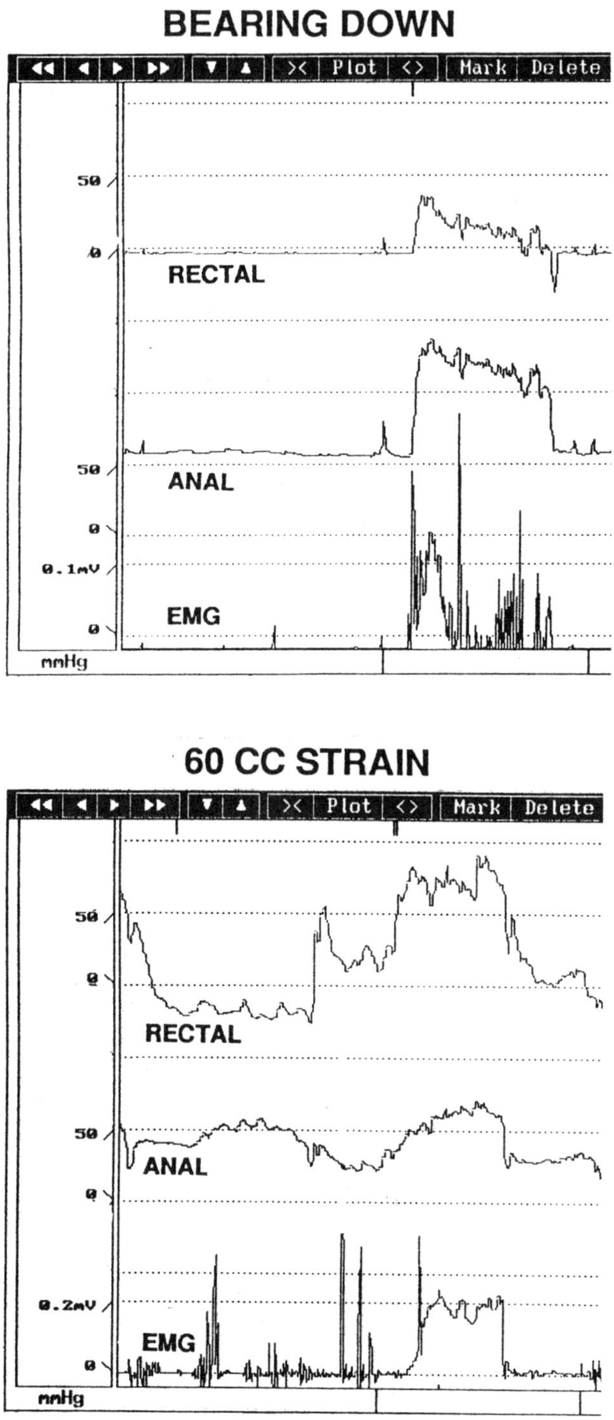

Figure 2. Changes in rectal and anal sphincter pressures and anal EMG changes when bearing down from the patient in the case study (refer to question 3).

Table 1. Common secondary causes of constipation

Anorectal and colonic disorders
Anal fissure
Hemorrhoids
Ulcerative proctitis
Diverticulitis
Colorectal Carcinoma
Inflammatory, post operative and radiation strictures
Drugs
Opioids and related agents
Anticholinergic drugs and antispasmodics
Antidepressants
Anti-hypertensive drugs, particularly calcium channel antagonists, Methyl Dopa
Anti Parkinsonian drugs
Anti-convulsants
Antihistamines
Diuretics
Metal ions such as antacids (aluminum or calcium), iron supplements, and calcium supplements.
Endocrine and metabolic disorders
Diabetes mellitus
Hypothyroidism
Hypokalemia
Hypercalcemia
Porphyria
Neuromuscular disorders
Spinal cord lesions
Parkinsons disease
Multiple sclerosis
Stroke/Cerebrovascular disease
Chagas disease
Hirschsprungs disease

bowel movements, c) a feeling of incomplete evacuation after at least 25% of bowel movements, and d) the passage of hard or pellet-like stool during at least 25% of bowel movements [10]. Because patients may complain of one or more of the aforementioned symptoms or may in fact use the generic term "constipation," it is the physician's responsibility to determine the exact nature of the problem.

ETIOLOGY

Constipation is often caused by a primary disturbance of colonic or anorectal neuromuscular function, as will be discussed. However, persistent constipation can also be caused by anatomic lesions of the colon or rectum, endocrine or metabolic disorders, neurologic diseases, or a variety of drugs. A list of common conditions that predispose to constipation is shown in Table 1. A few factors merit further discussion.

Diet

A high fiber diet increases stool weight and decreases colon transit time [11]. In contrast, a diet that is deficient of fiber may lead to constipation [11,12]. However, not every patient with constipation benefits with a high fiber diet [13]. Constipated patients with either slow colonic transit or pelvic floor dysfunction responded poorly to dietary supplementation with 30 grams of fiber/day, whereas those patients without an underlying motility disorder either improved or became symptom free [13]. Thus, fiber intake may not be a panacea for all patients. Furthermore, today, gastroenterologists are more likely to see patients who have failed a trial of high fiber diet than those who take insufficient fiber.

Psychosocial and Other Factors

For reasons that are not understood, constipation occurs in a subgroup of patients with psychosocial problems [14–16]. Patients with depression, anorexia nervosa, and other psychiatric conditions develop profound disturbances of gut motor function including constipation [15,16]. Some investigators report that 10–30% of patients with constipation have been sexually abused [17,18].

PATHOPHYSIOLOGY OF IDIOPATHIC CONSTIPATION

The diagnosis of idiopathic constipation is most often made after excluding secondary causes (Table 1). These patients suffer from what is essentially a colorectal motility disorder. At least two general subtypes of motor dysfunction, slow transit constipation and obstructive or dyssenergic defecation, have been identified. *Slow transit constipation* is characterized by delayed transit of food residues through the colon. This delay may be due to a primary dysfunction of colonic smooth muscle (myopathy) or colonic innervation (neuropathy) or secondary to obstructive defecation. *Obstructive defecation*, also known as anismus or pelvic floor dyssynergia, is characterized by difficulty or inability with expelling stools from the rectosigmoid region. Some patients with obstructive defecation may also have prolonged colonic transit [19]. There may be a third subtype, which is comprised of patients with constipation predominant irritable bowel syndrome.

CLINICAL EVALUATION

History

This should include an assessment of the duration, severity, nature of the problem, and any precipitating event [5]. The onset of constipation from birth or childhood suggests a congenital or a developmental disorder such as Hirschsprung's disease. It may also reflect poor toilet training and the presence of functional obstructive defecation. A long history of recurring problems which are refractory to dietary measures or laxatives, may suggest a functional colorectal disorder, whereas, the recent onset should alert the physician to seek and exclude an organic illness including neoplastic disease.

The history should also include an assessment of stool frequency, stool consistency (hard, formed or soft), stool size (small pellet-like or long column), degree of straining during defecation, whether there is a sensation of complete or incomplete evacuation, and the need for digital disimpaction of stool. A history of repeatedly ignoring "a call to stool," a history of travel or frequent changes in life style including shift work are all important. These issues are not discussed readily by patients and hence the onus is on the physician to elicit this information.

A dietary history should include an assessment of the amount of fiber and fluid intake, the number of meals and the time of day when meals are consumed. Many patients tend to skip breakfast as a result of the "early morning rush". This may prove to be a serious handicap because colonic propulsion primarily occurs during the "morning waking period" [5–7,20] and after a meal [20–23]. Thus, skipping breakfast may deprive the colon of an important stimulus. The history should also include the number and type of laxatives, and frequency of their use. A family history of using laxatives and a family history of bowel dysfunction are also important, because similar complaints are commonly present among family members. Obstetrical, surgical, and drug history are also useful. A history of back trauma or neurological problems may provide additional clues as to the etiology of constipation. In the elderly, fecal incontinence may be a presenting symptom of stool impaction with fecal overflow [24].

Physical Examination

A thorough physical examination that includes a detailed neurological examination should be performed to exclude systemic illnesses that may cause constipation. The abdomen must be carefully examined for the presence of stool in the left or right lower quadrant. A normal physical examination is not uncommon but it is important to exclude a gastrointestinal mass. Anorectal inspection may reveal skin excoriation, skin tags, anal fissure or hemorrhoids. Perineal sensation and the anocutaneous reflex can be assessed by gently stroking the perineal skin in all four quadrants with the help of a cotton bud (Q-tip) or with a blunt needle. Normally, stroking the perianal skin activates a reflex that contracts the external anal sphincter. If absent, one should suspect neuropathy.

A careful digital rectal examination should be performed to exclude the presence of a rectal stricture, stool or blood in the stool . During digital examination it is important to ask the patient to bear down as if to defecate. During this maneuver, the examiner should perceive relaxation of the external anal sphincter together with perineal descent. If absent, one should suspect functional obstructive or dyssynergic defecation.

DIAGNOSTIC PROCEDURES

General Measures

The first step in the evaluation of constipation is to exclude an underlying metabolic or pathologic disorder. A complete blood count, biochemical profile, serum calcium, glucose levels and thyroid function tests are usually sufficient. If there is a high index of suspicion, serum protein electrophoresis, urine porphyrins, serum parathyroid hormone and serum cortisol levels may be useful.

Flexible Sigmoidoscopy or Colonoscopy

This evaluation may help to identify mucosal lesions such as carcinoma, solitary rectal ulcer, diverticulosis, proctitis or anal fissure. Often, melanosis coli, a brownish-black or gray pigmentation can be recognized, but if absent it can be detected microscopically.

Radiographic Studies

A plain x-ray of the abdomen may provide evidence of an excessive amount of stool in the colon. If colonoscopy has not been performed, a *barium enema* may be useful for excluding significant colonic pathology. Patients with constipation may have a redundant sigmoid colon, a mega colon or mega rectum. The presence of Hirschsprung's disease can also be detected by barium enema, although manometry and histology are required to confirm the diagnosis.

Colon Transit Study. An assessment of colon transit time enables the physician to better understand the rate of stool movement through the colon, because the patient's recall of stool habit is often inaccurate [25]. Several techniques have been described for performing this test [3]. This includes the single capsule technique [26] and the multiple capsule technique [27]. The validity of multiple capsule technique has been questioned [28]. However, for routine clinical purposes a single capsule technique is sufficient. This test is performed by having the patient swallow a single sitzmarks capsule (Konsyl Pharmaceuticals, Fort Worth, Texas) containing 24 radio opaque markers on day 1 and by obtaining a plain radiograph of the abdomen on day 6 (i.e., 120 hours later). This transit study may reveal 3 patterns: *normal transit* consists of 0–5 markers remaining in the colon[3]; *slow transit* is manifest by >5 markers that are scattered throughout the colon; and *obstructive defecation pattern*, which is characterized by the hold up of > 5 markers in the rectosigmoid region with a near normal transit through the rest of the colon. Two-thirds of patients with obstructive defecation may also exhibit a mixed pattern consisting of both slow transit and obstructive delay [3]. In some patients with constipation, the colon transit time may be normal: in these subjects it is important to exclude pelvic floor dysfunction.

Defecography. is performed by placing approximately 150 ml of barium paste into the patient's rectum. This technique and its usefulness are discussed in a Chapter 9.

Colonic Manometry

The advent of solid state manometry probes and portable recorders now enables investigators to perform ambulatory colonic manometry over prolonged periods [20, 29, 30]. Preliminary studies have revealed that the colonic motor patterns are complex [6, 7, 20, 29, 30]. The significance of these complex patterns is not clearly understood. Additionally, colonic motor activity is intermittent, variable, and is influenced by many factors such as sleep, waking, meals [5–7, 20–23], and stressors [31–33]. Moreover, differences in regional colonic motor function have been observed [32]. Thus, in adults, at present colonic manometry is a research tool and may serve as an adjunct to current methods of assessing colonic function in some patients with severe constipation [20]. It has been suggested that colonic manometry may be more useful in children [34]. The absence of high amplitude propagating contractions or segmental loss of propagation is believed to represent colonic dysfunction. Such patients may benefit with total or segmental colonic resection.

Anorectal Manometry

This test provides a comprehensive assessment of pressure activity in the rectum and anal sphincters, and of rectal sensation, rectoanal reflexes, and rectal compliance [3]. Patients with constipation may demonstrate several manometric abnormalities that are summarized in Table 2 and discussed in detail in the chapter on anorectal manometry.

Normally, when someone bears down to have a movement, there is a rise in rectal pressure that is coordinated with a relaxation of the external anal sphincter [Figure 3]. This maneuver is under voluntary control and is primarily a learned response. Inability to perform this coordinated movement represents the chief pathophysiologic abnormality in patients with obstructive defecation and is a common cause of constipation [35, 36]. Poor rectoanal coordination may be due to impaired rectal contraction, paradoxical anal con-

Table 1. Colorectal abnormalities that are commonly seen in patients with constipation

Delayed colonic transit
Delayed segmental colonic transit
Impaired phasic contractions, particularly after meals and after waking
Obstructive defecation
Impaired rectal contraction
Paradoxical anal contraction
Impaired anal relaxation
Absent rectoanal inhibitory reflex (Hirschsprung's disease, dysganglionosis)
Impaired rectal sensation
Inability or prolonged difficulty with expelling a simulated stool

traction, or impaired anal relaxation [19]. Rectal sensation is also impaired in many of these patients [35] and may be associated with an increased rectal compliance that is thought to result from prolonged stool retention. Rectal compliance can be tested by sequentially inflating a balloon in the rectum and by simultaneously measuring the intraluminal pressures at each level of distention [37].

Balloon Expulsion Test

In order to assess patients with pelvic floor dysfunction, a simple test that simulates the act of defecation is desirable. Most normal subjects can expel a water filled balloon, within one minute [38]. If the patient is unable to expel the balloon within three minutes,

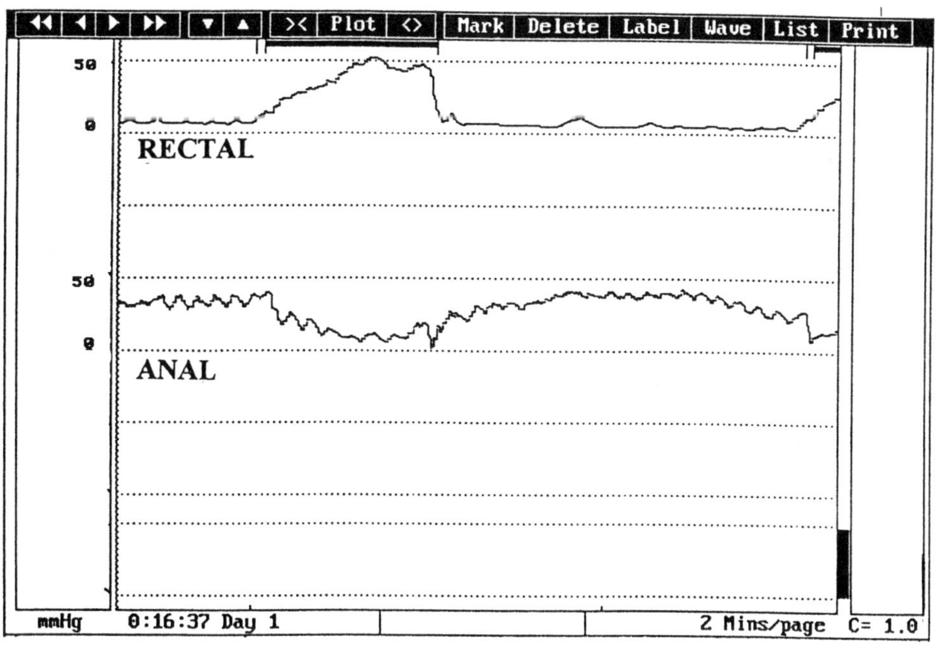

Figure 3. Changes in rectal and anal sphincter pressures when a normal healthy subject bears down as if having a bowel movement. Note that a rise in rectal pressure is coordinated with a drop in anal sphincter pressure or anal relaxation.

obstructive defecation should be suspected . A silicone-filled deformable device, (Fecom) can be used instead of a balloon to test defecation [19]. In a randomized study, the Fecom was felt to be more stool-like than the balloon [39].

CLINICAL USEFULNESS OF MANOMETRY

The clinical usefulness of manometry has been reported recently [35,36,40]. In one prospective study, manometric tests of anorectal function provided diagnostic and often clinically unsuspected information in more than 80% of patients with constipation. This information led to a change in management in 67% of patients. This study also showed that a majority of patients had more than one pathophysiological abnormality. Thus a combination of colon transit study, simulated defecation study and anorectal manometry can provide invaluable information regarding the underlying pathophysiology of constipation.

TREATMENT

The first step in the management of a patient with constipation is to elucidate an underlying cause or mechanism by performing appropriate tests as outlined above. An algorithmic approach to the diagnosis and treatment is shown in Figure 4. It is useful to reiterate that constipation is a common adverse effect of many *drugs* and that this is often overlooked. Some drugs have anticholinergic effects, others desiccate stool, and several drugs including analgesics cause constipation by reducing the awareness for stooling. Because depression and psychological distress are associated with constipation, it is important to identify and provide appropriate psychological or psychiatric support for these patients.

Education/General Advice

This should include information regarding normal bowel habit, diet, fluid intake, and exercise. Those with the obsession that their bowels must move every day should be informed that a bowel movement every other day or sometimes every third day is normal. Patients with poor eating habits, i.e., those who miss meals, should be encouraged to eat more regularly. Patients with inadequate fiber intake should be advised, with the help of a dietitian to increase their intake of natural fiber with fruits and vegetables. It is useful to emphasize that 20 to 30 gm of fiber/day is optimal.

Most individuals with normal bowel function have a bowel movement at approximately the same time of day [41]. This suggests that the initiation of defecation is in part a conditioned reflex that is learned. Therefore, it is advisable to encourage patients to establish a regular pattern of bowel movement. Colonic motor activity is more active after waking or after a meal [6,7,20–22]. Hence, the optimal time for a bowel movement is usually within the first two hours after waking and after breakfast.

Slow Transit Constipation

Before labeling a patient as having colonic inertia or slow transit constipation, it is important to exclude pelvic floor dysfunction. Ideally, slow transit constipation should be treated with a prokinetic agent that selectively stimulates colonic peristalsis. Unfortu-

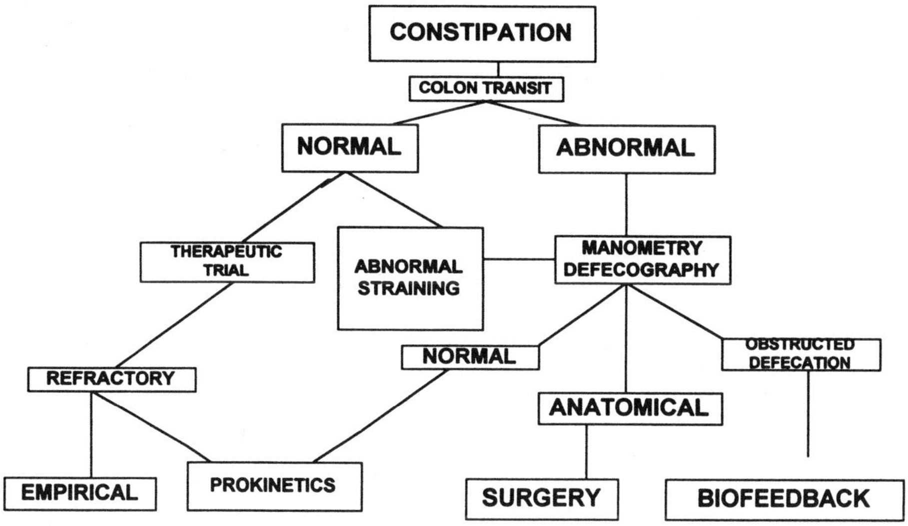

Figure 4. Algorithm for the evaluation and treatment of patients with chronic constipation.

nately, such an agent is not available. Prokinetic drugs such as cisapride (Propulsid®) [42] and the prostaglandin E analogue misoprostol [Cytotec®] [43] have been tried with some success, but the long-term outcome of their use is not known. Other agents that have been tried include, colchicine [44], and ursodeoxycholic acid [45]. However, laxatives remain the main stay of treatment for slow transit constipation.

Based on their mode of action, laxatives can be classified into at least four groups. *Bulk forming laxatives* consist of natural substances such as psyllium or synthetic polysaccharides such as methylcellulose. *Emollients* are anionic surfactants that lower the surface tension of stool and thereby allow easy mixing of its fatty and aqueous components. This interaction may potentially soften stool and permit easy defecation. The two most popular emollients are sodium docusate [Colace®] or calcium docusate. [Surfak®]. *Saline & Osmotic Laxatives* include agents such as milk of magnesia, magnesium citrate, magnesium sulfate, sodium phosphate, and those agents that consist of a mixture of sodium and potassium salts and polyethylene glycol. Other agents include nonabsorbable sugars such as lactulose (Cephulac,® Chronulac®), Lactitol, Sorbitol, Mannitol, and PEG solutions [COLYTE® or GoLYTELY®]. These salts retain water and exert an osmotic effect in the small bowel and colon. *Stimulant Laxatives.* This group includes anthraquinones (Casacara, Sagrada), senna (Senokot®), bisacodyl (Dulcolax®) castor oil and phenolphthalein. The net effect of these agents is to stimulate intestinal motility and secretion.

Obstructive or Dyssynergic Defecation

This problem is best treated by neuromuscular conditioning and biofeedback therapy [19,46,47]. The goal of biofeedback therapy is to restore a normal pattern of defecation by methods discussed in detail in the chapter on biofeedback therapy. Because biofeedback therapy effectively corrects the underlying pathophysiology that causes obstructive defecation (19), it should produce long term benefits. However, it must be emphasized that biofeedback therapy should be performed in conjunction with other supportive measures addressed earlier.

Stool Impaction

Patients with stool impaction or those with hard stools that are difficult to expel may require physical removal of the stool by digital disimpaction or through a rigid sigmoidoscope. This can be painful and may require sedation or anesthesia. Once the hard pellets have been removed, these patients require a bowel conditioning regime of laxatives and suppositories. In order to prevent a recurrence of this problem, Glycerin or bisacodyl suppositories are usually sufficient. Rarely patients require phosphate or saline enemas. In addition, patients may require saline or osmotic laxatives or PEG solution [48].

Refractory Constipation and Surgery

In patients with refractory slow transit constipation, surgery may be an option [49]. However, before considering surgery, it is important to establish that the problem is confined to the colon and not a generalized neuromuscular dysfunction of the gut. For this purpose, a normal gastric emptying study and a normal antroduodenojejunal manometry is required. Surgical options for refractory constipation include colectomy with ileostomy or ileorectal anastomosis. In a large series of carefully selected patients, the results of surgery were quite favorable [49]. However, surgery should be a last resort for patients with constipation. It is important to emphasize that colectomy does not improve symptoms in patients with obstructive defecation. Similarly, colectomy will not relieve chronic abdominal pain or psychosocial problems. On the other hand, surgery is often beneficial for patients with Hirschsprung's disease, rectal prolapse, or colorectal obstruction due to stricture or neoplasia.

CONCLUSIONS

Constipation is a common malady that is often misdiagnosed and poorly managed. Recent advances in technology together with a better understanding of its underlying pathophysiology have led to real progress in the diagnosis and treatment of this problem. Prokinetic drugs that are colon-selective are the optimal treatment for patients with slow transit constipation, but these are not yet available. In contrast, biofeedback therapy has revolutionized the management of patients with dyssynergic or obstructive defecation.

ANSWER KEY AND EXPLANATIONS FOR CASE STUDY

1. Having obtained a good history, the next step in the evaluation of a patient with constipation is to perform a detailed physical and rectal examination. Based on these assessments one should embark on a systematic plan of investigation(s) as outlined in the algorithm in Fig 4. All of the tests included in the question such as colon transit study or colonoscopy may be required but only after eliciting appropriate clues from the history and examination. The best option is **C**.

2. This plain abdominal x-ray taken on day 6 shows 23 markers which are all retained within the rectosigmoid region, i.e the findings are compatible with the inability to empty stools from the distal colon and a normal transit through the rest of the colon. Thus, this patient does not have a normal transit study but has features suggestive of obstructive defecation. The upper limit for the 95% confi-

dence interval for normal colon transit has been estimated to be 66 hours. Because most markers are present in the colon at 120 hours, in this patient, the transit time is likely to be much longer and hence response C is incorrect. It is possible to assess the mean transit time by plotting the number of remaining markers on a normogram that was obtained from a large group of healthy volunteers (Evans et al –reference 26). Answer D is technically correct since more than 5 markers are present on this x-ray but it is not the most appropriate answer. Diabetic colopathy is a nonspecific term that describes colonic neuropathy secondary to diabetes and may manifest as either diarrhea or constipation or with pelvic floor dysfunction. The best option is **B**.

3. The normal response to bearing down as if to defecate is shown in Figure 3. In contrast, in Figure 2 when asked to bear down, the patient shows a rise in rectal pressure that is associated with a rise in anal sphincter pressure and in the anal EMG voltage. This is typical for an obstructive or dyssynergic pattern i.e the patient is unable to coordinate the movements of the rectal and anal muscles to facilitate defecation. Hence response A is incorrect. In the second set of maneuvers, when a 60 cc balloon was distended in the rectum, there is a rise in intrarectal pressure and a fall in the resting anal sphincter pressure-the normal rectoanal inhibitory reflex. The presence of this reflex excludes Hirschsprung's disease. Hence responses B and D are incorrect. Furthermore, when the patient was once again asked to bear down, she again shows an obstructive pattern consisting of a paradoxical rise in anal sphincter pressure and an increase in the anal EMG voltage. Thus the correct response is C. Patients with spinal cord injury usually loose their ability to evacuate voluntarily. In these patients, usually there is no change in anal sphincter pressures when bearing down but the rectoanal inhibitory reflex is intact and normal. The best option is **C**.

REFERENCES

1. Sonnenberg A, Koch TR. Physician visits in the United States for constipation. 1958–1986. Dig Dis Sci 1989;34:606–611.
2. Drossman DA, Li Z, Andruzzi E, et al. US householder survey of functional gastrointestinal disorders; prevalence, sociodemography and health impact. Dig Dis Sci 1993;38:1569–1580.
3. Rao SSC, Manometric evaluation of constipation – Part I. The Gastroenterologist 1996;4:145–154.
4. Sonnenberg, A, Everhart, JE, Brown, DM. The economic cost of constipation. In: Kamm, MA & Lennard-Jones, JE eds. Constipation. Petersfield, UK: Wrightson Biomedical Publishing, Ltd., 1994:19–29.
5. Rao, SSC, Leistikow J, Ackerson K, Hatfield R, Welcher K, Chamberlain M. Colonic motor patterns in healthy humans. A 24-hour ambulatory study. Gastroenterology 1998:114-A829.
6. Narducci F, Bassotti G, Gavurri M, Morelli A. Twenty-four hour manometric recordings of colonic motor activity in healthy man. Gut 1987; 28:17–25.
7. Bassotti G, Gavurri M. Manometric investigation of high amplitude propagated contractile activity of the human colon. Am J Physiol 1988; 255:G660-G664.
8. Rao SSC, Sadeghi P, Leistikow J. Colonic pressure activity during spontaneous defecation. Dig Dis Sci 1998; A-12:1594.
9. Rao SSC, Hatfield R, Leistikow J. Does the rectum generate a pushing force during defecation. Gastroenterology 1998;114:A824–825.
10. Whitehead WE, Chaussade S, Corazziari E, Kumar D. Report of an international workshop on management of constipation. Gastroenterol Intl 1991; 4:97–113.
11. Burkitt DP, Walker ARP, Painter NS. Effect of dietary fibre on stool and transit times and its role in the causation of disease. Lancet 1972;ii:1408–1411.

12. Tucker DM, Sandstead HH, Logan GM, et al. Dietary fiber and personality factors as determinants of stool output. Gastroenterology 1981;81:879–883.
13. Voderholzer WA, Schtke W, M†hldorfer BE, Klauser AG, Birkner B, M†ller-Lissner SA. Clinical response to dietary fiber treatment of chronic constipation. Am J Gastroenterol 1997;92:95–98.
14. Wald A, Hinds JP, Caruanna BJ. Psychological and physiological characteristics of patients with severe idiopathic constipation. Gastorenterology 1989;97:932–937.
15. Devroede G, Girand G, Bouchoucha M, et al. Idiopathic constipation by colonic dysfunction: Relationship with personality and anxiety. Dig Dis Sci 1989;34:1428–1433.
16. Kamal N, Chami T, Andersen A, et al. Delayed gastrointestinal transit times in anorexia nervosa and bulimia nervosa. Gastroenterology 1991;101:1320–1324.
17. Drossman DA, Leserman J, Nachman G, Li Z, Gluck H, Toomey TC, Mitchell CM. Sexual and physical abuse in women with functional or organic gastrointestinal disorders. Ann Intern Med 1990;113:828–833.
18. Walker EA, Katon WJ, Roy-Byrne PP, Jemelka RP, Russo J. Histories of sexual victimization in patients with irritable bowel syndrome or inflammatory bowel disease. Am J. Psychiatry 1993; 150:1502–1506.
19. Rao SSC, Welcher KD, Leistikow JS. Obstructive Defecation (OD): A failure of rectoanal coordination. Am.J Gastroenterol 1998;93:1042–1050.
20. Rao S, Chamberlain M, Hatfield J, Leistikow J, Zimmerman B. Ambulatory 24-hour colonic motility in health and in patients with constipation (abstract). Gastroenterology 1997;112:A809.
21. Rao SSC, Hatfield R, Chamberlain M, Stumbo P. Is the gastrocolonic response influenced by fat or carbohydrate content of a meal? Gastroenterology 1996;110:A742.
22. Bazzocchi G, Ellis J, Villanueva-Myer J, et al. Postprandial colonic transit and motor activity in chronic constipation. Gastroenterology 1990;98:686–693.
23. Snape WJ Jr, Matrazzo SA, Cohen SM. Effect of eating and gastrointestinal hormones on human colonic myoelectrical motor activity. Gastroenterology 1978;75:373–378.
24. Read NW, Abouzekry L, Read MG, Howell P, Ottewell D, Donnelly TC. Anorectal function in elderly patients with fecal impaction. Gastroenterology 1985;89:959–966.
25. Ashraf W, Park F, Quigley EMM, Lof J. An examination of the reliability of reported stool frequency in the diagnosis of idiopathic constipation. Am J Gastroenterol 1996, 91:26–32.
26. Evans RC, Kamm MA, Hinton JM, Lennard-Jones JE. The normal range and a simple diagram for recording whole gut transit time. Int J Colorectal Dis 1992;7:15–17.
27. Metcalf AM, Phillips SF, Zinsmeister AR, et al. Simplified assessment of segmental colonic transit. Gastorenterology 1987;92:40–47.
28. Ehrenpreis ED, Jorge JMN, Schiano TD, Zaitman D, Kane SV. Why colonic marker studies don't measure transit time. Gastroenterology 1997; 110A 728 (Abstract).
29. Rao SSC, Welcher K. Periodic rectal motor activity: Intrinsic colonic gatekeeper. Am J Gastroenterol 1996;91:890–897.
30. Soffer EE, Scalabrini P, Wingate DL. Prolonged ambulant monitoring of human colonic motility. Am J Physiol 1989;257:G601-G606.
31. Chaudhary NA, Truelone SC. Human colonic motility: a comparative study of normal subjects, patients with ulcerative colitis and patients with irritable bowel syndrome. I Resting pattern of motility. Gastroenterology 1961; 40:11–17.
32. Rao SSC, Suls J, Hatfield R, Summers RW. Psychological and physical stress induce differential effects on human colonic motor activity (abstract). Am J Gastroenterol 1998;93:985–990.
33. Narducci F, Snape WJ, Battle WM, London RL, Cohen S. Increased colonic motility during exposure to a stressful situation. Dig Dis Sci 1985; 30:40–44.
34. Lorenzo DiC, Flores AF, Reddy SN, Hyman PE. Use of colonic manometry to differentiate causes of intractable constipation in children. J Pediatr 1992; 120:690–695.
35. Rao SSC, Patel RS. How useful are manometric tests in the evaluation of defecation disorders. Am J. Gastroenterol 1997;92:469–475.
36. Wexner SD, Jorge JMN. Colorectal physiological tests. Use or abuse of technology? Eur J Surg 1994;160:167–174.
37. Wald A: Colonic and anorectal motility testing in clinical practice. Am J. Gastroenterol 1994;89:2109–2115.
38. Rao SSC, Rao S, Hatfield R, Chamberlain M. Is the simulated defecation test influenced by position, presence of artificial stool and its consistency? Gastroenterology 1996;110:A741.
39. Pelsang RE, Rao SSC, and Welcher K.: FECOM: A new artificial stool for assessing defecation. Am J Gastroenterol 1999; 94:183–186.
40. Sun WM, Donnelly TC, Read NW. Utility of a combined test of anorectal manometry, electromyography, and sensation in determining the mechanism of idiopathic fecal incontinence. Gut 1992;33:807–813.

41. Heaton KW, Wood N, Cripps HA, Philipp R. The call to stool and its relationship to constipation: a community study. Eur J Gast Hepaol 1994;6:145–149.
42. Krevsky B, Mauer A, Malmud LS, Fisher RS. Cisapride accelerates colonic transit in constipated patients with colonic inertia. Am J Gastroenterol 1989;84:882–887.
43. Soffer EE, Metcalf A, Launspach J. Misoprostol is effective treatment for patients with severe chronic constipation. Dig Dis Sci 1994;39:923–933.
44. Sninsky CA, Verne GN, Gordon JM, Eaker EY, Davis RH. Double-blind, placebo-controlled, cross-over study evaluating the effectiveness of colchicine in the treatment of severe idiopathic constipation (abstract). Gastroenterology 1998;114(4):A839.
45. Zamboni L, Talarico R, Morselli Labae AM, Malavolti M, Cicognani C, Sama C, Barbara L. Urosodeoxy cholic acid in the management of primary constipation. A cross over study. Gastroenterology 1995;108:A1207.
46. Rao SSC, Loening-Baucke V, Enck P. Biofeedback therapy for defecation disorders. Dig Dis 1997; 15 (Suppl 1):78–92.
47. Enck P. Biofeedback training in disordered defecation: A critical review. Dig Dis Sci 1993; 38:1953–1960.
48. Andorsky RI, Goldner F. Colonic lavage solution (polyethylene glycol electrolyte lavage solution) as a treatment for chronic constipation: a double-blind placebo-controlled study. Am J Gastroenterol 1990;85:261–265.
49. Pemberton JH, Rath DM, Ilstrup DM. Evaluation and surgical treatment of severe chronic constipation. Ann Surg 1991;214: 403–413.

21

FECAL INCONTINENCE

John H. Pemberton*

Mayo Clinic
Rochester, Minnesota

INTRODUCTION

Idiopathic "fecal incontinence" is the most common type of fecal incontinence. It is usually caused by denervation injury to the pelvic floor, although its manifestations vary greatly. In young women, incontinence occurs most commonly after direct injury to the anal sphincter during childbirth. This is usually recognized immediately and repaired successfully by sphincter re-approximation. In older women, fecal incontinence may occur several years after childbirth without an apparent cause. In such patients, a progressive denervation injury to the striated muscles of the pelvic floor has occurred and continues to occur during straining to defecate. It is associated with an abnormal pudendal nerve terminal motor latency and increased muscle fiber density. Often, there is no visible evidence of sphincter trauma.

When a similar clinical picture occurs *in childless women*, the etiology of pelvic floor denervation is chronic straining at defecation, excessive pelvic floor descent and concomitant stretch injury to the pudendal nerves. Associated with this denervation injury is extensive histologic damage to the external anal sphincter, puborectal muscle and levator ani muscles. Moreover, the smooth muscle cells of the internal sphincter are often destroyed and replaced by fibrous tissue.

CASE STUDY

1. A 52-year-old woman presents with a four year history of incontinence, anal seepage, urgency and passage of liquid stools without blood. Which of the following information is most important?

   A. Caffeine and alcohol intake.
   B. Obstetric Hx.
   C. Hx of laparoscopic cholecystectomy.
   D. Medications.

* Phone: 507-284-2359, Fax: 507-284-1794, E-mail: pemberton.john@mayo.edu

Gastrointestinal Motility, edited by Rao *et al.*
Kluwer Academic / Plenum Publishers, New York, 1999.

E. Back problems.

2. Further inquiry revealed that she was a gravida 3, para 3, and had one forceps delivery. She had laparoscopic cholecystectomy 1 year ago and stress urinary incontinence for 10 years. In this patient, which test is most useful in the further management?

   A. Routine blood tests with thyroid profile.
   B. Anorectal manometry.
   C. Defecography.
   D. Anal endosonography.
   E. Saline infusion into the rectum.

3. Routine blood tests including thyroid function tests were normal. Anorectal manometry showed; Resting sphincter pressure of 40 mm Hg, Squeeze sphincter pressure of 82 mm Hg, Squeeze duration of 25 seconds, Threshold for first sensation was 35 ml and threshold for pain was 110 ml. Saline continence test showed that the patient retained 540 ml out of a total infusion of 800 ml of saline. Defecography showed obtuse anorectal angle with perineal descent.

   Which of the following therapeutic options would you advice her ?

   A. Kegel exercises
   B. Loperamide
   C. Cholestyramine
   D. Neuromuscular conditioning with biofeedback
   E. Anal sphincter repair

4. Biofeedback therapy for fecal incontinence can reliably improve all of the following parameters except:

   A. Increase rectal reservoir capacity.
   B. Increase the duration of squeeze.
   C. Eliminate sensory delay.
   D. Improve sensory threshold
   E. Increase voluntary squeeze pressure to 110 mmHg.

COMMON CAUSES OF FECAL INCONTINENCE

Obstetric injury to the perineum and pelvic floor structures is usually due to two mechanisms. One is direct and the other is indirect trauma.

Fecal Incontinence Associated with Direct Trauma

Here, there is an outright tear or disruption of the pelvic floor, puborectal muscle or the sphincter complex which results in a loss of anal barrier function and fecal incontinence.

Fecal Incontinence Associated with Indirect Trauma (Neurogenic Fecal Incontinence)

Here, the incontinence is neurogenic i.e. a result of indirect trauma. Nerves to the puborectal muscle, pudendal nerve and to a lesser extent nerves to the puborectal muscle are stretched either by prolonged childbirth or difficult labor. Such stretching causes a neuropraxic injury leading to denervation of the external anal sphincter. This in turn causes an inability to augment an already relatively low resting pressure, and in stressful situations (loose or watery stool), fecal incontinence can occur.

Several specific neurologic problems have been found in patients complaining of fecal incontinence and thus the term neurogenic rather than "idiopathic" incontinence is more accurate.

Most patients with neurogenic fecal incontinence have low resting and squeeze pressures, and abnormal external anal sphincter (EAS) electromyograms, suggesting a neuropathy. Elevated rectal pressures, like those seen during a valsalva maneuver may exceed anal canal pressures and cause leakage of stool.

Many patients with neurogenic incontinence also have an obtuse anorectal angle. If anal canal pressures (resting and squeeze) are adequate, however, incontinence does not occur. If, on the other hand, the sphincter apparatus is incompetent, incontinence is inevitable. If the angle is normal (puborectalis is competent) but the anal sphincter is inadequate, continence to *solid stool* is usually maintained.

The etiology of neurogenic fecal incontinence is likely to be progressive neuropathic damage to the striated musculature of the pelvic floor, the EAS and of the internal anal sphincter (IAS).

EMG abnormalities of the EAS and puborectal muscle are characterized by decreased activity and the motor unit potentials are of larger amplitude and longer duration. The fiber density is increased on single fiber examination. These findings are consistent with a denervation—reinnervation pattern of injury of the striated muscle fibers of the EAS. Other studies stimulating the pudendal nerve as they exit the pudendal canal have shown that pudendal nerve terminal motor latency (PNTML) is increased significantly in patients with incontinence compared with controls. Transcutaneous electrical stimulation of the spinal cord can be used to measure spinal nerve motor latencies so that cauda equina lesions can be differentiated from those due to neurogenic incontinence. Most patients with neurogenic fecal incontinence have a prolonged PNTML. These neuropathic changes of the EAS and puborectal muscles in women are associated with childbirth. However, excessive straining to defecate over prolonged periods may also increase PNTML.

Severe incontinence does not occur unless the EAS and puborectal muscles are both damaged, and the IAS is weak. Stretch injury to the pudendal nerve affects both the somatic afferents to the EAS and the autonomic fibers to the IAS. Consequently, in these patients the IAS degenerates, the fiber density and PNTML increase and the sensation is decreased. Thus, abnormal descent of the pelvic floor, whether caused acutely by childbirth or chronically by excessive straining is associated with damage to the EAS, puborectal muscles, and weakness of the IAS. Incontinence also occurs in about 50% of patients with complete rectal prolapse, and in 1/3 of patients with rectal intussusception (occult rectal prolapse). Finally, disturbed sensation in the rectum and anal canal, especially in women who have sustained a perineal laceration may also play a role in the genesis of incontinence, but this awaits clear documentation

EVALUATION OF PATIENTS WITH FECAL INCONTINENCE

The primary question to be addressed in this section is: "Do tests that quantify anatomy and physiology 'add value' to the management of patients with disorders of fecal continence?" In other words, is the digital exam with a practiced finger adequate to delineate the different etiologies of fecal incontinence, and thus lay the groundwork for a rational treatment plan? Fecal incontinence is a complex disorder. If extensive testing adds little value, then its usefulness is limited.

What does it mean to "add value"? Tests that add value in *a*) establishing diagnoses, *b*) helping categorize and delineate diagnoses such that a plan of treatment is facilitated, *c*) establishing what is normal (which is often incompletely understood), and *d*) redirecting the approach to patient management.

The disorders of continence which might prompt testing beyond a digital exam include disorders of function, such as constipation, disordered defecation and fecal incontinence. Moreover, disorders of anatomy such as disrupted sphincters, descending perineum, occult prolapse, complete prolapse, Hirschprung's disease and solitary rectal ulcer syndrome should also prompt testing. The etiology of fecal incontinence, as noted above, can sometimes be simplified into (1) anatomical disruption of the sphincter, such as obstetrical injury, surgical misadventure, accidents, and (2) neurogenic injury to the pudendal nerve caused by child birth and straining. Moreover, aging, impaired rectal compliance, sensory disturbances, irritable bowel syndrome and combination of these disorders may each exacerbate fecal incontinence.

Perhaps the biggest advance in our understanding of fecal incontinence has come via the work of the neurologists who have described neurogenic fecal incontinence. Previous to this, neurogenic incontinence was lumped in to something called "idiopathic fecal incontinence," but since Swash and Henry delineated the abnormalities associated with neuropraxic injury to the pelvic floor, our understanding of pelvic floor function and dysfunction has increased dramatically.

Tests Available

These are several tests available to delineate the etiology of fecal incontinence. Obviously, the first of these tests should be a physical exam and a sigmoidoscopy. A detailed assessment of the tests of anorectal function can be found in Chapter 8.

Investigations

A careful, thorough history and physical examination will identify the majority of causes of fecal incontinence. There may be a history of prolonged or difficult childbirth. A history of forceful straining at defecation suggests denervation of the pelvic floor. True fecal incontinence must be differentiated from conditions that cause seepage such as grade 3 or 4 hemorrhoids, fistulas, low rectal or anal tumors, and poor perineal hygiene.[1]

The severity of fecal incontinence is classified as minor if fecal seepage occurs less than once a month, moderate if there is incontinence to solids more than once a month or to liquids more than once a week, and severe if there is loss of control of solids several times a week or liquids on a daily basis. Another classification that is used for grading continence[2] is shown in Table 1.

The physical examination includes inspection of the perianal skin for irritation, fistulous tracks, and hygiene. Scars from previous episiotomies or obstetric tears should be noted. A gaping anus suggests a marked loss of function. *Digital rectal examination* includes an assessment of anal sphincter tone at rest and during squeeze. Moreover, masses and impaction are easily identified digitally. Function of the puborectalis muscle which wraps around the posterior anal canal at the anorectal junction is assessed by asking the patient to squeeze the sphincter; the puborectalis should push the examiner's finger anteriorly. Next, anoscopy and

Table 1. This shows a method of grading fecal incontinence

Grade 1:	Complete continence
Grade 2:	Incontinence of flatus
Grade 3:	Incontinence of flatus and liquid stools
Grade 4:	Incontinence of flatus, liquids stools, and solid stools

rigid proctoscopy should be performed to exclude anorectal pathology. Finally, flexible sigmoidoscopy and/or colonoscopy should be performed if colonic pathology is suspected.

Physiologic and radiologic tests are used to confirm clinical suspicions and provide objective data about the function of the anorectum. Pelvic floor dysfunction is a complex problem and multiple tests may be needed.[3] The finger is a fundamental tool for understanding pelvic floor anatomy, but because disorders can sometimes be complex, other studies are often needed to establish the diagnosis and determine management strategies. A suggested algorithm for the investigation of fecal incontinence is shown in Figure 1.

Anorectal Manometry

Anal manometry maps the anal sphincter pressure profile. It measures resting and squeeze pressure and the length of the anal canal, and demonstrates the presence of the rectoanal inhibitory reflex.[4] Resting anal pressure reflects largely the functional status of the internal anal sphincter, while squeeze pressure quantifies the functional status of the external sphincter. Finally, the rectoanal inhibitory reflex demonstrates the neuromuscular integrity of the anorectum. Interestingly, the length of the anal canal is often the only detectable change in anorectal function tests after successful overlapping sphincteroplasty.

The data generated by multiport anorectal manometry can be used to create a three-dimensional resting and squeeze pressure map of the anal canal. Several different catheter arrays have been used, such as air- or water-filled microballoons, water-perfused catheters, sleeve catheters, or solid-state microtransducers. A station pull-through technique is most commonly used, with resting and squeeze pressures being recorded at 0.5 cm intervals.[5] Directional manometry describes a pressure vector that can identify the site of a sphincter defect. Importantly, simple digital examination by an experienced clinician gives a very good estimation of anal sphincter pressures.[6]

Unfortunately, there is considerable overlap between pressure profiles of normal and incontinent patients. For example, patients with normal resting pressures but reduced squeeze pressures may be incontinent because the external sphincter cannot be recruited to occlude the anal canal at times of stress. Although a significant number of *continent* patients have *impaired* pressure profiles, some *incontinent* patients generate completely *normal* pressure profiles! Unfortunately, manometric studies of the anal canal do not always interpret the patient's clinical status reliably.

Anorectal Sensation

To measure rectal sensation, a balloon is attached to the manometry catheter. The balloon is placed in the rectum and inflated slowly. The patient is asked to report when the sensation of a fecal mass is first noticed and when they wish to defecate. Both blunted and hyperacute rectal sensation can be implied from this test.

Sensation of the anal canal can be quantified by means of an electrical probe inserted into the anal canal and connected to a constant current stimulator.[7] The amplitude of current is gradually increased until the patient experiences a tapping or pricking sensation.

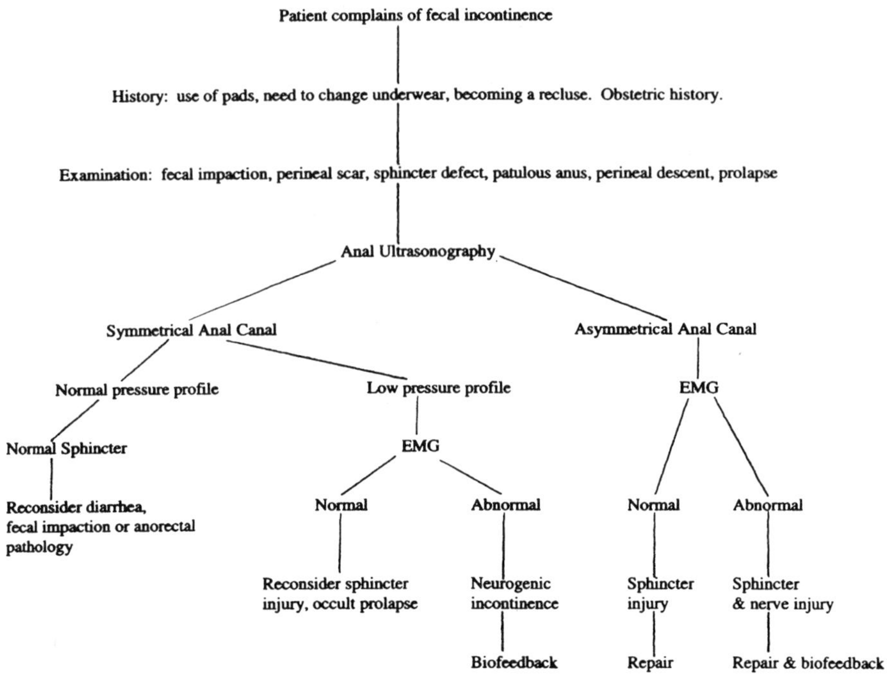

Figure 1.

This threshold to electrostimulation is increased in patients with incontinence but the test seems of limited value in clinical practice.

Electromyography of the Pelvic Floor and Pudendal Nerve Terminal Motor Latency

Electromyography (EMG) of the anal sphincter demonstrates electromuscular activity of the puborectalis and will identify reinnervation of the sphincter typical of pelvic neuropathy. The test is done using single-fiber or concentric needle techniques. Single-fiber electrodes detect potentials that originate from a single motor unit. Fiber density may be determined from the mean number of spikes recorded from the uptake area of a single-fiber EMG electrode sited in 20 different positions.[8] Increased fiber density, suggested by increased number of spike potentials detected by the single-fiber electrode, implies denervation followed by reinnervation. Incontinent patients usually show small and prolonged motor unit potentials. EMG is uncomfortable and unpopular with patients. Moreover, objective interpretation of concentric EMG can be at times very difficult. An alternative, less painful technique is to use a sponge EMG electrode, which is inserted into the anal canal. Two electrodes, placed 180 degrees apart, record electrical activity of the puborectalis at rest and during squeezing and pushing (which simulates defecation). Normally, the puborectalis generates the greatest signal during contraction and generates little signal during normal defecation which requires relaxation of the puborectalis, permitting straightening of the anorectal angle. Incontinent patients show diminished activity in all phases of activity.

Pudendal nerve terminal motor latencies (PNTML) are measured by using a glove-mounted, stimulating electrode that is inserted through the anal canal and positioned on the nerve as it exits the pudendal canal. A recording electrode, adjacent to the external sphincter, detects the evoked potential associated with muscle contraction.[9] The process is carried out bilaterally. The normal delay between stimulating and recording is about 2.0 (+/- 0.5 (SD) ms). Increased delay suggests damage to the pudendal nerves.

Although no medical or surgical treatment has been shown to be effective for pudendal nerve damage, it is an important diagnosis to make. First, the outcome of surgical intervention (such as overlapping sphincteroplasty) may be poor. Second, if the surgeon is persuaded to attempt a procedure, then the surgeon can prepare the patient for the possibility of failure. Third, some surgeons argue that a neosphincter may be more appropriate for this group of incontinent patients.

Defecating Proctogram

Defecography visualizes rectal emptying and indeed, the entire, complex process of defecation dynamically. The patient is positioned upright on a radiolucent commode and barium paste are instilled into the rectum and distal sigmoid. Lateral fluoroscopic images are taken as the patient squeezes the sphincter, while deferring defecation, and while straining to defecate. Several processes can be seen and assessed. Defecating proctograms assess *i* function of puborectalis, *ii* efficiency of emptying, *iii* length of anal canal, *iv* movement of anorectal angle, and *v* extent of perineal descent.[10] Proctograms are difficult to interpret correctly, however, because 50% of *normal* subjects will show some abnormality. Pouchograms also demonstrate occult or overt rectal intussusception, as well as the presence or absence of a rectocele, sigmoidocele, or enterocele. It should be noted that although a careful history and physical examination may lead to a diagnosis of rectocele, the role it plays in fecal incontinence remains unclear until clarified by defecography.

Endoanal Ultrasound

Endoanal ultrasound (EAUS) has only recently been used for the investigation of patients with fecal incontinence. The morphology of the internal anal sphincter, part of the external anal sphincter, the puborectalis, and the rectovaginal septum can be seen quite well.[11] EAUS is particularly useful if there is a history suggestive of a sphincter defect. Defects, evident as a break in the ring of the sphincter deep to the submucosa are identified and mapped in a straightforward relatively unambiguous fashion. EAUS has shown that 85% of patients with fecal incontinence of traumatic origin have external sphincter defects and 40% of these patients also have disruption of the internal sphincter. Moreover, defects are often associated with a weak puborectalis muscle that can be relatively inert on dynamic testing (when patient is asked to squeeze the sphincter). Interestingly, incontinent patients with low resting anal pressures have a thinner internal anal sphincter than do controls. Finally, EAUS appears to be superseding electromyography in the investigation of patients with fecal incontinence.

MEDICAL AND BIOFEEDBACK THERAPY FOR FECAL INCONTINENCE

Medical Treatment

Many patients with fecal incontinence have intermittent diarrhea, which greatly exacerbates the problem. Simple constipating agents such as codeine phosphate or loperamide should be tried. Loperamide slows transit and is particularly useful if there has been a previous small bowel resection. Stool bulking agents improve the consistency of liquid stool. Successful treatment of fecal incontinence with antidiarrheal agents may be limited to only 15% to 20% of patients but this represents a subgroup of patients for whom subsequent aggressive treatment may not be necessary.

Fecal impactions require removal and should be dislodged manually. General anesthesia is sometimes required. Subsequent impaction should be prevented with stool softeners, bulk agents, water, laxatives, and enemas.

If the rectum is kept empty, fecal incontinence can be controlled. The use of glycerin suppositories or bisacodyl may help but they rarely empty the rectum completely. Disposable phosphate enemas are more effective but patients' compliance is not as good. A daily tap water enema can induce a bowel action and keep the rectum empty.

Biofeedback

Biofeedback means "learning through reinforcement".[12] Feedback signals have been used to influence behavior of a variety of organs of the gastrointestinal tract such as to control lower esophageal sphincter pressure[13] and gastric acid secretion.[14]

Biofeedback reduces episodes of fecal incontinence in 50% to 90% of patients.[15] It is especially helpful for patients who have forgotten how to contract the pelvic floor. A balloon distended in the rectum gives a sensation similar to that provided by stool and the patient is trained to coordinate this sensation with contraction of the external anal sphincter. Contraction of the sphincter is measured by an anal balloon or electrical plug and displayed visually. A detailed description has been provided in Chapter 22.

The balloon system similar to that described by Schuster is usually used. This consists of two intra-anal balloons and one rectal balloon. The patient receives visual feedback from the tracings of changes in sphincter pressure and verbal feedback from the therapist. Biofeedback training is in three parts. The first part is sensory training with the patient taught to recognize decreasing volumes of rectal distension. The second part involves training how to contract the external anal sphincter in response to rectal distension, and the third part involves weaning of the visual feedback. The program consists of sessions that last for up to 1 hour at weekly intervals for four weeks. Patients continue their program at home with the help of home-training units.

Patients need to be motivated and able to understand the instructions. Some rectal sensation is needed as well as the ability to contract the anal sphincter voluntarily. To be carried out correctly and effectively, biofeedback is time consuming and requires a dedicated therapist.

Indications for biofeedback are debatable. Patients in whom incontinence is related to previous anorectal surgery do well.[16] The greater the number of incontinent episodes, the less likelihood of success with biofeedback, because it suggests more extensive denervation injury. Nevertheless, the published results are encouraging with success rates (variably defined as either complete continence or a 75% to 90% reduction in incontinent episodes) of between 63% and 90% of patients.[15]

Biofeedback is an attractive therapeutic option as it is painless and harmless. Most patients maintain their improved state although some require refresher courses. The precise technique does not appear to influence results as long as it is carried out correctly. Studies on the mechanisms of the improvement have produced conflicting results. Sphincter pressure may[17,18] or may not[19] increase and rectal sensation threshold may decrease.[20]

CONCLUSIONS

Combining the history and physical exam with endoanal ultrasound and pudendal nerve terminal latency should provide a powerful approach to the diagnosis and categorization of patients with fecal incontinence. These investigations can facilitate the development of rational treatment strategies that could improve the clinical outcome in a majority of these patients.

ANSWER KEY AND EXPLANATIONS

1. All of the historical features posed in this question are important in the evaluation of a patient with fecal incontinence. In any given patient, one or more factors may be responsible for precipitating incontinence. However, the most important factor in the development of fecal incontinence in a female patient is obstetric trauma. Hence, obstetric history with particular reference to the use of forceps, prolonged labor, breech presentation and history of perineal tears or excessive stitches following delivery may provide useful clues regarding the underlying pathophysiology. The best option is **B.**
2. Among the various tests that are available for the evaluation of a patient with fecal incontinence, anorectal manometry provides functional and clinically useful information. Routine blood tests and thyroid function tests can be useful in identifying a metabolic dysfunction or asthenia that may contribute to incontinence. Similarly, defecography may provide useful information regarding the presence of a rectal prolapse and a wide anorectal angle that may contribute to incontinence. Anal endosonography provides very useful morphological evaluation regarding sphincter defects but there are many patients with anal sphincter defects without functional impairment. Hence, this test has a lower specificity in the evaluation of a patient with incontinence. Saline infusion into the rectum (see chapter 8) can objectively demonstrate incontinence in the laboratory but does not provide precise diagnostic information regarding the underlying pathophysiology. The best option is **B**.
3. The results of anorectal manometry reveal that this patient has a weak external anal sphincter and a weak internal anal sphincter with impaired rectal sensation, reduced rectal compliance and reservoir capacity. All of these factors may contribute to her incontinence. Modified Kegel exercises can help to strengthen the pelvic floor but often many of these patients have poor perception of their anatomy and the correct maneuvers that they ought to perform. Loperamide may help to control diarrhea temporarily but is not useful in the long term. Cholestyramine can be a useful adjunct to the treatment of diarrhea in many patients with incontinence who may have coexisting bile salt malabsorption. Anal sphincter repair can be useful in selected patients with fecal incontinence without significant neuropathy. Because this patient has reasonable sphincter muscle, as a first step, biofeedback training should be tried. If unsuccessful and if the patient does not have neuropathy, surgery may be another useful option. The best option is **D.**
4. Recently, it has been shown that biofeedback therapy not only improves subjective parameters but also objective parameters of anorectal function in patients with fecal incontinence. This includes the squeeze sphincter pressure, the duration of squeeze, the sensory delay, the sensory thresholds and the rectal reservoir capacity. However, it is not possible to ensure that the voluntary squeeze pressure increases to 110 mm Hg in every patient with incontinence. The best option is **E.**

REFERENCES

1. Madoff RD, Williams JG, Caushaj PF: Fecal incontinence. N Engl J Med 1992:326:1002–1007.
2. Fleshman JW, Dreznick Z, Fry RD, et al: Anal sphincter repair for obstetric injury: Manometric evaluation of functional results. Dis Colon Rectum 1991:34:1061–1067.
3. Pemberton JH: Anorectal and pelvic disorders: Putting physiology into practice. J Gastroenterol Hepat 1990: 5:127:143.
4. Barkel DC, Pemberton JH, Pezim ME, et al: Scintigraphic assessment of the anorectal angle in health and following ileal pouch-anal anastomosis. Ann Surg 1988: 208:42–49.
5. Johnson GP, Pemberton JH, Ness J, Samson M, Zinsmeister AR: Transducer manometry and the effect of body position on anal canal pressures. Dis Colon Rectum 1990: 33 (6):469–75.
6. Hallan RI, Marzouk DEMM, Waldron DJ, et al: Comparison of digital and manometric assessment of anal sphincter function. Br J Surg 1988:76:973–975.
7. Miller R, Bartolo DCC, Cervero F, et al: Differences in anal sensation in continent and incontinent patients with perineal descent. Int J Coloretal Dis 1989:4:45–59.
8. Neill ME, Swash M: Increased motor unit fibre density in the external anal sphincter muscle in ano-rectal incontinence: a single fibre EMG study. J Neurol Neurosurg Psychiatry 1980:43:343–347.
9. Snooks SJ, Barnes PRH, Swash M: Damage to the innervation of the pelvic floor musculature in chornic constipation. Gastroenterology 1985:89:977–981.
10. Bartram CI, Turnbull GK, Lennard-Jones JE: Evacuation proctography: An investigation of rectal expulsion in 20 subjects without defecatory disturbance. Gastrointest Radiol 1988:13:72–80.
11. Law PJ, Bartram CI: Anal endosonography: Technique and normal anatomy. Gastrointest Radiol 1989:14:349–353.
12. Skinner BF: Science and human behaviour. New York, Macmillan, 1953.
13. Nikoomanesh P, Wells D, Schuster MM: Biofeedback control of lower esophageal sphincter contraction in man. Clin Res 1973:21:521.
14. Welgan PR: Learned control of gastric acid secretion in ulcer patients. Psychosom Med 1974:36:411–419.
15. Enck P: Biofeedback training in disordered defecation: A critical review. Dig Dis Sci 1993:38:1953–1960.
16. McHugh S, Walma K, Diamant NE: Fecal incontinence: A controlled trial of biofeedback. Gastroenterology 1986:90:1545.
17. Whitehead WE, Burgio KL, Engel BT: Biofeedback treatment of fecal incontinence in geriatric patients. J Am Geriatr Soc 1985:33:320–324.
18. Riboli RB, Fraseco M, Pitt OG: Biofeedback conditioning for fecal incontinence. Arch Phys Med Rehabil 1988 :68:29–31.
19. Loening-Bauke V: Biofeedback for fecal incontinence. Dig Dis 1990:112–124.
20. Cerulli MA, Nikoomanesh P, Schuster MM: Progress in biofeedback conditioning for fecal incontinence. Gastroenterology 1979:76:742–746.

22

BIOFEEDBACK THERAPY FOR CONSTIPATION AND FECAL INCONTINENCE

Satish S. C. Rao*

Department of Medicine
University of Iowa College of Medicine
Iowa City, Iowa

INTRODUCTION

Constipation with obstructive defecation and fecal incontinence are problems that are frequently encountered in the gastrointestinal clinic (1). Until recently, most patients were treated with supportive or empirical therapy or surgery. However, over the last decade, biofeedback therapy has emerged as a useful adjunct for the treatment of these problems. Biofeedback therapy is a learning process that is based on "operant conditioning" techniques. The governing principle is that any behavior—be it a complex maneuver such as eating or a simple task such as muscle contraction—when reinforced, its likelihood of being repeated increases (2).

CASE STUDY

A 30-year-old heterosexual, male engineer was referred with a 20 year history of constipation. He had no urge to defecate and in spite of excessive efforts at straining he could not pass stool. Recently, his bowel symptoms became refractory to laxatives. His active problems included insulin dependent diabetes mellitus (2 years) and bulimia nervosa (1 year). At presentation, he was taking a high fiber diet with 60g of fiber/day, sodium docusate 2 tablets tid, metamucil 4 tablespoons tid and frequent enemas. With this regime, his bowels moved once every two or three weeks. Digital examination revealed inability to bear down and paradoxical anal sphincter contraction. Occult blood was negative.

1. What would be your approach to the further management of this patient?

   A. Order a colonoscopy.
   B. Change his laxative regime.

* Satish S.C. Rao, Tel. No.: (319) 353-602, Fax No.: (319) 353-6399

Gastrointestinal Motility, edited by Rao *et al.*
Kluwer Academic / Plenum Publishers, New York, 1999.

C. Advice better control of his diabetes.
D. Order a colon transit study.

2. A hematology and biochemical profile, and thyroid function test was ordered and were normal. His diabetes was well controlled with a HbA1C of 6%. A barium enema was normal. A colon transit study showed that 22/24 markers were retained in the colon at 120 hours indicating slow colon transit. Results of anorectal manometry are summarized below:

	Patient	Normal (95% C.I.)
Max Resting Pressure mm Hg	65	67(7)
Max Squeeze Pressure mm Hg	140	138(14)
Max Rectal pressure when straining mmHg	45	61(9)
Anal residual pressure -strain mm Hg	80	47(8)
Defecation index	0.6	1.7(0.5)
Threshold vol. for first sensory perception cc	240	20(3)
Threshold vol. for Desire to defecate cc	320	98(27)

How would you interpret these manometric findings?

A. Impaired rectal sensation.
B. Obstructive defecation.
C. Hirschsprung's disease.
D. Obstructive defecation with impaired rectal sensation.
E. Normal sphincter with abnormal rectal compliance.

3. Based on your evaluations, you have made a definitive diagnosis. Which of the following treatments would you now recommend ?

A. A better laxative regime.
B. Surgery for refractory constipation.
C. Neuromuscular conditioning using biofeedback therapy.
D. Botox injection of anal sphincter.
E. Misoprostol

The patient elected to undergo biofeedback therapy. After 4 sessions of retraining, his defecation pattern changed from that of an obstructive pattern to a normal coordinated pattern (Figure 1). Subsequently, there was marked improvement in his constipation. One year later, he was able to pass stools without laxatives and has remained well. The manometric changes are summarized below.

	Patient's baseline	After therapy	Normal (95% C.I.)
Max resting pressure mm Hg	65	66	67(7)
Max squeeze pressure mm Hg	140	140	138(14)
Max rectal pressure when straining mmHg	45	70	61(9)
Anal residual pressure—strain mm Hg	80	20	47(8)
Defecation index	0.6	3.5	1.7(0.5)
Threshold vol. for first sensory perception cc	240	30	20(3)
Threshold vol. for desire to defecate cc	320	60	98(27)
50 cc balloon expulsion time s	>300	35	57(34)

4. Which of the following is not true?
A. Biofeedback therapy is effective both in the short term as well as in the long term.

Table 1. Indications for biofeedback bowel retraining

Constipation—obstructive defecation (Anismus)
Fecal incontinence
Solitary rectal ulcer syndrome

B. There is subjective improvement in bowel function but no objective improvement in anorectal function.
C. This mode of treatment is reimbursed by most insurance carriers.
D. Biofeedback therapy is safe and effective but labor intensive.

INDICATIONS FOR BIOFEEDBACK THERAPY

The indications for biofeedback therapy are shown in Table 1. Several techniques of performing biofeedback therapy have been developed (2, 3). Although, some of the components of neuromuscular conditioning are similar, there are distinct differences in the methods that are used for training patients with obstructive defecation (anismus) and for training patients with fecal incontinence. Also, the principles and goals of training are different. Hence, the treatment methods for each condition are discussed separately.

BIOFEEDBACK THERAPY FOR CONSTIPATION

Obstructive defecation accounts for 30–50% of patients with chronic constipation (4). Other terms that have been used to describe this condition include anismus (5), spastic pelvic floor syndrome (6), paradoxical anal contraction (7,8) and outlet obstruction (9) or pelvic floor dyssynergia (10). Several studies have reported the usefulness of biofeedback therapy in the management of patients with obstructive defecation and these are summarized in Table 2.

Principles of Therapy

Although several pathophysiological disturbances may lead to obstructive defecation, it is likely that the primary mechanism is a failure of rectoanal coordination (11). This may be associated with one or more of the following changes; impaired or inadequate rectal contraction, paradoxical anal contraction (4, 11), absent or inadequate anal relaxation (9, 11), or impaired rectal sensation (11). Thus, the goals of retraining are to correct each of these pathophysiological disturbance(s) so that a patient learns to expel stool without using laxatives, suppositories or digital manipulations and to restore normal bowel movement. Generally, visual, auditory or verbal feedback techniques are used for neuromuscular conditioning. These methods help to reinforce a particular task as it is being performed. For example, the display of the pressure or EMG changes of the anal sphincter—on a monitor screen—provides instant visual feedback to the patient regarding their performance. Similarly, when bearing down, the intensity (pitch) of the electrical activity of the anal sphincter muscle provides auditory feedback to the patient regarding their performance. Throughout the training session, the therapist complements the patient and reinforces their behavior when it is appropriate or corrects the behavior when it is inappropriate. This verbal feedback to the patient is just as important as the visual or the auditory feedback.

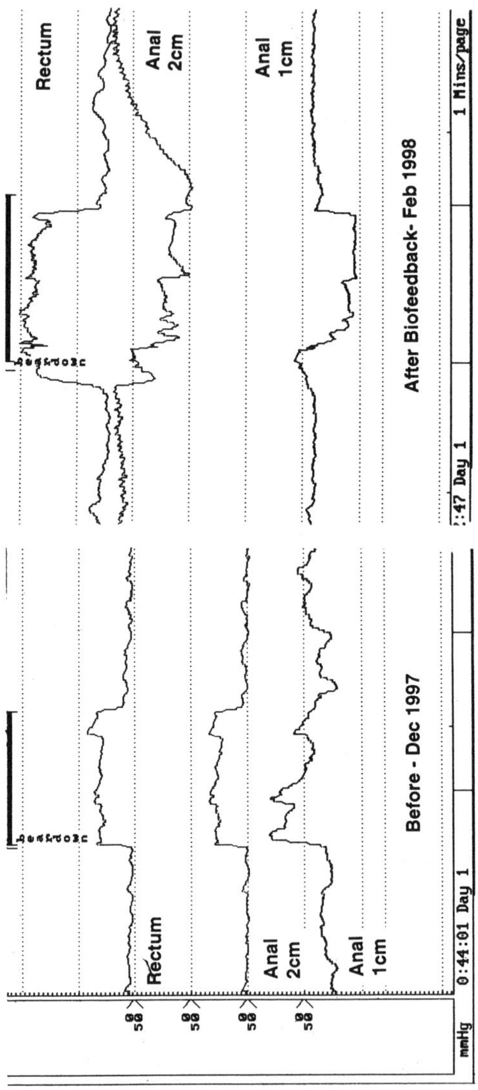

Figure 1. Effects of biofeedback therapy on the rectal and anal sphincter pressures during simulated defecation in a normal subject and in a patient with obstructive defecation before and after treatment.

Table 2. Biofeedback studies in adults with obstructive defecation. [Adapted with permission from reference 7.]

Author	Year	Ref	N	Age range	F/M	Training method	Setting	Sessions $\bar{x}$ (range)	Follow up $\bar{x}$ or range	Outcome assessment	% Improved
Bleijenberg	1988	6	10	10–48	?	EMG + B.D	In patient	Daily	1–14 mo	Interview	70%
Weber	1988	19	22	18–55	22/0	Manometry (W)	Outpatient	2–4 ?	?	Interview	18%
Kawimbe	1991	9	15	22–76	12/3	EMG	Home	2/day	3–6 mo.	Interview	87%
Dahl	1991	7	9	20–60	9/0	EMG	Inpatient	5	2–6 mo.	Interview	78%
Wexner	1992	8	18	10–84	13/5	EMG	Outpatient	9	1–12 mo.	Diary	89%
Fleshman	1992	15	9	35–62	8/1	EMG + B.D.	Outpatient	6/day´2	6 mo.	Interview	100%
Turnbull	1992	16	7	29–42	7/0	Manometry (W)	Outpatient	5	3 mo–3 yrs.	Diary/Phone	71%
Papachrysostomou	1993	17	22	32–50	17/5	EMG+ B.D. + H.T.	Outpatient	3	1.5 mo.	Manometry/ defecography/ interview	59%
Koutsomanis	1994	12	20	18–53	?	EMG	Outpatient	2–6	2–12 mo.	Diary/Transit	50%
Keck	1994	32	12	17–82	10/2	Manometry(W)	Outpatient	3 (1–14)	8 mo.	Phone	33%
Bleijenberg	1994	18	20	20–50	15/5	EMG or B.D.	Outpatient	8	9–14 mo.	Diary	B.D.—22% EMG—73%
Rao	1997	14	26	21–87	17/9	Manometry (S) + B.D.	Outpatient	4(2–8)	2–4 mo.	Manometry/ Diary	88%
Koutsomanis	1995	12	59	20–66	52/7	EMG or B.D.	Outpatient	2	3 mo.	EMG/Diary	44%
Siproudhis	1995	33	27	21–77	20/7	Manometry (W) + B.D.	Outpatient	6 (1–10)	14 mo.	Interview	52%
Total	—	—	276	10–87	202/44	6:2 1:2:3*	—	2–14	1 mo–3 yrs.	—	58%

B.D. = Balloon defecation, Manometry (W) = Water perfused manometry, Manometry (S)= Solid state manometry probe system, H.T.= Home trainer.
* EMG alone - 6, B.D. alone - 2, Manometry alone - 1, EMG + B.D. - 2, Manometry + B.D. - 3.

Abdominal Muscle Coordination Exercises. This exercise improves the abdominal effort that is required to facilitate defecation (12–14). Here, patients are taught to distend the abdomen by inhaling slowly and then to hold their breath for at least 10–15 seconds. They are asked to practice this maneuver for 20 minutes, three times a day.

Recto-Anal Coordination. The goal of this maneuver is to produce a coordinated movement (Figure 1) that consists of a rise in the intra rectal pressure which is synchronized with a relaxation of the anal sphincter (7–9, 11–13, 15–19). This training is performed by either placing an EMG device or a manometry probe into the anorectum. Next, the patient is asked to sit on a commode, in front of a monitor, and to take a deep abdominal breath and bear down as if to defecate, while observing the display on the monitor. During this maneuver, the patient is asked to titrate the degree of abdominal and anal effort that may be required to produce a coordinated movement of the pelvic floor (Figure 1).

Sensory Conditioning. The objective is to teach the patient to perceive a particular distention with the same intensity as they had experienced previously with a larger volume (14, 20–23). The first step here is to induce an urge to defecate by inflating a balloon in the rectum. Thereafter, with each subsequent inflation the balloon volume is decreased stepwise by about 10–15%. During each distention, the subject is encouraged to observe the rectal pressure changes on the monitor and to use this cue, particularly at volumes that are not readily perceived (14, 20–24). If the patient fails to perceive a particular volume, or reports a change in the intensity of perception, the balloon inflation is repeated after a 5 second warning either by using the same volume or by using the previously perceived (higher) volume. Thus, by repeated inflation and deflation and through a process of trial and error, newer thresholds for rectal perception are established.

Duration and Frequency

The number of biofeedback training sessions should be tailored for each patient based on their needs. On average, three to six, one hour training sessions are required. After retraining, periodic reinforcement may provide additional benefit, but there is no data to support this.

Devices and Techniques for Biofeedback

Because biofeedback therapy is an instrument based learning technique, several devices and methods have been described and newer techniques continue to evolve.

The *EMG biofeedback system* consists of surface EMG electrodes that are either mounted on a probe or affixed to the surface of the external anal sphincter muscle (6-9, 13). A few studies (6, 9, 25), have used *home training kits* such as *a*n EMG home trainer or a silicon probe device attached to a hand-held monitor with an illuminated color LCD.

The *balloon defecation method* consists of placing a water-filled or air-filled balloon in the rectum and assisting the patient to expel this distended balloon (6, 12, 13, 14, 17, 18) During the maneuver, the patient is instructed to relax the external anal muscles and to coordinate the movements of the abdominal and pelvic floor muscles. Preferably, this technique should be performed in the sitting posture (2, 14), though some centers have used the lying posture (6, 13).

A *solid state manometry probe with microtransducers and a latex balloon* (Figure 2) can also be used for biofeedback therapy (14). In this method, the pressure activity from the rectum and anus that is displayed on a monitor is used to provide the visual feedback.

Table 3. Selection criteria for patients with constipation and obstructive defecation

Paradoxical anal contraction or inability to relax the anal sphincter during attempts to strain as if to defecate.
Failure or extreme difficulty with expelling a simulated stool (>3 minutes).
Retention of markers in the rectosigmoid region with prolonged colonic transit.
Inability to expel > 50 % of barium paste during defecography.

Subsequently, in order to simulate the act of defecation, the distended balloon can also be pulled through the anus. Thus, this system may serve as a comprehensive device for neuromuscular conditioning, since it appears to incorporate many of the useful features from each of the aforementioned devices.

Table 3 shows criteria that are used in our center for selecting patients for biofeedback therapy. Because pelvic floor dysfunction is multifactorial (3,14), and some of the manometric features described in patients may also be seen in healthy subjects (11) several criteria are required. In addition to the *first* criteria, *two or more* criteria must be fulfilled before a definitive diagnosis of obstructive defecation can be made and before a patient is considered for enrollment into a neuromuscular conditioning program (14).

The number of biofeedback training sessions should be customized for each patient, as discussed previously. On average, three to five, one hour training sessions may be required. After retraining, additional benefit may be provided by periodic reinforcement, although there is no data to support this.

Treatment Efficacy

The symptomatic improvement rate has varied from 44%–100% (Table 2). One should exercise caution when interpreting clinical outcome of these studies because the end point for successful treatment has not been clearly defined and the duration of follow up has been quite variable. Controlled trials have not yet been performed and are difficult to organize. Despite these shortcomings, it is gratifying to note that a simple, noninvasive and a relatively inexpensive technique which has no adverse effects can ameliorate symptoms and improve the quality of life in a majority of patients with this affliction.

BIOFEEDBACK THERAPY FOR FECAL INCONTINENCE

Principles of Therapy

The goals of biofeedback therapy are to a) improve the strength of the anal sphincter, b) improve the coordination between the rectum and anal sphincter and c) improve the sensory awareness for stool perception. Because each goal requires a specific training method, the treatment protocol for a given patient will depend on the underlying dysfunction (20, 22–23, 26, 27). All three maneuvers are complimentary. The indications for biofeedback therapy and the criteria for selecting patients is shown in Table 4. There is no uniform consensus nor have these criteria been validated. But these could serve as guidelines.

Components and Methods of Neuromuscular Conditioning

This could involve one or more of the following three components. Again, visual, audio or verbal feedback techniques are used to perform the following maneuvers.

Table 4. Indications for biofeedback therapy and selection criteria for patients with fecal incontinence

Indications for therapy
Adults
Sphincter dysfunction secondary to obstetric trauma
Sphincter dysfunction secondary to surgical trauma
Diabetes with fecal incontinence
Idiopathic fecal incontinence
Selection criteria
Manometric criteria
Weak resting or squeeze sphincter pressure but with some ability to squeeze
Impaired rectal sensation
Impaired rectal compliance
Abnormal saline continence test
Mild or moderate pudendal neuropathy
At least two criteria must be present
Subjective criteria
Age: Patients from any age group are eligible provided their cognitive function is intact.
Motivation
Ability to follow commands and pursue exercise strengthening

Adapted with permission, Reference #2.

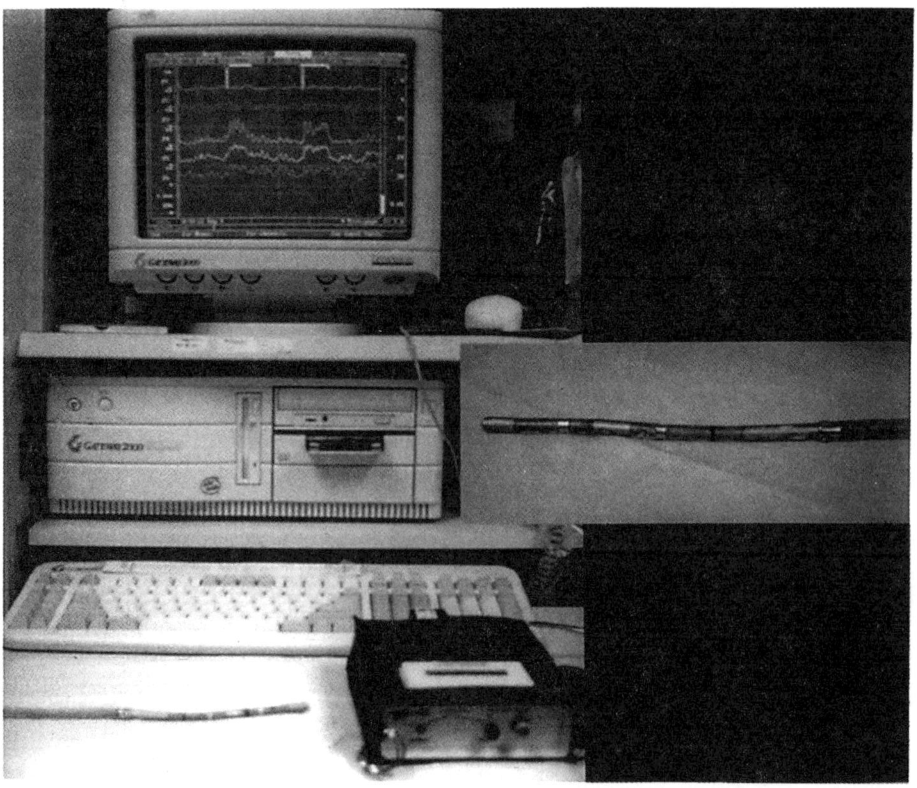

Figure 2. A solid state manometry probe, (inset), a data logger and a computer system that can be used for biofeedback therapy.

Conditioning the External Anal Sphincter. In order to help the patient to understand the correct maneuvers and what is expected of them during the training session, at the outset, a normal pressure recording is shown (Figure 3). Next, a manometric device is placed in the anorectum and the patients are instructed to squeeze the anus and to maintain the squeeze for 20 seconds. During the squeeze maneuver, by observing the pressure or EMG changes on a monitor or by listening to the pitch of the auditory signals generated by the EMG activity the patients receive instant feedback of their performance. They are advised to selectively squeeze the anal muscles and not to use their abdominal, thigh, or gluteal muscles. After two or three sessions, the patients are encouraged to perform these maneuvers without visual or auditory feedback (11, 20–23, 27–30). Also, they are instructed to practice the anal muscle strengthening exercises, at home, for at least 20 minutes, three times a day (26).

Conditioning Rectal Sensory Perception. The goal is to improve the thresholds for rectal perception by intermittently inflating a balloon in the rectum, as described under sensory training for obstructive defecation.

Conditioning Rectoanal Coordination. The aim of this training is to achieve a maximum voluntary squeeze in less than 1 second, after a balloon is inflated in the rectum (17–19, 31–33). This maneuver mimics the arrival of stool in the rectum and prepares the patient to react appropriately by using the right group of muscle. After each inflation, the patient is asked to signal when they have perceived rectal distention and to react to this distention by promptly squeezing their anal sphincter and to maintain the squeeze for at least 20 seconds.

Which of the three components namely, sphincter training, sensory conditioning or rectoanal coordination is most useful remains unclear. In general, most experts, believe that all three components are useful and that the treatment program should be customized for a given patient depending on the underlying dysfunction. The key element however, is to condition the external anal sphincter response so as to improve the squeeze profile and the ability to respond to small volumes of rectal distention.

Duration and Frequency

Each training session can take 1 to 1 ½ hours. Patients are asked to visit the motility laboratory once or twice a week. At the outset, it is often difficult to predict how many sessions are required for successful training. On average, 3–6 training sessions are required. After successful training, in our center, we provide periodic reinforcement. A recent study has shown that reinforcement of biofeedback therapy at 6 weeks, 3 months, and 6 months and 12 months can improve long-term outcome of these patients (26).

Devices Used for Biofeedback

Several types of manometry probes and technical systems are available (34). This includes the Schuster balloon system which was used in the early studies (23, 29), the EMG devices (35), (Myotron; Dantec EMG, Dantec Ltd, Copenhagen, Denmark; Perry Meter[tm] anal EMG sensor, EPS-21, Perry Meter systems, Stafford, PA; Pelvimed, Essen, Germany), water perfused manometry probe system (20) and the solid state manometry system with a latex balloon (11, 14, 26). The advantages of any one device has not been scientifically tested. A solid state probe system (Figure 2), is preferable to a water per-

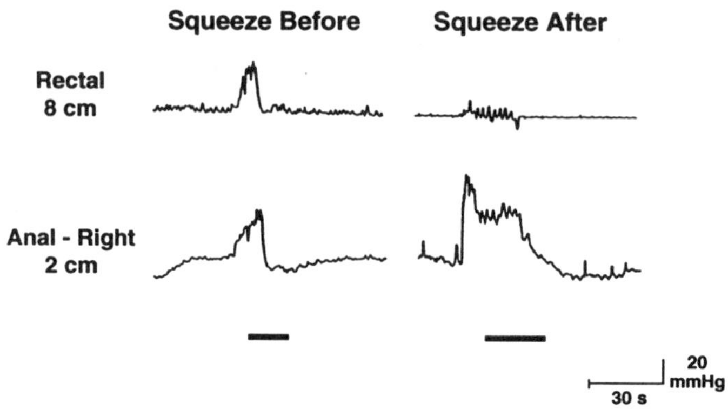

Figure 3. The squeeze anal sphincter pressures, the duration of squeeze and the recto anal coordination in a patient with fecal incontinence before and after biofeedback therapy [Modified with permission from Rao, et al (26)].

fused system because there is no distraction or embarrassment from leakage of fluid, a problem often seen in patients with a weak or incompetent anus. Also, if required the subject can be moved to a sitting position without adversely affecting calibration (26). However, a solid state manometry system is expensive and the sensors are fragile. Although, each system has inherent advantages and disadvantages, most systems provide reproducible and useful measurements (34). The choice of any one system depends on many factors that include cost, goals of training and facilities which are already available.

Treatment Efficacy

After biofeedback therapy, symptomatic improvement has been reported in 80% of patients (3). However, no uniform criteria has been used for defining improvement. A few studies have used symptom diaries for assessing improvement (3, 20, 22-24, 26, 27) whereas, others have relied on interviews or telephone surveys (29–31). Because patients' recall of stool habit is unreliable (36) the high success rate that has been reported in some studies (3) remain questionable. Some authors have argued that the therapeutic efficacy of biofeedback training cannot be predicted by manometric results (20–22, 29). However, a recent prospective study reported that one year after starting therapy, 80% of patients showed improvement in subjective and objective parameters of anorectal function (26).

Neuromuscular conditioning not only improves symptoms (3, 20, 26) but also objective parameters of anorectal function such as squeeze sphincter pressure (26), rectal sensation (20, 23, 26), and rectal reservoir capacity (26). The number of training sessions that are required to achieve clinical improvement is variable and depends not only on the severity of the problem but also on the patient's ability to learn new techniques. Hence, the number of training sessions should be customized for each patient based on their needs. Furthermore, periodic reinforcements and follow up assessments using objective parameters may help to improve the long-term outcome of biofeedback therapy (26).

CONCLUSIONS

Neuromuscular conditioning is an instrument based learning technique. Several devices and methods are available for performing this training. No single method is either superior or universally popular. The three modalities that are commonly used for neuromuscular conditioning are visual, verbal and audio feedback. Ideally, the training program should be customized for each patient, based on the underlying dysfunction(s). Following biofeedback therapy, clinical trials have reported symptomatic improvement in 70–80% of patients with either incontinence or obstructive defecation. Recent studies have also demonstrated objective improvement in anorectal function.

Biofeedback therapy is a useful method of treatment for patients with refractory anorectal disorders. It is safe, relatively inexpensive, and easy to administer, and has been shown to improve symptoms and objective parameters of anorectal function. However, this therapy is not widely available. This problem can be overcome by further refinements in technology, both the tools and techniques of therapy, coupled with a genuine desire to learn and implement this training program in clinical practice.

ANSWER KEY AND EXPLANATIONS

1. The history, physical findings particularly digital rectal examination is highly suggestive of constipation associated with obstructive or dyssenergic defecation. There are no alarm symptoms such as weight loss, GI bleeding or recent change in bowel habit. Also, constipation is one of the most frequent complications of diabetes although in this patient it started before the onset of his diabetes. Hence, a colonoscopy is unlikely to be helpful. Yes, it may exclude inflammatory disease or colon cancer, but these are low priorities given the history and age of this patient. He has already become refractory to laxatives and changing this regime further would be an exercise in vain. His diabetic control is pretty good. The best option would therefore be to perform further investigations to identify a colorectal dysfunction that may be responsible for his constipation and to identify if this is a generalized colonic problem or more localized to the pelvic floor. The best option is **D**.
2. The manometry shows that the patient has normal anal sphincter pressure, but impaired rectal pressure when bearing down together with an abnormally high anal residual pressure which is compatible with paradoxical anal contraction. These features are typically seen in patients with dyssynergic defecation. The defecation index which is the ratio between the maximum rectal pressure during straining and the minimum anal residual pressure during straining was 45/80 = 0.6. Normal values are >1.5. In addition the patient has impaired rectal perception and desire to defecate. Hence the best option which describes all of these features is Option D. Option A and option B are also correct but only partially describe the findings. Hirschsprung's disease cannot be diagnosed based on the information provided. In this condition the rectoanal inhibitory reflex is absent. The values for rectal compliance have not been given. The best option is **D**.
3. The patient has already failed laxatives and a better regime is unlikely to be beneficial, at least in the long term. Surgery is an option because he has life-long constipation that is now refractory to treatment. But, he is a young man with dyssynergic defecation that can be treated without surgery. Neuromuscular conditioning has been shown to be extremely useful and effective in treating this problem. Botox injections have been tried with rather mixed results. Although this may work temporarily, clearly until the patient learns to improve rectoanal coordination, botox by itself would not work. Surgical myotomy has also been

tried with disappointing results. Misoprostol can be useful but is not a long term solution. The best option is **C**.

4. Recent studies have shown that biofeedback therapy not only improves subjective but also objective parameters of anorectal function. Although there are limited and poorly controlled studies, current experience is that this treatment modality is effective in the long term. Most insurance carriers will reimburse this therapy, but the word "biofeedback" is taboo in insurance literature. Hence, when billing, codes for bowel retraining or physical rehabilitation must be used. Option D is correct. The best option is **B**.

REFERENCES

1. Drossman, D.A., Zhiming, L., Andruzzi, E., Temple, R.D., Talley, N.J., Thompson, W.G., Whitehead, W.E., Janssens, J., Funch-Jensen, P., Corazziari, E., Richter, J.E., Koch, G.G. US householder survey of functional gastrointestinal disorders: Prevalence, sociodemography, and health impact. *Dig Dis Sci* 1993; 38:1569–1580.
2. Rao, S.S.C., Enck, P., Loening-Baucke, V. Biofeedback therapy for defecation disorders. *Dig Dis* 1997; 15; 78–92.
3. Enck, P. Biofeedback training in disordered defecation: A critical review. *Dig Dis Sci* 1993; 38:1953–1960.
4. Rao, S.S.C., Patel, R.S. How useful are manometric tests of anorectal function in the management of defecation disorders. *Am J Gastroenterol* 1997; 92:469–475.
5. Preston, D.M., Lennard-Jones, J.E. Anismus in chronic constipation. *Dig. Dis. Sci.* 1985;30:413–418.
6. Bleijenberg, G., Kuijpers, H.C. Treatment of spastic pelvic floor syndrome with biofeedback. *Dis Colon Rectum* 1987;30:108–111.
7. Dahl, J., Lindquist, B.L., Tysk, C., Leissner, P., Philipson, L., Jarnerot, G. Behavioral medicine treatment in chronic constipation with paradoxical anal sphincter contraction. *Dis Colon Rectum* 1991; 34:769–776.
8. Wexner, S.D., Cheape, J.D., Jorge, J.M.N.,, Heymen S., Jagelman, D.G. Prospective assessment of biofeedback for the treatment of paradoxical puborectalis syndrome. *Dis Colon Rectum* 1992;35:145–150.
9. Kawimbe, B.M., Papachrysostomou, M., Binnie, N.R., Clare, N., Smith, A.N. Outlet obstruction constipation (anismus) managed by biofeedback. *Gut* 1991;32:1175–1179.
10. Whitehead WE, Devrode G, Habib F, Meunier P, Wald A. Functional disorders of the anus and rectum. In: Drossman DA et al. (Editors), The Functional Gastrointestinal Disorders. Little Brown & Company, Boston 1994; pp. 217–263.
11. Rao, S.S.C., Welcher K, Leistikow J. Obstructive Defecation (O.D.): A failure of rectoanal coordination. *Am J Gastroenterol* 1998;93:1042–1050.
12. Koutsomanis, D., Lennard-Jones, J.E., Kamm, M.A. Prospective study of biofeedback treatment for patients with slow and normal transit constipation. *Eur J Gastroenterol. Hepatol.* 1994;6:131–137.
13. Koutsomanis, D., Lennard-Jones J.E., Roy, A.J., Kamm, M.A. Controlled randomized trial of visual biofeedback versus muscle training without a visual display for intractable constipation. *Gut* 1995;37:95–99.
14. Rao, S.S.C., Welcher, K.D, Pelsang, R.E. Effects of biofeedback therapy on anorectal function in obstructive defecation. *Dig Dis Sci* 1997;42:2197–2205.
15. Fleshman, J.W., Dreznik, Z., Cohen, E., Fry, R.D., Kodner, I.J. Balloon expulsion test facilitates diagnosis of pelvic floor outlet obstruction due to nonrelaxing puborectalis muscle. *Dis Colon Rectum* 1992;35:1019–1025.
16. Turnbull, G.K., Ritvo, P.G. Anal sphincter biofeedback relaxation treatment for women with intractable constipation symptoms. *Dis Colon Rectum* 1992;35:530–536.
17. Papachrysostomou, M., Smith, A.N. Effects of biofeedback on obstructive defecation - Reconditioning of the defecation reflex? *Gut* 1994;35:242–245.
18. Bleijenberg, G., Kuijpers, H.C. Biofeedback treatment of constipation: A comparison of two methods. *Am J Gastroenterol* 1994;89:1021–1026.
19. Weber, J., Ducrotte, P.H., Touchais, J.Y., Roussignol, C., Denis, P.H. Biofeedback training for constipation in adults and children. *Dis Colon Rectum* 1987;30:844–846.
20. Miner, P.B., Donelly, T.C., Read, N.W. Investigation of the mode of action of biofeedback in treatment of fecal incontinence. *Dig Dis Sci* 1990;35:1291–1298.
21. Latimer, P.R., Campbell, D., Kasperski, J. A component analysis of biofeedback in the treatment of fecal incontinence. *Biofeedback Self-Regul* 1984;9:311–324.

22. Loening-Baucke, V. Efficacy of biofeedback training in improving fecal incontinence and anorectal physiologic function. *Gut* 1990;31:1395–1402.
23. Wald, A., Tunuguntla, A.K. Anorectal sensorimotor dysfunction in fecal incontinence and diabetes mellitus. *N. Engl. J Med* 1984;310:1282–1287.
24. Wald, A. Biofeedback for neurogenic fecal incontinence: Rectal sensation is a determinant of outcome. *J. Pediatr. Gastroenterol. Nutr.* 1983;2:302–306.
25. Heymen, S., Vickers, D., Weiss, E.G., Nogueras, J.J., Wexner, S.D. A prospective randomized trial comparing biofeedback techniques for patients with constipation. *Gastroenterology* 1996;110:A678.
26. Rao, S.S.C., Welcher, K. and Happel, J. Can biofeedback therapy (BT) improve anorectal function in fecal incontinence? *Am J Gastroenterol* 1996;91:2360–2366.
27. Whitehead, W.E., Burgio, K.L., Engel, B.T. Biofeedback treatment of fecal incontinence in geriatric patients. *J Am Geriatr Soc* 1985;33:320–324.
28. MacLeod, J.H. Management of anal incontinence by biofeedback. *Gastroenterol* 1987;93:291–294.
29. Wald, A. Biofeedback therapy for fecal incontinence. *Ann Intern Med* 1981;95:146–149.
30. Cerulli, M.A., Nikoomanesh, P., Schuster, M.M. Progress in biofeedback conditioning for fecal incontinence. *Gastroenterol* 1979;76:742–746.
31. Engel, B.T., Nikoomanesh, P., Schuster, M.M. Operant conditioning of rectosphincteric responses in the treatment of fecal incontinence. *New Engl J Med* 1974; 290:646–649.
32. Keck, J.O., Staniunas, R.J., Coller, J.A., Barrett, R.C., Oster, M.E., Schoetz, D.J., JR, Roberts, P.L., Murray, J.J., Veidenheimer, M.C. Biofeedback training is useful in fecal incontinence but disappointing in constipation. *Dis Colon Rectum* 1994;37:1271–1276.
33. Siproudhis, L., Dautreme, S., Ropert, A., Briand, H., Renet, C., Beusnel, C., Juguet, F., Rabot, A.F., Bretagne, J.F., Gosselin, M. Anismus and Biofeedback: Who benefits? *Eur J Gastro Hepatol* 1995; 7:547–552.
34. Rao, S.S.C. Manometric evaluation of constipation – Part I. *The Gastroenterologist* 1996;4:145–154.
35. Haskell, B., Rovner, H. Electromyography in the management of the incompetent anal sphincter. *Dis Colon Rectum* 1976;10:81–84.
36. Manning, A.P., Wyman, J.B., Heaton, K.W. How trustworthy are bowel histories? Comparison of recalled and recorded information. *BMJ* 1976; 2:213–4.

23

PEDIATRIC COLORECTAL DISORDERS

Carlo Di Lorenzo*

University of Pittsburgh
Children's Hospital of Pittsburgh
Pittsburgh, Pennsylvania

INTRODUCTION

Approximately 5% of all visits to pediatricians and up to 30% of referrals to pediatric gastroenterologists involve children with a disorder of defecation. Abnormal stooling is invariably a source of great anxiety for parents and may be quite distressing for the affected child. Chronic fecal incontinence is particularly devastating for the child's self esteem and for the parent-child interaction. One must remember that constipation and fecal incontinence are symptoms that can be caused by many different diseases. A plethora of organic disorders has been associated with constipation and/or fecal incontinence, but in the majority of cases children presenting with these symptoms suffer from functional constipation which may lead to fecal soiling. Childhood functional constipation differs significantly from constipation in adults. The diagnostic evaluation, differential diagnosis and treatment should be tailored to the pediatric population, because tests or therapies commonly used in adults may be unhelpful or even counterproductive in a constipated child (1).

After discussing a clinical problem, I will review the differential diagnosis and the clinical approach to the child with a colorectal disorder.

CASE STUDY

A 4-year-old boy presents with a history of intractable constipation. The past history reveals that he failed to pass meconium at birth. He had a suction rectal biopsy and a diagnosis of Hirschsprung's disease was made. He underwent a diverting colostomy at 6 days of life. He did well until 9 months of age when the colostomy was closed and a pull-through Duhamel procedure was performed. Since that time he has been constipated. There is no history of soiling or withholding behavior.

* Tel: (412) 692-5180, Fax: (412) 692-7355, e-mail: dilorec@chplink.chp.edu

Gastrointestinal Motility, edited by Rao *et al.*
Kluwer Academic / Plenum Publishers, New York, 1999.

1. What is the next step in the evaluation of this patient?

   A. Digital rectal and perianal neurological examination
   B. Barium enema
   C. Rectal biopsy
   D. Anal ultrasonography

Physical examination reveals a distended abdomen without palpable stool. Bowel sounds are normal. Neurological examination is normal and the examination of the back shows no evidence of spinal dysraphism. Laboratory tests rule out hypothyroidism and celiac disease. A sweat test is normal. Digital rectal exam shows no evidence of stricture. The anal tone is normal. There is a small amount of soft stool within the reach of the examining finger. There is no explosive diarrhea following the rectal examination. Initially, he is treated with anal dilatations and subsequently with stool softeners. Because persistent symptoms continue, a barium enema and a rectal suction biopsy are performed. Both of these tests are normal.

2. What is your next diagnostic step?

   A. Defecography
   B. Anorectal manometry
   C. Colon transit study
   D. Colonic manometry
   E. Laparotomy with full thickness biopsies

The anorectal manometry shows a non-relaxing internal anal sphincter with a basal sphincter pressure of 45 mmHg. The child receives a single sitzmarks capsule containing 24 radioopaque markers. After 3 days the markers are scattered throughout the colon. On the 4th day, the child receives an enema and the markers are expelled. A colonic manometry is performed.

3. Figure 1 shows an excerpt from the patient's colonic manometry in comparison with a colonic manometry from a normal child. How do you interpret the manometric findings?

   A. Normal study suggestive of a behavioral problem
   B. Colonic inertia
   C. Abnormal colonic motility in the distal colon
   D. Not enough is known about colonic physiology to interpret colonic manometry studies

We felt that the study was indicative of abnormal motor function of the distal colon and the child underwent a revision of the Duhamel procedure with resection of the distal 35 cm of colon. The surgery was curative. The pathologic specimen showed findings suggestive of neuronal intestinal dysplasia (NID). This condition is characterized by hyperplasia of the enteric nerve plexuses and heterotopia of neural elements (2, 3). It has been suggested that the incidence of NID is similar to Hirschsprung's disease and that it can be associated with Hirschsprung's disease, proximal to the aganglionic segment (4).

THE CONSTIPATED CHILD

Between 5% and 10% of all children at some time seek medical attention because of constipation. Although the list of organic diseases potentially responsible for childhood con-

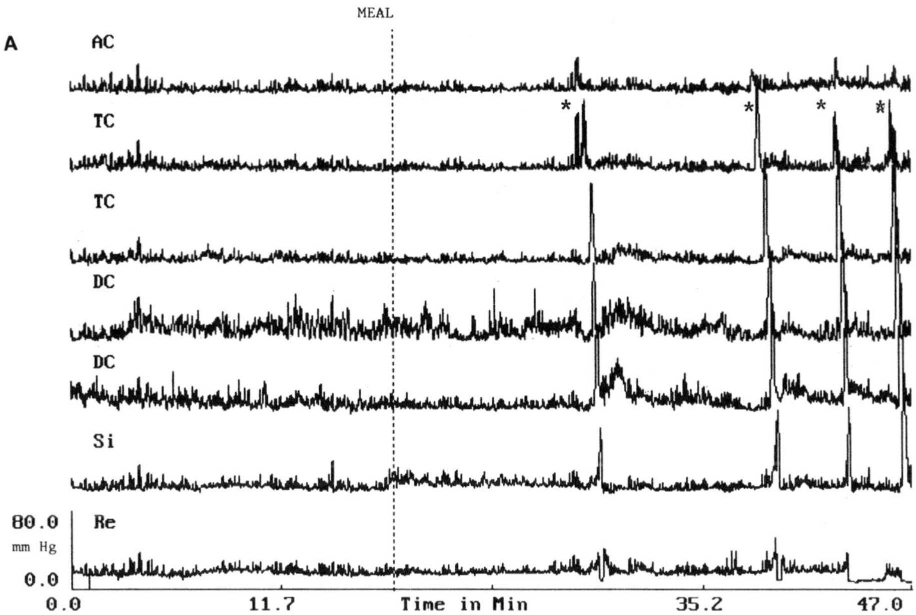

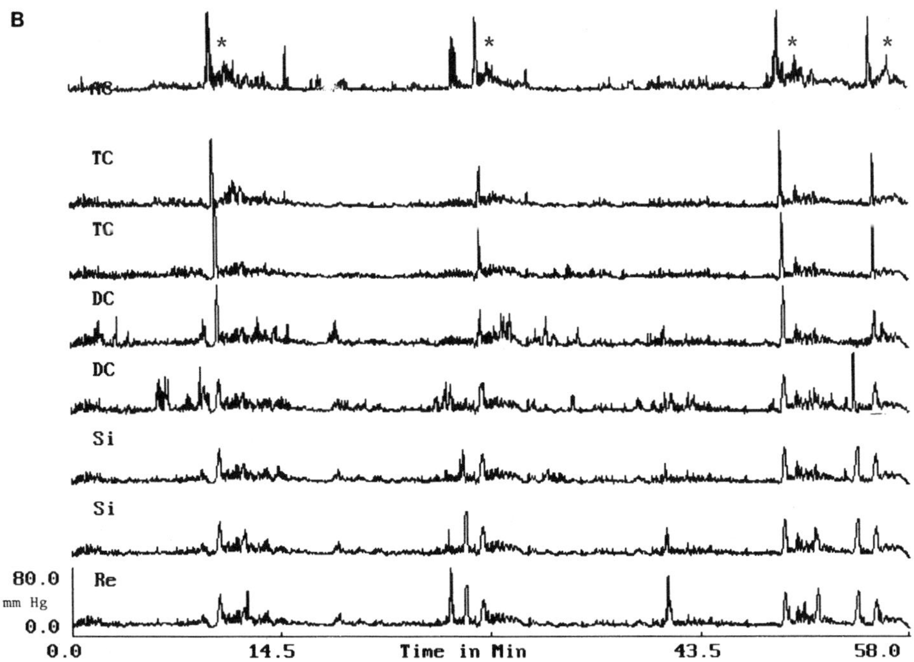

Figure 1. Examples of colonic manometry from a child with functional constipation (A) and from the child described in the case study (B). It is possible to notice that the HAPCs (*) migrate to the sigmoid region in panel A and only propagate to the beginning of the descending colon in panel B. AC = Ascending colon; TC = Transverse colon; DC = Descending Colon; Si = Sigmoid colon; Re = Rectum

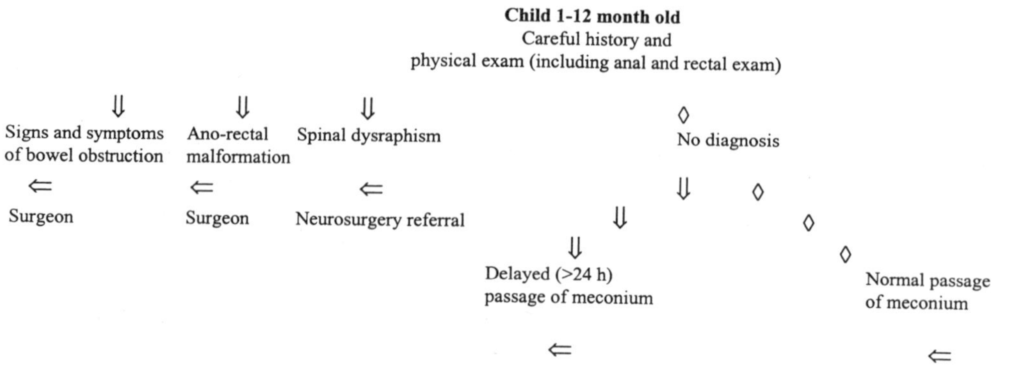

Figure 2. Suggested algorithm for the evaluation of an infant 1–12 month old presenting with constipation.

stipation is quite extensive, most of them are rare and in most cases are associated with other signs and symptoms besides constipation (5). Hirschsprung's disease occurs only in approximately 1:5000 live births, with approximately 700 new cases diagnosed every year in the United States. Thus, fewer than 1% of toddlers and school-age children who present for the first time with constipation will have Hirschsprung's disease. Anorectal malformations are even less common and most cases are diagnosed at birth. Other organic causes of childhood constipation include neurologic diseases (spinal cord disorders, static encephalopathy, hypotonia), endocrine and metabolic disorders (hypothyroidism, hypercalcemia, cystic fibrosis), celiac disease, and the use of constipating drugs. Beyond the neonatal period, the most common cause of constipation is functional constipation, which has also been called functional fecal retention, idiopathic constipation, or psychogenic megacolon (6). Two algorithms for the evaluation of the infant and the older child presenting with constipation are proposed in Figure 2 and Figure 3. The newborn who fails to pass stools often has a surgical condition responsible for his symptom and will not be discussed here.

FUNCTIONAL CONSTIPATION

Functional constipation in children starts almost always as a behavioral problem (7). Infants and toddlers become constipated when they develop a fear of defecation after experiencing unpleasant stool evacuations. Children have a very concrete way of thinking

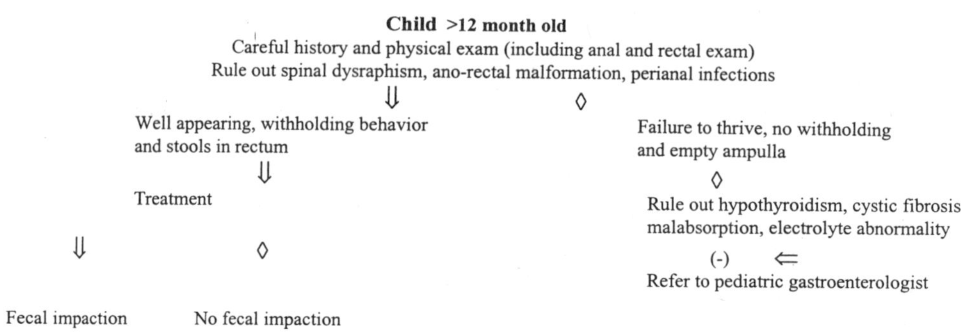

Figure 3. Suggested algorithm for the evaluation of a child >1 year old presenting with constipation.

and avoiding painful activities is part of their learning process. The fear of pain may be due to past experiences with the passage of hard stools or even just stools with different consistency. The stool characteristics change when children introduce new types of foods into their diet and this modification may be frightening for the infant or the toddler. Anal fissures or perianal infections, such as those due to a group A beta hemolytic streptococcus can also cause painful defecation (8, 9). Sexual abuse with anal penetration should also be suspected in a child with perianal lesions who presents with a sudden onset of severe constipation. Children with functional constipation respond to the urge to defecate with repeated attempts to withhold. The longer they are able to postpone the defecation, the harder and bigger the stool becomes. Once they finally have a bowel movement the anus painfully stretches, reinforcing the conviction that defecation is an act to be avoided at all cost. When a large fecal mass occupies the rectum, fecal soiling is quite common, triggering angry responses from the parents who are not aware that the child has lost control of his ability to maintain continence. The psychological consequences of chronic soiling in a school age child may be enormous.

Successful treatment of functional fecal retention depends on the ability of the clinician to establish a therapeutic relationship with the parents *and* the child (10). It is important to reassure the child, who often is brought to see the physician against his or her will and is too embarrassed and depressed even to make eye contact, that this is a common condition with a very favorable outcome. Common myths surrounding constipation need to be dispelled. The child does not have a "bowel blockage", childhood constipation does not cause cancer, failure to thrive, or "leakage of toxins into the body"! The physician then proceeds to prescribe stool softeners to facilitate the passage of stools. Behavioral modifications emphasizing positive reinforcement (e.g. stickers on calendars, diaries, rewards after a number of successful evacuations) are extremely helpful especially with younger children (11). It is essential to empower the child to take an active role in the treatment plan. The child will have to respond to the urge by attempting to defecate instead of withholding. No medical or behavioral treatment will be beneficial until the child will agree that "it feels better to let the poop go" instead than to withhold it. Enemas and suppositories are despised by children and should be avoided whenever possible.

Treatment should be continued for several months. After the child experiences months of pain-free and accident free defecation, a slow weaning from the medications should be attempted. Relapses are common. Changes in the daily routine, such as vacations, new school schedules, and weekend visits to a divorced parent disrupt routines and may result in non-compliance. Relapses tend to be handled more easily and quickly than the initial presentation.

For children older than 6 years of age who can cooperate, biofeedback can be used with success. Children with functional constipation develop a learned behavior responding to rectal distension with a paradoxical contraction of the external anal sphincter response to rectal distension. This is the manometrical equivalent of "withholding" in response to the urge. It is possible to teach children to inhibit this contraction and normalize their manometric pattern with resolution of symptoms (12). Biofeedback accelerates the successful treatment of children with functional constipation without changing the long term outcome (13).

HIRSCHSPRUNG'S DISEASE AND OTHER NEUROMUSCULAR DISEASES

Hirschsprung's disease is the most common colonic neuromuscular disease in children. It is characterized by the presence of an aganglionic segment beginning at the inter-

Table 1. Comparison of different causes of childhood constipation

Signs and symptoms	Functional constipation	Hirschsprung's disease	Other colonic neuromuscular diseases
Onset of symptoms	Toilet training, school	Birth	Birth or infancy
Stool withholding	Common	Never	Rare
Obstructive symptoms	Rare	Common	Possible
Fecal soiling	Common	Only in ultra-short segments	Rare
Upper GI symptoms	Never	Unusual	Common
Enterocolitis	Never	Common	Rare
Urinary symptoms	Frequent infections in children with fecal soiling	None	Poor bladder emptying in hollow visceral diseases
Anorectal manometry	Normal sphincter relaxation	No sphincter relaxation	Abnormal sphincter relaxation
Colon manometry	Normal	Varies	Abnormal
Treatment	Medical	Surgical	Medical or surgical
Prognosis	In most cases resolves before puberty	Common long term sequelae	Favorable outcome if disease is limited to the colon

nal anal sphincter and extending proximally for a variable distance. In 75% of cases the disease is limited to the rectosigmoid area and it rarely extends beyond the colon. The manometric hallmark of Hirschsprung's diseases is the absence of the rectoanal inhibitory reflex. The internal anal sphincter does not relax upon rectal distension and the aganglionic segment is in a state of constant contraction. The ganglionated bowel proximal to the aganglionic zone becomes dilated due to the distal functional obstruction.

A variety of other histopathologic abnormalities of the colonic nervous system ranging from hypoganglionosis to hyperganglionosis have been described in association with constipation (14). The pathologic diagnosis of these conditions is often difficult due to the lack of normal control tissues matched for age and colonic location (15). Pseudo-Hirschsprung's, neuronal intestinal dysplasia (NID), colonic pseudo-obstruction and visceral neuropathy are some of the labels used to describe these colonic neuromuscular diseases. They represent a heterogeneous and still poorly understood group of disorders which are only now beginning to be recognized. In NID there is proliferation of the myenteric and submucosal plexuses, and neurons are found in the lamina propria and within the smooth muscle layers. Incomplete relaxation of the internal anal sphincter has been described in NID (16). Thus, it is possible that NID is responsible for some cases previously believed to represent "ultrashort" segment Hirschsprung's disease and in which no aganglionosis could be demonstrated.

There are enough differences between functional constipation and colonic neuromuscular diseases (Table 1) to make rectal biopsy, radiological tests, and manometry studies unnecessary in most children presenting with constipation. A more detailed discussion of Hirschsprung's disease and other neuromuscular colonic diseases is beyond the scope of this chapter.

ANORECTAL MANOMETRY

The main indication to perform anorectal manometry in a child with chronic constipation is to rule out Hirschsprung's disease. Anorectal manometry represents a sensitive and specific diagnostic test for Hirschsprung's disease but only when performed in cooperative children. In crying infants or when restraint is needed to perform the test, artifacts are common and caution should be used in interpreting the results. Manometry is particularly useful when the aganglionic segment is short and radiologic or pathologic studies are equivocal. In Hirschsprung's disease the absence of neurons in the myenteric plexus causes a failure of the rectoanal inhibitory reflex and there is lack of relaxation or paradoxical contraction of the internal anal sphincter in response to rectal distention. In the presence of normal internal anal sphincter relaxation in response to rectal distention, Hirschsprung's disease is ruled out. It is important to note that when the rectum is dilated, the rectal ballon must be inflated with larger volumes to stretch the rectal wall and elicit a normal sphincter relaxation. In the child with retentive behavior, the voluntary contraction of the external anal sphincter and the gluteal muscles often confound the interpretation of anorectal manometry. Sedation does not interfere with the rectoanal inhibitory reflex and may be used in newborns and uncooperative children.

Anorectal motility has been investigated extensively in childhood functional constipation, often with conflicting results. There seems to be a higher threshold for rectal sensation and for internal sphincter relaxation in response to balloon distention (17). Different investigators have reported increased, decreased or normal basal anal pressure (18–20). The external anal sphincter is described as abnormally contracted during straining in a subgroup of constipated children. The manometric abnormalities may represent a primary disorder or be secondary to the chronic fecal retention. Anorectal manometry should not be performed in a child with typical features of functional constipation, unless biofeedback treatment is being planned.

COLONIC MANOMETRY

Colonic manometry is currently considered a useful clinical tool in the evaluation of selected children with intractable constipation. Children have a more predictable colonic response to the meal and more frequent high amplitude propagated contractions (HAPCs) than do adults making colonic manometry studies easier to interpret in pediatric patients. Colonic motility in children has been measured with both water perfused catheters and catheters with implanted transducers (21). A colonoscopy is used to position the manometry catheter with the tip in the proximal colon. It is essential to perform colonic manometry studies only in unsedated and awake children because both sedation and sleep substantially alter colonic motor activity. We routinely study children on the day after the placement of the catheter. During the past few years, colonic manometry has been used to clarify the pathophysiology of constipation and incontinence in children with colonic neuromuscular disorders. Characteristics of normal colon motility in children of different ages have been defined (22). In cases of children with atypical presentations or a lack of response to treatment, colonic manometry differentiates functional from organic causes of constipation (23). It has also been helpful in clarifying the pathophysiology of constipation in children with intestinal pseudo-obstruction (24). In functional constipation, colonic motility is normal. The characteristics of colonic motility in children with functional constipation include an increase in motility after a meal (the "gastrocolonic reflex"), and the presence of HAPCs after waking, after ingestion of a large meal or after infusion of bisacodyl, and absence of discrete motor abnormalities. During colonic manometry, HAPCs are easily identified, based upon their characteristic high amplitude, their prolonged duration and the propagation over most of the colon (25). Usually, HAPCs stop at the rectosigmoid junction, carrying the stools to the rectum where they are "stored" until a choice is made to evacuate them. They are considered the manometric equivalent of the "mass movements" described by radiologists. In children, there is an age related decrease in the frequency of HAPCs (22). The increased number of HAPCs probably correlates with the increased number of bowel movements commonly observed in infants and children.

The absence of HAPCs has been reported in adults with colonic inertia (26). In children, absence of the gastrocolonic response and the lack of HAPCs after ingestion of a meal or stimulation with bisacodyl have been associated with a colon neuropathy (24, 27). A myopathy is more likely in children with consistently low amplitude contractions or complete absence of contractions. We believe that colonic manometry, a technique well tolerated by children and less invasive than surgical full thickness biopsies, should be considered part of the evaluation of children with intractable constipation prior to considering them for surgery. Colonic manometry also identifies the segment of colon with abnormal motor function, helping to plan a possible surgical intervention. Colonic manometry is reported to be beneficial in clarifying the pathophysiology of persistent symptoms in children who continue to be constipated or develop fecal incontinence after surgery for Hirschsprung's disease (28).

COLONIC TRANSIT STUDIES

Total and segmental colonic transit times using radioopaque markers (29) or scintigraphic techniques (30) have been studied in children. These techniques have not reached the same degree of popularity achieved in the evaluation of adults with constipation. There are no well defined normal values for segmental colonic transit in children of different ages. Diet is difficult to standardize in pediatrics. Unless an aggressive clean out is performed before the ingestion of the radioopaque markers, the stool filled rectosigmoid always causes accumulation of the markers in the more proximal colon. Healthy children excrete swallowed markers within 48 hours. In functional constipation, the markers accumulate in the rectum without slowing of colonic transit time. In generalized colonic neuromuscular diseases the markers may be scattered throughout the colon for many days. When that happens, colonic manometry can differentiate between neuropathy and myopathy and assess with more precision the length of colonic segment exhibiting abnormal motor function.

CONCLUSION

Children frequently suffer from functional constipation. A clear understanding of the pathophysiology of childhood constipation is critical to develop a successful diagnostic and therapeutic plan. A variety of congenital and acquired colonic neuromuscular disorders have recently been described and should be suspected in children with atypical presentation or failure to improve with conventional management. In these children anorectal and colonic manometry may be extremely helpful in clarifying pathophysiology and extent of the disease.

ANSWER KEY AND EXPLANATIONS FOR CASE STUDY

1. The work up of a child presenting with constipation after surgery for Hirschsprung's disease is outlined in Figure 4. It is important to perform a complete

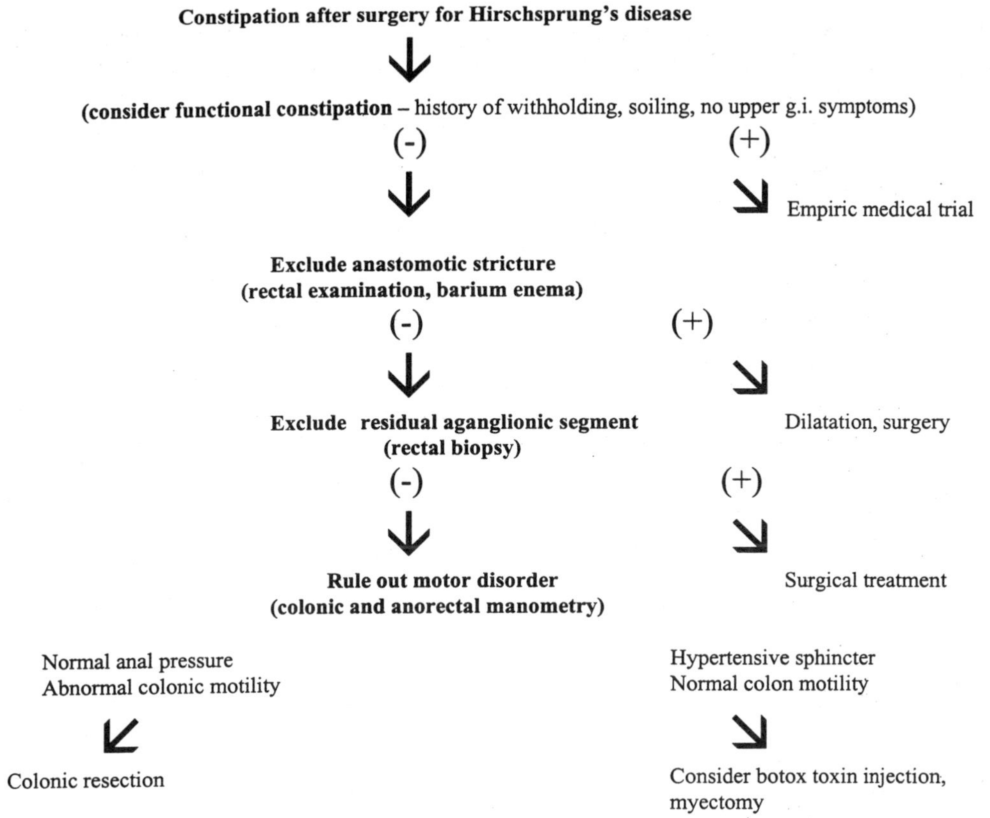

Figure 4. Suggested algorithm for the evaluation of a child presenting with constipation after surgery for Hirschsprung's disease.

physical examination in any child presenting with constipation, including a careful inspection of the perianal area and a rectal examination. Children with chronic defecation problems are often terrified by anally invasive procedures and the perianal area should be inspected only after giving reassurance that there will be no pain. Important findings on physical examination include the presence of fecal soiling, perianal dermatitis and pseudoverrucous papules and nodules (often a consequence of chronic soiling), a rectoperineal fistula, an anteriorly placed anus, and vascular, pigmented or hairy patches in the lumbosacral region, suggestive of occult spinal dysraphism (31). When the child is old enough to cooperate, light touch sensibility in the sacral dermatomes should be tested. The integrity of the sensorimotor apparatus of fecal continence can be assessed by examining the reflex contraction of the external anal sphincter and pelvic floor in response to stroking of the perianal skin. Abnormal plantar or Achilles tendon reflexes may also be suggestive of spinal dysraphism.

In children who have had anorectal surgery it is important to assess the size and tone of the anus. This can be accomplished by a thorough rectal examination, although in many cases a barium enema is needed to rule out more conclusively anastomotic strictures. In the case of a child who carries a diagnosis of Hirschsprung's disease and who is not doing well after surgery, suction rectal biopsies should be performed to rule out a residual aganglionic segment. There is no experience with the use of anal ultrasonography in toddlers and this test would probably not be helpful in this scenario. The best option is **A** although B and C are also correct and these tests should be performed in that order.

2. Defecography is impossible to perform in uncooperative children and a laparotomy should not be done prior to a more extensive work up. Anorectal manometry in a child with Hirschsprung's disease will continue to demonstrate a non relaxing internal anal sphincter even after a pull-through procedure, but it may be useful in order to evaluate the internal anal sphincter basal pressure (32). If the sphincter is hypertensive ("internal sphincter achalasia"), an intrasphincteric botox toxin injection (33) or a myectomy (34) may solve the symptoms. Colonic manometry, as previously discussed, is a reliable mean to study colonic motor function. It may demonstrate the physiologic abnormality and assess extent of the disease. Colonic transit studies with radioopaque markers are non invasive and easy to perform. They are often performed as screening tests in children with defecation disorders prior to more invasive testing. Overall, they are not felt to be as helpful in children as they are in adults. Normal colonic transit time values are not well established in children of different age. It is the author's opinion that no surgical intervention should be based entirely on a radioopaque markers transit study. The best option is **B**, although C and D are reasonable options.
3. The colonic manometry study shows high amplitude propagated contractions (HAPCs) migrating from the ascending colon to the descending colon. They become simultaneous and non propagating in the distal colon. This pattern has been associated with colonic neuropathy (24). Resection of the affected segment is usually curative. The best option is **C.**

REFERENCES

1. Di Lorenzo C, Hyman PE. Gastrointestinal motility in neonatal and pediatric practice. In: Gastroenterology Clinics of North America: Gastrointestinal motility in clinical practice. New York W.B. Saunders, Philadelphia, PA, 1996, 203–25.
2. Scharli AF. Neuronal intestinal dysplasia. Pediatr Surg Int 1992; 7: 2–7.
3. Schofield DE, Yunis EJ: Intestinal neuronal dysplasia. J Pediatr Gastroenterol Nutr 1991; 12: 182–189.
4. Fadda B, Pistor G, Meier-Ruge W, et al. Symptoms, diagnosis, and therapy of neuronal intestinal dysplasia masked by Hirschsprung's disease. Pediatr Surg Int 1987; 2: 76–80.
5. Di Lorenzo C. Constipation. In: Hyman PE, Di Lorenzo C eds. Pediatric Gastrointestinal Motility Disorders. New York, Academy Professional Information Services, 1994: 129–144.
6. Hyman PE, Fleisher D: Functional fecal retention. Practical Gastroenterology 1992; 16: 29–37.
7. Clayden G: Constipation as a behavioural problem in children. In: Kamm M, Lennard-Jones JE (eds.): Constipation, pp. 117–122. Wringhtson Biomedical Publishing Ltd, Petersfield, UK, 1994
8. Kokx NP, Comstock JA, Facklam RR. Streptococcal perianal disease in children. Pediatrics 1987; 80: 659–63.
9. Spear RM, Rothbaum R, Keating J, Biaufuss MC, Rosemblum JL. Perianal streptococcal cellulitis. J Pediatr. 1985; 107: 557–59.
10. Fleisher D. Diagnosis and treatment of disorders of defecation in children. Pediatric Annals. 1976; 5: 71–101.
11. Stark LJ, Owens-Stively J, Spirito A, Lewis A. Guevremont D. Group behavioral treatment of retentive encopresis. J Ped Psy 1990; 15: 659–71.
12. Loening-Baucke VA, Younoszai MK. Effect of treatment on rectal and sigmoid motility in chronically constipated children. Pediatrics 1984; 73: 199–205.
13. Loening-Baucke VA. Biofeedback training in children with functional constipation. A critical review. Dig Dis Sci 1996; 41: 65–71.

14. Milla PJ, Smith VV: Aganglionosis, hypoganglionosis and hyperganglionosis: clinical presentation and histopathology. In: Kamm M, Lennard-Jones JE (eds.): Constipation, pp. 183–192. Wrightson Biomedical Publishing Ltd, Petersfield, UK, 1994.
15. Smith VV: Intestinal neuronal density in childhood: a baseline for the objective assessment of hypo- and hyperganglionosis. Pediatr. Pathol. 1993; 13: 225–37.
16. Koletzko S, Ballauff A, Hadziselimovic F et al: Is histological diagnosis of neuronal dysplasia related to clinical and manometric findings in constipated children? Results of a pilot study. J Pediatr Gastroenterol Nutr 1993; 17: 59–65.
17. Loening-Baucke VA: Sensitivity of the sigmoid colon and rectum in children treated for chronic constipation. J Pediatr Gastroenterol Nutr 1984; 3: 454–9.
18. Loening-Baucke VA, Younoszai MK: Abnormal anal sphincter response in chronically constipated children. J Pediatr 1982; 100: 213–8.
19. Wald A, Chandra R, Chiponis D et al.: Anorectal function and continence mechanisms in childhood encopresis. J Pediatr Gastroenterol Nutr 1986; 5: 346–51
20. Meunier P, Marechal JM, deBeaugeu MJ: Rectoanal pressures and rectal sensitivity studies in chronic childhood constipation. Gastroenterology 1979; 77: 330–6.
21. Di Lorenzo C. Colonic manometry. In: Hyman PE, Di Lorenzo C eds. Pediatric Gastrointestinal Motility Disorders. New York, Academy Professional Information Services, 1994: 215–30.
22. Di Lorenzo C, Flores AF, Hyman PE: Age-related changes in colon motility. J Pediatr 1995; 127: 593–6.
23. Di Lorenzo C, Flores AF, Reddy SN et al.: Colonic manometry differentiates causes of intractable constipation in children. J Pediatr 1992; 112: 690–5.
24. Di Lorenzo C, Flores AF, Reddy SN, et al.: Colonic manometry in children with chronic intestinal pseudo-obstruction. Gut 1993; 34: 803–7.
25. Bassotti G, Gaburri M: Manometric investigation of high amplitude propagated contractile activity of the human colon. Am J Physiol 1988; 255: G660–4.
26. Bassotti G, Imbimbo BP, Betti C et al: Impaired colonic response to eating in patients with slow transit constipation. Am J Gastroenterol 1992; 87: 504–8.
27. Rivera A, Di Lorenzo C, Flores AF, Reddy S, Idries S, McCloud E, Hyman PE. Colonic manometry differentiates three causes of persistent symptoms following surgery for Hirschsprung's disease. Gastroenterology 1996; 110: A746.
28. Hamid SA, Di Lorenzo C, Reddy SN, Flores AF, Hyman PE. Bisacodyl and high amplitude propagating contractions (HAPCs) in children. J Pediatr Gastroenterol Nutr 1998; 27: 398–402,.
29. Scaillon M, Cadranel S. Transit tests. In: Hyman PE, Di Lorenzo C eds. Pediatric Gastrointestinal Motility Disorders. New York, Academy Professional Information Services, 1994: 265–276.
30. Vattimo A, Burroni L, Bertelli P, et al. Total and segmental colon transit time in constipated children assessed by scintigraphy with 111In-DTPA given orally. J Nucl Biol Med 1994; 37: 218–22.
31. Anderson, F. Occult spinal dysraphism: a series of 73 cases. Pediatrics, 1975; 55: 826–35.
32. Mishalany HG, MM Wooley: Postoperative functional and manometric evaluation of patients with Hirschsprung's disease. J Pediatr Surg 1987; 22: 443–6.
33. Langer JC, Birnbaum E. Preliminary experience with intrasphincteric botulinum toxin for persistent constipation after pull-through for Hirschsprung's disease. J Pediatr Surg 1997; 32: 1059–62.
34. Abbas Banani S, Forootan H. Role of anorectal myectomy after failed endorectal pull-through in Hirschsprung's disease. J Pediatr Surg 1994; 29: 1307–9.

24

BLOATING AND GAS

William J. Snape, Jr.

University of California, Irvine
Director, Bowel Disease and Motility Center
Long Beach Memorial Medical Center

INTRODUCTION

The symptom of abdominal bloating is extremely common (1; 2). Clinically, patients with abdominal bloating complain of a generalized abdominal discomfort not localized in a specific quadrant of the abdomen. They will describe their abdomen as markedly distended, often using the term "similar to my pregnancies". Patients avoid tight-fitting clothes because of discomfort from the clothes on their skin. Often patients have associated increases in both belching and flatulence, which may or may not relieve the discomfort.

Abdominal bloating is a non-specific symptom and can be associated with multiple gut disorders (2–4). It is the physician's job first to determine the cause of the abdominal bloating and then to initiate a treatment regimen for the patient's symptoms. Even though therapy often does not completely relieve the symptom, the patient may resume full activity after understanding the pathophysiology of the underlying condition and its overall significance.

CASE STUDY

1. A 40-year-old woman presents in your office with post-prandial bloating for the past 6 months. The discomfort is relieved with the passage of flatus. There is no significant past history and system review was unremarkable. Physical exam was normal.

   What would be the next step in the management of this patient?

   A. Upper gastrointestinal endoscopy
   B. Restrict milk and/or wheat products
   C. CT-scan of the abdomen

Gastrointestinal Motility, edited by Rao *et al.*
Kluwer Academic / Plenum Publishers, New York, 1999.

D. Serum test for Helicobacter pylori

2. The intake of milk and wheat products was restricted but she continued to have both bloating and increased flatus and returns to your office 3 months later. The next test you would do in the work up of this patient is?

   A. Upper gastrointestinal barium x-ray with small bowel follow through.
   B. CT-scan of the abdomen
   C. Initiate anticholinergic therapy
   D. Initiate therapy with an antidepressant

3. The small bowel series showed delayed transit through the small intestine. The next step in the management of this patient is?

   A. Obtain a breath hydrogen study
   B. Initiate treatment with metronidazole
   C. Octreotide 50mg subcutaneously at night
   D. Cisapride

ABDOMINAL BLOATING

Pathophysiology

Abdominal bloating occurs in approximately 10% to 20% of the population (2). Except for the bloating, these patients are often otherwise healthy. Bloating often increases throughout the day and recedes at night or during sleep. Commonly bloating increases after eating a meal, especially following larger meals. Chewing gum or drinking carbonated beverages further exacerbates the bloating. Belching or flatus often relieves the symptoms.

Other bowel complaints should be sought in patients with abdominal bloating. Patients with gastroparesis, non-ulcer dyspepsia, or irritable bowel syndrome may have abdominal bloating as one of their major complaints (5,6–10). In these patients the abdominal bloating appears to be a component of the visceral hypersensitivity, since the amount of abdominal gas is similar in bloated and non-bloated patients (11). Lactose intolerance can cause abdominal bloating (12), secondary to the breakdown of lactose by enteric bacteria to hydrogen gas and short chain fatty acids. The increased osmolality due to the short chain fatty acids in the small intestinal lumen increases fluid accumulation further accentuating the bloating. Psyllium ingestion may cause bloating due to the bacterial breakdown of cellulose with the generation of by-products similar to the digestion of lactose. Changes in estrogen and progesterone concentrations associated with the perimenstrual period, also may accentuate abdominal bloating in affected women. The bloating is more often seen prior to the onset of menses, but can also be seen throughout the perimenstrual period. The physician should not have tunnel vision when approaching the patient with increased abdominal bloating. Even though most often the symptoms are associated with functional bowel disorders or increased gas production by food ingestion, partial small bowel obstruction may also cause abdominal bloating. When appropriate the physician should search for organic causes for bowel dysfunction.

In rare patients exaggerated lumbar lordosis, usually associated with a psychological disorder, can present as bloating (2). When the lordotic position is removed in these patients, the abdominal bloating recedes. Bloating also can occur in patients with extremely poor muscle tone of the abdominal muscles.

Investigations

Diagnostic tests are needed to identify the underlying cause of abdominal bloating. The extent of testing must be determined by the clinician's impression as to the most likely pathophysiologic cause for the patient's symptoms. In the patient who presents for the first time with abdominal bloating, an abdominal series, including decubitus x-rays looking for air-fluid levels, are a bare minimum. If the abdominal series appears normal, further tests examining for small bowel obstruction appear unwarranted. Recent studies have demonstrated that patients with bloating associated with functional bowel disorders do not have an increase in the amount of gas within the abdomen, but rather have an increased sensitivity to a normal amount of intraluminal gas (11, 13, 14).

If, the abdominal series is equivocal or positive for a potential partial small bowel obstruction, a small bowel follow-through or even enteroclysis may be necessary. In these patients often a CT scan of the abdomen will provide further information as to the underlying cause of the obstruction. Rarely antroduodenal motility studies may be necessary to identify a subtle small bowel obstruction (15). The causes of partial small bowel obstruction often revolve around previous abdominal surgery, but a lengthy discussion of bowel obstruction will not be included in this brief review.

Additional studies to further understand the cause of bloating may include a radionuclide gastric emptying. Slow gastric emptying associated with gastroparesis will give a generalized sense of bloating, although often the bloating appears to be more localized to the left upper quadrant and mid-epigastrium. In the patient in whom gastroparesis is suspected, a history should be examined for potential causes of slow gastric emptying including diabetes mellitus (16–19).

A breath hydrogen study may be helpful in the diagnosis of increased intestinal gas secondary to bacterial overgrowth within the bowel. If lactose is used as the administered sugar, an increased amount of breath hydrogen is produced as the lactose reaches the colon, in lactase deficient patients. Lactulose is used to test for increased intestinal bacteria leading to increased gas and volatile fatty acid production to exclude confusion with lactase deficiency. If the patient has an elevated breath hydrogen which is greater than 20 parts per million at baseline or if the breath hydrogen increases greater than 12 parts per million above baseline over the first 30 minutes, intestinal bacterial overgrowth should be suspected. In some patients rapid intestinal transit may increase breath hydrogen level more rapidly than in healthy subjects. The rapid transit may contribute to bloating. The normal transit time is approximately 75 minutes.

GASEOUSNESS

Physiology

Gastrointestinal gas associated with belching or flatus is generally composed of varying quantities of nitrogen, oxygen, hydrogen, carbon dioxide and methane. Similar to the surrounding air, nitrogen is the most common gas within the gut and most often comes from swallowed air. Once nitrogen is within the gut, it diffuses poorly across the mucosal barrier. Oxygen also comes from swallowed air. However, oxygen levels within the gut are altered by free diffusion across the small bowel mucosa. Twenty percent of the gut hydrogen can come from diffusion into the gut lumen. However, hydrogen within the gut comes predominantly from bacterial fermentation of carbohydrate. For this reason, hydrogen is measured when testing for degradation of complex carbohydrates in the small intestine. Carbon dioxide may come from a neutralization reaction between acid in the stomach and bicarbonate produced by the pancreas or ingestion of carbonated beverages. There is also free diffusion of carbon dioxide from the body into the gut. A minority of patients produce methane through bacterial fermentation. Colonic methane production is inherited (20, 21). Often the transmission occurs within families. Methane can be measured within expired air since it diffuses across the mucosal barrier.

Pathophysiology

When the patient presents either with bloating or increasing flatus and/or belching, the origin of the increased gastrointestinal gas should be determined. Belching is most often associated with air swallowing, which can be associated with normal eating and drinking, or with a nervous habit. If a patient has increased gas during periods of increased anxiety, they should be encouraged to watch for signs of increased air swallowing. Patients with gastroesophageal reflux disease, even if they recognize minimal heartburn, may have increased belching associated with increased swallowing, often subconsciously to clear the esophagus of acid.

Increased flatus is most often due to intestinal production of excess gas. This can be secondary to increased amounts of fermentation of non-absorbed carbohydrate. Wheat products, as well as milk, have been implicated in the increased production of flatus (13, 14). In some patients intestinal bacterial overgrowth may lead to bloating and flatus.

TREATMENT OF BLOATING AND GAS

Figure 1 demonstrates the algorithm I use for the clinical management of bloating and gaseousness. First, one must assess the overall importance of the symptoms and how they affect the patient's well being and performance of life activities. The presence of alarm symptoms (leucocytosis, increased abdominal pain, vomiting, or gastrointestinal blood loss) suggests a more serious organic cause of the symptoms. A thorough dietary history as well as the potential life activities that lead to gaseousness should be identified. At some time each patient should have an abdominal series, looking for dilated bowel and/or air-fluid levels in the bowel.

After excluding an obvious organic cause of the bloating, foods that may exacerbate symptoms should be eliminated. Patients should eliminate milk and other non-absorbable carbohydrates including fiber, wheat and most beans. If this elimination diet demonstrates that certain foods are a potential cause of the patient's symptoms, supplemental lactase (lactaid) or alpha-galactosidase (Beano) can be tried for long-term therapy. If the symptoms resolve after dietary manipulation, no further work-up is necessary. If the symptoms remain, simethicone or activated charcoal provides some benefit to patients. Generally the patients do not get complete relief from these agents and some patients may get no relief at all.

If the symptoms remain, further tests are appropriate. These may include investigation of the upper gut with either an upper endoscopy or upper GI and small bowel series. Although studies are as yet inconclusive, gastritis due to Helicobacter pylori may be associated with abdominal discomfort and even symptoms of bloating (22–25). If on examination of the upper GI tract a specific anatomic disease is identified, specific therapy should be initiated. If no organic disease is identified, the patient may have visceral hypersensi-

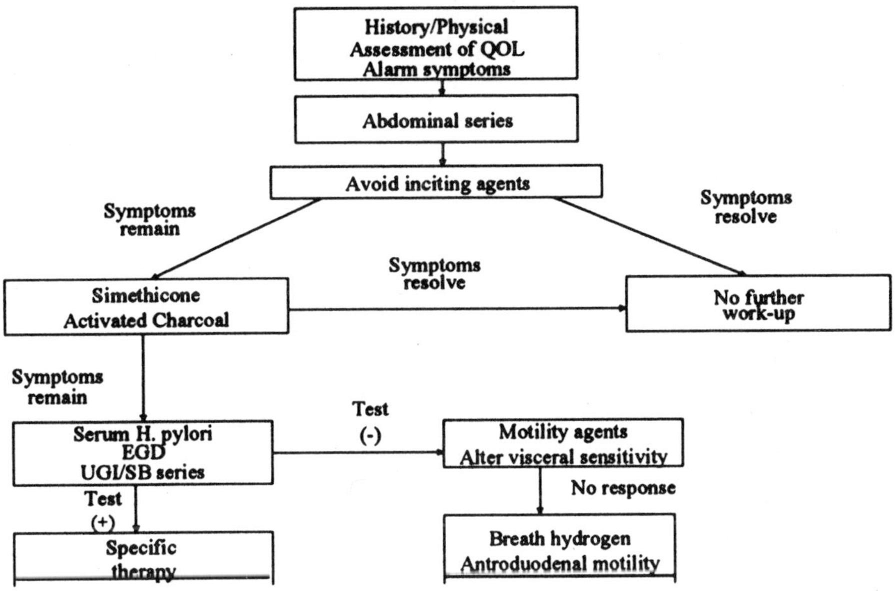

Figure 1. Clinical management of bloating and gaseousness.

tivity and this should be treated with drugs that modify abdominal pain, which will be discussed in another chapter.

If the patient does not respond to the treatment of visceral hypersensitivity, further studies of the small bowel should be undertaken. Breath hydrogen studies to evaluate the speed of intestinal transit or bacterial overgrowth may help to identify a motility disorder that could contribute to bloating (5). Antroduodenal motility may also be necessary to identify abnormalities that produce abdominal symptoms.

CONCLUSIONS

In summary, it is common for patients to complain of abdominal bloating and gaseousness. Most often these patients can be treated by dietary manipulation and/or enzyme supplements such as eliminating lactose from their diet. Occasionally a motility disorder or visceral hyperalgesia is identified which will require specific therapy.

ANSWER KEY AND EXPLANATIONS

1. This otherwise fit and healthy woman has no alarm symptoms and nothing in the history or exam to suggest a serious organic disorder. Hence, further investigations can be avoided and an empiric dietary trial would be quite appropriate. A detailed dietary history and more importantly asking the patient to maintain a daily food and symptom diary can be very useful and could provide insights regarding their problem. The best option is **B**.

2. Often dietary manipulation is successful, because the patient may have developed a transient or permanent intolerance for a specific food item(s) due to viral enteritis, infections such as giardia or loss of mucosal disaccharidases through aging or genetic factors. Because she has now developed refractory symptoms that have not responded to dietary changes, it would be reasonable to perform further investigations. Further empirical therapy should be avoided at this stage. Given the choices, a barium x-ray would be the most appropriate option although a lactose hydrogen breath test may be equally useful. The best option is **A.**
3. A breath hydrogen study at this time documented bacterial overgrowth. Because of coexisting delayed transit, patient may be prone to recurrent attacks of bacterial overgrowth. Hence, having made the diagnosis of bacterial overgrowth, treatment with prokinetic agent (cisapride or octreotide) or antibiotic should be considered. Also at this stage it would be appropriate to consider performing a small bowel manometry study to identify if the patient has a myopathy or neuropathy. The best option is **A.**

REFERENCES

1. Sullivan SN. A prospective study of unexplained visible abdominal bloating. N.Z.Med.J. 1994;107:428–30.
2. Sullivan SN. Functional abdominal bloating. J.Clin.Gastroenterol. 1994;19:23–7.
3. Drossman DA. Chronic functional abdominal pain. Am.J.Gastroenterol. 1996;91:2270–81.
4. Greenbaum DS, Greenbaum RB, Joseph JG, Natale JE. Chronic abdominal wall pain. Diagnostic validity and costs. Dig.Dis.Sci. 1994;39 :1935–41.
5. Di Lorenzo C, Hyman PE, Flores AF, et al. Antroduodenal manometry in children and adults with severe non- ulcer dyspepsia. Scand.J.Gastroenterol. 1994;29:799–806.
6. Maxton DG, Martin DF, Whorwell PJ, Godfrey M. Abdominal distension in female patients with irritable bowel syndrome: exploration of possible mechanisms. Gut. 1991;32:662–4.
7. Jones R. Dyspeptic symptoms in the community. Gut. 1989;30:893–8.
8. Kaneko H, Mitsuma T, Uchida K, Furusawa A, Morise K. Immunoreactive-somatostatin, substance P, and calcitonin gene- related peptide concentrations of the human gastric mucosa in patients with nonulcer dyspepsia and peptic ulcer disease. Am.J.Gastroenterol. 1993;88:898–904.
9. Talley NJ, Phillips SF. Non-ulcer dyspepsia: potential causes and pathophysiology. Ann.Intern.Med. 1988;108:865–79.
10. Talley NJ. Non-ulcer dyspepsia: myths and realities. Aliment.Pharmacol.Therap. 1991;5:145–62.
11. Lasser RB, Bond JH, Levitt MD. The role of intestinal gas in functional abdominal pain. N.Engl.J.Med. 1975;293:524–6.
12. Christopher NL, Bayless TM. Role of the small bowel and colon in lactose-induced diarrhea. Gastroenterology. 1971;60:845–52.
13. Levitt MD, Lasser RB, Schwartz JS, Bond JH. Studies of a flatulent patient. N.Engl.J.Med. 1976;295(5):260–2.
14. Sutalf LO, Levitt MD. Follow-up of a flatulent patient. Dig.Dis.Sci. 1979;24(8):652–4.
15. Camilleri M. Jejunal manometry in distal subacute mechanical obstruction: significance of prolonged simultaneous contractions. Gut. 1989;30:468–75.
16. Farrell FJ, Keeffe EB. Diabetic gastroparesis. Dig.Dis. 1995;13:291–300.
17. Horowitz M, Fraser RJ. Gastroparesis: diagnosis and management. Scand.J.Gastroenterol.Suppl. 1995;213:7–16.
18. Kendall BJ, McCallum RW. Gastroparesis and the current use of prokinetic drugs. The Gastroenterologist. 1993;1:107–14.
19. Parkman HP, Harris AD, Krevsky B, Urbain JL, Maurer AH, Fisher RS. Gastroduodenal motility and dysmotility: an update on techniques available for evaluation. Am.J.Gastroenterol. 1995;90:869–92.
20. Florin TH, Woods HJ. Inhibition of methanogenesis by human bile. Gut. 1995;37:418–21.
21. Wolever TM, ter Wal P, Spadafora P, Robb P. Guar, but not psyllium, increases breath methane and serum acetate concentrations in human subjects. Am.J.Clin.Nutr. 1992;55:719–22.

22. Holtmann G, Talley NJ, Goebell H. Association between H. pylori, duodenal mechanosensory thresholds, and small intestinal motility in chronic unexplained dyspepsia. Dig.Dis.Sci. 1996;41:1285–91.
23. Scolapio JS, Camilleri M. Nonulcer dyspepsia. The Gastroenterologist. 1996;4:13–23.
24. Talley NJ. Functional dyspepsia-Should treatment be targeted on disturbed physiology? Alim Pharmacol Therap 1995; ((2): 107–15.
25. Veldhuyzen van Zanten SJ, Cleary C, Talley NJ, et al. Drug treatment of functional dyspepsia: a systematic analysis of trial methodology with recommendations for design of future trials [see comments]. Am.J.Gastroenterol. 1996;91:660–73.

25

NEUROPHYSIOLOGY OF VISCERAL PAIN

G. F. Gebhart*

Department of Pharmacology
College of Medicine
University of Iowa
Iowa City, Iowa 52242-1109

INTRODUCTION

Visceral pain differs significantly from the more commonly experienced and better understood somatic pain that arises from cutaneous structures. Whereas somatic pain is generally easy to localize, visceral pain is diffuse in nature, poorly localized and typically referred to cutaneous and/or other somatic structures. Most importantly, visceral pain is not linked to tissue injury, as is cutaneous pain. Indeed, early researchers concluded that injury to visceral organs did not produce pain, advancing the erroneous postulate that the viscera were insensate. While it has long been appreciated that the principal sensations that arise from the viscera are discomfort and pain, we have only recently begun to understand the function of the sensory innervation of the viscera. Because few clinical or basic scientists have focused efforts on the study of visceral pain, knowledge specific to visceral pain (as opposed to cutaneous pain, for which models were readily available) has been slow to accumulate. However, recent developments have led to a better understanding of visceral pain (1).

VISCERAL HYPERALGESIA

Historically, clinical observations have led the way towards a change in the understanding of visceral pain. In 1973, Ritchie (2) documented that irritable bowel syndrome (IBS) patients reported pain at lower volumes of balloon distension of the colon than did normal subjects (Figure 1).

* Phone: 319/335-7946, Fax: 319/335-8930, E-mail: gf-gebhart@uiowa.edu

Gastrointestinal Motility, edited by Rao *et al.*
Kluwer Academic / Plenum Publishers, New York, 1999.

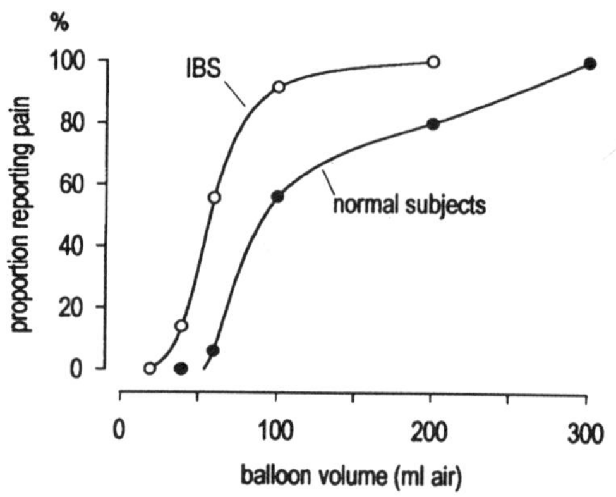

Figure 1. Proportion of normal subjects and of irritable bowel syndrome (IBS) patients reporting pain from balloon distension in the pelvic colon (adapted from Ritchie, 1985).

As illustrated, less than 10% of normal subjects reported pain at a distending volume of about 60ml whereas more than 50% of irritable bowel patients reported pain at the same distending volume. This leftward shift in the psychophysical function of IBS patients suggests the presence of hyperalgesia in IBS. Qualitatively similar results have been confirmed subsequently by others and also extended to other hollow viscera in patients with other so-called functional bowel disorders (non-cardiac chest pain, non-ulcer dyspepsia, and irritable bowel syndrome). Similar results may be obtained in other circumstances such as pain associated with interstitial cystitis. Accordingly, it is considered that the altered sensations (bloating, discomfort and abdominal pain) associated with these functional disorders represent visceral hyperalgesia (3, 4). If this hypothesis is correct, then visceral afferent fibers should exhibit sensitization (primary hyper algesia) and the spinal neurons on which they terminate should undergo a change in excitability; mechanisms for primary and secondary hyperalgesia, respectively (4, 5). Evidence in support of central mechanisms contributing to visceral hyperalgesia includes the documentation that when the colon is distended, the area of referred sensation in irritable bowel patients is significantly greater in size than in normal subjects (Figure 2).

As illustrated, the areas of referred sensation in IBS patients were documented by Dawson (6) to be significantly greater in size in IBS patients. Because visceral and somatic inputs converge on to spinal cord neurons (considered to be the basis of referral of

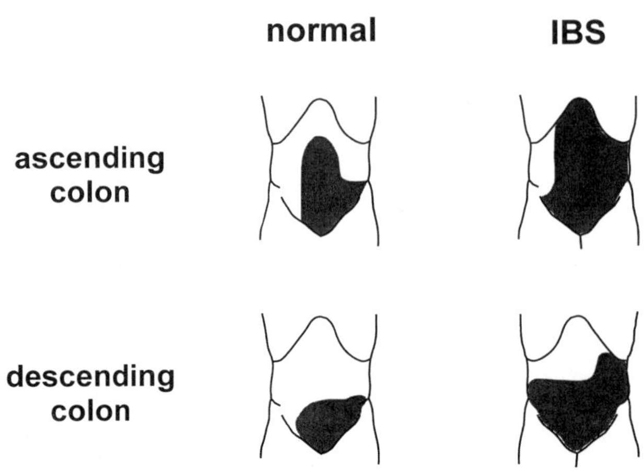

Figure 2. Areas of referred sensation (shaded) associated with balloon distension of the colon in normal subjects and irritable bowel syndrome (IBS) patients (adapted from Dawson, 1985).

sensations from the viscera to somatic structures), the changes in size of areas of referral in IBS patients represents a central mechanism likely reflective of an increase in excitability of spinal neurons. Interestingly, even in the absence of either an inflammatory or functional bowel disorder, central mechanisms can contribute to the size of areas of referred sensation following as few as ten repetitive colonic distensions in normal subjects (7).

BASIC ANATOMY

Innervation of the viscera is unlike innervation of any somatic structure. The viscera can be considered to receive a dual sensory innervation, classically referred to as sympathetic and parasympathetic, but more appropriately designated by nerve name. The viscera are innervated by vagal and spinal afferent pathways comprised principally of thinly myelinated Aδ-fibers and unmyelinated C-fibers (Figure 3). Although the focus here is visceral pain and on hyperalgesia, it should be understood that activity in most visceral sensory neurons, whether in vagal afferent fibers or in spinal afferent fibers, does not reach consciousness. For example, there is regular input into the central nervous system from gastric and hepatic chemoreceptors, aortic baroreceptors, etc. but these are not consciously perceived.

The sensory innervation of the gastrointestinal tract involves all layers of a viscus (mucosa, muscle and serosa), and visceral receptors exhibit chemosensitivity, thermosensitivity and mechanosensitivity. It has become clear in recent years that visceral receptors

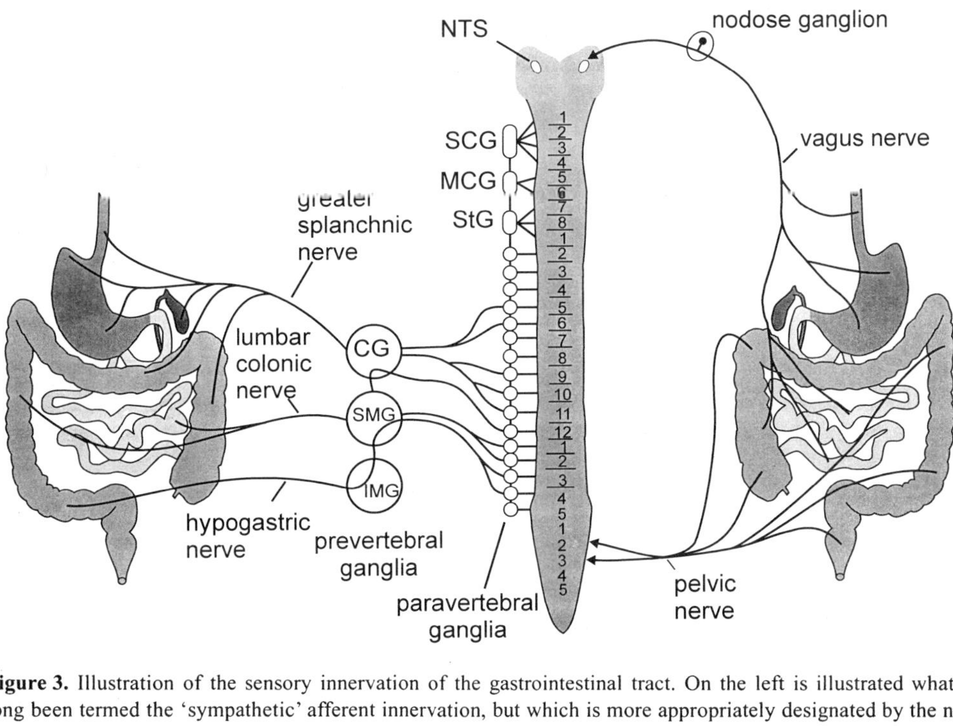

Figure 3. Illustration of the sensory innervation of the gastrointestinal tract. On the left is illustrated what has long been termed the 'sympathetic' afferent innervation, but which is more appropriately designated by the nerve name. On the right, vagal and pelvic nerve innervation is illustrated. Abbreviations: CG, celiac ganglion; IMG, inferior mesenteric ganglion; MCG, middle cervical ganglion; NTS, nucleus tractus solitarius; SCG, superior cervical ganglion; StG, stellate ganglion; SMG, superior mesenteric ganglion; spinal segments are numbered.

are generally polymodal in character (i.e., respond to multiple modalities of stimulation) and thus may represent a phylogenetically older sensory apparatus than the well-defined receptors associated with cutaneous structures. The cell bodies of visceral afferent neurons, like the cell bodies of somatic afferent neurons, are located in the dorsal root ganglia. However, the route visceral afferent neurons take to the spinal cord typically involves passage through or near pre-vertebral ganglia (where they can give off collateral axons to influence autonomic ganglion cell bodies and, accordingly, secretory and motor function) and para-vertebral ganglia.

In contrast to afferent fibers arising from somatic structures, the number of spinal visceral afferent fibers is estimated to be only 2–10% of the total spinal afferent input from all sources. Some compensation for this low number is provided by the significantly greater rostral-caudal intraspinal spread of visceral afferent fiber terminals. Sugiura (8) has examined the spinal organization of somatic and visceral C-fiber afferent fiber terminals and documented that the central terminals of somatic C-fibers were circumscribed and concentrated rostro-caudally in the guinea pig. In contrast, visceral C-fibers were found to have many more terminal swellings (suggestive of synapses) and terminal areas spread over several spinal cord segments. As indicated above, but not illustrated in Figure 3, the spinal neurons upon which visceral afferent fibers terminate typically also receive convergent cutaneous and/or other deep input (e.g., from muscle or other viscera).

VISCERAL PAIN

Experimentally, balloon distension is the most reliable stimulus for hollow visceral organs and is the most widely used stimulus in both human and non-human animal experiments (9). Accordingly, we know most about the response properties for mechanosensitive visceral afferent fibers, usually those innervating the muscle layer. The recent past has documented that the hollow viscera are innervated by two populations of mechanosensitive afferent fibers: a larger group (70–80%) of fibers have low thresholds for response and a smaller group (20–30%) of afferent fibers have high thresholds for response (e.g., >30mmHg distending pressure) (Figure 4).

Those visceral mechanoreceptors that have low thresholds for response (e.g., <5mmHg) obviously respond within the physiological range. Interestingly, as illustrated in figure 4, these afferent fibers also respond to distending pressures in the noxious range and in fact encode stimulus intensity fairly well. Note also that response magnitude in the

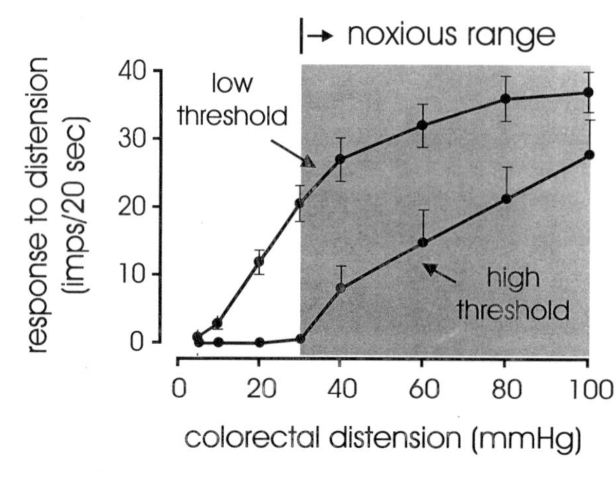

Figure 4. Summary of responses of pelvic nerve afferent fibers in the rat to graded intensities of colorectal distension. Two populations of pelvic nerve afferent fibers, low threshold and high threshold, innervate the colon. Low threshold fibers typically have response thresholds <5mmHg; high threshold fibers typically have response thresholds ≥30mmHg. Both low and high threshold fibers encode distending pressures in the noxious range.

noxious range is typically greater than response magnitude of high threshold fibers at the same intensity of stimulation. Those visceral mechanoreceptors that have high thresholds for response do not begin to respond until the distending pressure is at or exceeds pressures that are likely noxious. It was long argued that a homogenous population of afferent fibers innervated the viscera. All had response thresholds in the physiological range and visceral pain was considered to be encoded in the discharge frequency of the afferent neurons. We now know that all hollow viscera studied (esophagus, gall bladder, stomach, urinary bladder, colon, uterus), in a number of different species, are innervated by a population of afferent fibers that do not respond until the distending stimulus intensity is in the range of pressures associated with pain. Whether these afferent fibers are nociceptors, which have been extensively studied in skin, is currently a matter of debate (10).

SENSITIZATION AND SILENT NEURONS

Cutaneous nociceptors, but not *non*-nociceptors, are characterized by their ability to be sensitized. By sensitization is meant an increase in response magnitude to a noxious intensity stimulus. Sensitized nociceptors also often develop increased spontaneous activity and/or a reduction in the threshold at which they respond. Sensitization, which is a key component in the development of primary hyperalgesia, is produced by a wide variety of chemical mediators that are released at the site of tissue injury, synthesized at the site of tissue injury, or attracted to the site of injury. These chemical mediators of nociceptor activation and/or sensitization include products of arachidonic acid (prostaglandins and leukotrienes), histamine, serotonin, cytokines, bradykinin, protons, peptides such as substance P, neurotrophins, etc. It was at one time considered that some of these putative mediators (e.g., bradykinin) individually sensitized responses of peripheral mechano-nociceptors, but it has since been documented that many of these mediators likely act in concert and synergistically to sensitize nociceptors. By far most of our understanding of nociceptor sensitization has been derived from studies of cutaneous receptors, but it has been documented that visceral afferent fibers can also be sensitized experimentally (Figure 5).

As illustrated in Figure 5, both low threshold and high threshold mechanosensitive visceral afferent fibers can become sensitized during experimental visceral inflammation. Note in the example that the spontaneous activity of the low threshold fiber increases significantly, as does the response magnitude, particularly at the greater distending pressures. It is clear that response threshold decreases significantly in the example of the high thresh-

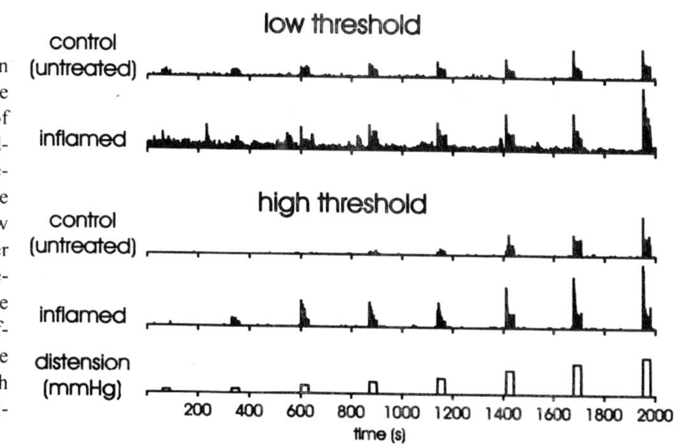

Figure 5. Examples of sensitization of pelvic nerve afferent fibers in the rat after intracolonic instillation of 2.5% acetic acid. Colorectal distending pressures are given below and responses presented as peristimulus time histograms (1 sec bin width) for a low threshold and a high threshold fiber are illustrated above. Both spontaneous activity and response magnitude of the low threshold fiber increase after colonic inflammation; response threshold decreases for the high threshold fiber while response magnitude increases after inflammation.

old fiber (from about 30mmHg to 10mmHg). As indicated above, there is debate whether the viscera are innervated by both nociceptors and non-nociceptors (as opposed to the concept that all sensory fibers that innervate the viscera can convey noxious information). Because both low and high threshold visceral afferent fibers encode distending stimuli well into the noxious range (>30 mmHg; as illustrated in figure 4), and both low and high threshold fibers can be sensitized by experimental inflammation (e.g., figure 5), the position that there exists a specific class of visceral nociceptors is not supported.

Sensitization of visceral afferent fibers results in increased afferent input to the spinal cord and contributes to altered sensation. Tissue injury or visceral inflammation can also contribute to increased afferent input to the spinal cord by activation of previously silent or "sleeping" afferent fibers. It has been shown that mechanically insensitive cutaneous or visceral afferent fibers acquire spontaneous activity and mechanical sensitivity after tissue injury or inflammation (Figure 6). There is some consideration, but no evidence, that such silent fibers are chemo-nociceptors that respond only to noxious chemical stimuli, after which they become sensitized to mechanical stimulation and lose their selectivity for chemical stimuli. Experimentally, this is a difficult proposition to investigate and the issue is unresolved at present. If correct, however, it suggests that there may be families of chemo-nociceptors innervating the viscera. Activity in these newly discovered afferent fibers adds to the total afferent barrage entering the spinal cord, contributing to what is called central sensitization.

CENTRAL SENSITIZATION

One consequence of an increased afferent input into the spinal cord is an excessive release of neurotransmitters such as glutamate and substance P in the spinal dorsal horn.

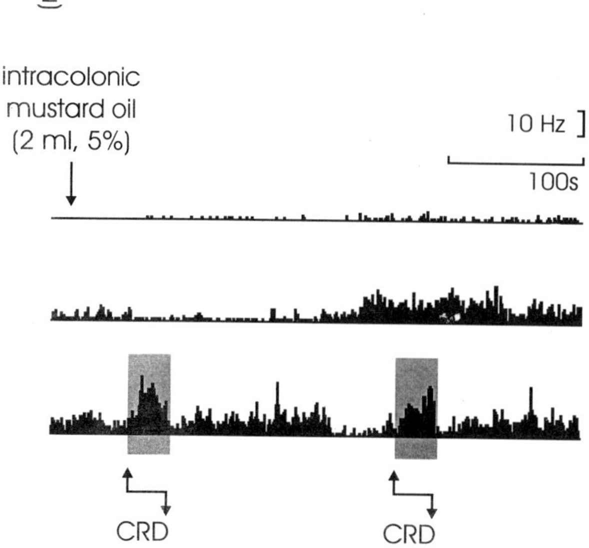

Figure 6. Example of a mechanically-insensitive ("silent") pelvic nerve afferent fiber that innervates the colon of the rat. The fiber did not respond to noxious colorectal distension (CRD) or noxious urinary bladder distension (UBD) (top record). After intracolonic instillation of the irritant mustard oil, this fiber acquired spontaneous activity and mechanosensitivity to colorectal distension. Data are illustrated as 1 sec bins of counted action potentials; the record is continuous left to right and top to bottom.

These neurotransmitters increase the excitability of second order spinal neurons that receive visceral (and somatic) input. These spinal neurons not only become easily excitable, but also show an increase in the sensory receptive field (Figure 2). In addition to an increase in the size of peripheral receptive fields, there is an increase in the sensitivity to perceiving previous sub-threshold intensities of stimulation (e.g., Figure 1). This phenomenon has been termed central sensitization (in analogy to nociceptor sensitization) and can lead to altered sensations that may outlive the period of increased afferent input.

Thus, visceral afferent fibers and the spinal neurons upon which they terminate undergo similar changes in sensitivity that have been better documented to date in models of cutaneous nociception. It would appear that visceral hyperalgesia is analogous to cutaneous hyperalgesia and is mediated by similar, if not the same mechanisms. Accordingly, central sensitization as representative of secondary hyperalgesia may persist in the absence of continued peripheral input. This "memory" in the spinal cord and at supraspinal sites has been suggested as the basis of some chronic pain states, including those arising from the viscera.

CLINICAL IMPLICATIONS

Although the anatomy and organization of visceral sensory fibers differs from their cutaneous counterpart, the central (principally spinal cord, which has been the focus of investigation) consequences of excessive peripheral input are similar for both viscera and skin. This should not be surprising, given that spinal cord neurons typically receive convergent somatic and visceral inputs. Accordingly, the central mechanisms of visceral hyperalgesia do not appear to differ significantly from the central mechanisms of cutaneous hyperalgesia. This information suggests that management of visceral hyperalgesia, such as associated with functional bowel disorders, would be similar to the management of cutaneous hyperalgesia (e.g., post-incisional hyperalgesia).

Because the peripheral sensory apparatus innervating the viscera differs in several ways from that associated with somatic tissue, an alternative strategy for the management of the bloating, discomfort and/or pain that characterizes some aspects of functional bowel disorders would focus on the sensory innervation of the viscera. While a potentially attractive strategy, at present, a selective pharmacological treatment that would achieve this objective is not available.

There remains much to be learnt about the sensory innervation of the viscera, not the least of which is the interaction between the extrinsic, sensory and the intrinsic (enteric) innervations. Information about such interactions is important in order to selectively manage visceral hyperalgesia.

REFERENCES

1. Gebhart, G.F. (ed.) *Visceral Pain*, IASP Press, Seattle, 1995, pp. 516.
2. Ritchie, J. Mechanisms of pain in the irritable bowel syndrome. In: *Irritable Bowel Syndrome* (Read, ed.), Grune and Stratton, Philadelphia, 1985, pp. 163–170.
3. Gebhart, G.F. Visceral pain mechanisms. In: *Current and Emerging Issues in Cancer Pain: Research and Practice* (Chapman and Foley, eds.), Raven Press, New York, 1993, pp. 99–111.
4. Mayer, E.A. and Gebhart, G.F. Basic and clinical aspects of visceral hyperalgesia. *Gastroenterology* 1994, 107:271–293.
5. Bonica, J.J. Clinical importance of hyperalgesia. In: *Hyperalgesia and Allodynia* (Willis, ed.), Raven Press, New York, 1992, pp. 17–43.

6. Dawson, A.M. Origin of pain in the irritable bowel syndrome. In: *Irritable Bowel Syndrome* (Read, ed.), Grune and Stratton, Philadelphia, 1985, pp. 155–162.
7. Ness, T.J., Metcalf, A.M. and Gebhart, G.F. A psychophysiological study in humans using phasic colonic distension as a noxious visceral stimulus. *Pain* 1990, 43:377–386.
8. Sugiura, Y. Spinal organization of C-fiber afferents related with nociception or non-nociception. In: *The Polymodal Receptor: A Gateway to Pathological Pain* (Kumazawa, Kruger and Mizumura, eds.), Elsevier, Amsterdam, 1996, pp. 319–339.
9. Ness, T.J. and Gebhart, G.F. Visceral pain: A review of experimental studies. *Pain* 1990, 41:167–234.
10. Cervero, F. Sensory innervation of the viscera: Peripheral basis of visceral pain. *Physiol. Rev.* 1994, 74:95–138.

26

CLINICAL EVALUATION OF VISCERAL PAIN

Satish S. C. Rao*

University of Iowa
College of Medicine
Department of Internal Medicine
Iowa City, Iowa 52242

INTRODUCTION

Because of rigorous efforts to characterize visceral pain, today, visceral hyperalgesia has emerged as the key mechanism for unexplained chest pain, unexplained abdominal pain, functional dyspepsia, and irritable bowel syndrome (1,2). Consequently, several innovative techniques have been developed both for evoking visceral pain and for identifying the locus of this problem. This chapter will focus on the methods that are currently available for the evaluation of visceral sensation and pain.

TECHNIQUES FOR ASSESSING VISCERAL SENSATION

In spite of a wide array of tests and techniques that are available for evaluating visceral pain, it is unclear which tool to use, how to use these tools effectively, and how useful are these techniques in clinical practice. Table 1 shows a list of techniques that have been used for evoking visceral pain.

What follows is a brief discussion regarding the merits and shortcomings of some of the most popular tools.

Balloon Distention Test

In its simplest form, this test consists of placing into the lumen of the gut either a latex balloon or a highly compliant polyvinyl chloride (PVC) balloon that is mounted on a plastic probe. Subsequently, the balloon is inflated with either air or water, using phasic or ramp distentions. This simple test has proved extremely useful for developing the concept of visceral hyperalgesia and for facilitating research on this topic. In patients with unex-

* Phone: 319/353-6602, Fax: 319/353-6399, E-mail: satish-rao@uiowa.edu

Gastrointestinal Motility, edited by Rao *et al.*
Kluwer Academic / Plenum Publishers, New York, 1999.

Table 1. Techniques of evoking visceral sensation

• Balloon distention test	• Electrical stimulation
• Impedance planimetry	• Chemical/pharmacological stimulation
• Barostat balloon distention	

plained gastrointestinal symptoms, balloon distention studies have shown that an abnormality of visceral sensation is more likely to be present than a motor dysfunction (1).

By distending a balloon in the rectum, it has been shown that patients with irritable bowel syndrome (IBS) have lower sensory thresholds for pain (3,4). Others have shown that pain is referred over a much wider area in IBS patients, for example, balloon distention of the hepatic flexure produced pain in the left upper and left lower quadrants (5).

An early study suggested that a balloon volume of 8 ml could distinguish patients with visceral hypersensitivity from those with normal esophageal sensation (6). But other investigators could not reproduce this result (7,8). Also, threshold volumes between 5 to 30 cc have been reported to induce esophageal pain (9). In one study, the same number of normals and patients reported chest pain with a balloon volume of 30 cc (7). These discrepancies may be due to the type of balloon, the size of balloon, the location of balloon in the esophagus, the rate of distention, and the volumes used for balloon distention (9,10). In order to overcome some of these shortcomings, newer techniques of balloon distention such as impedance planimetry and barostat have been developed.

Impedance Planimetry

This technique provides a method of measuring the cross-sectional area of the lumen in a selected plane at a range of distending pressures as well as arousing visceral sensation (11,12). The impedance planimetry system consists of two components, the impedance probe and the signal processing system (Figure 1). The probe consists of four ring electrodes, two outer excitation electrodes and two inner detecting electrodes which are enclosed within a latex balloon. The probe has three perfusion side holes that are located 2.5 cm, proximal and distal to the balloon, and one inside the balloon. Additionally, there is a large infusion channel for balloon distention. The balloon is distended by infusing 0.018% NaCl. Stepwise graded balloon distentions are performed using incremental pressures ranging from 5–70 cm H_2O. During balloon distention, the subject reports first sensation at a particular pressure threshold and moderate discomfort or pain at higher thresholds (11–13). Balloon distention also changes the balloon cross-sectional area which is measured using the field gradient principle (8–10). From these measurements, other biomechanical properties such as tension/strain relationships can be calculated (11–14).

Impedance planimetry has been used to evaluate the sensory and biomechanical properties of the esophagus (11), both in healthy subjects (9) and in patients with unexplained chest pain (13), in the duodenum (15), and in the rectum (16). In one study, twenty-four consecutive patients with recurrent chest pain for more than six months and with negative cardiac workup, normal endoscopy, and either normal 24-hour esophageal pH study or failure to respond to omeprazole (20 mg per day) were evaluated with impedance planimetry (13). Balloon distention reproduced typical chest pain in 84% of these patients. Also, the thresholds for first sensation, moderate discomfort and chest pain, were significantly lower in the patient group compared to healthy normal subjects (Figure 2). In addition, the esophagus was poorly compliant and exhibited a lower threshold for reactive contractions (13). These findings indicated that the esophagus was hypersensitive, poorly compliant and hyper reactive in patients with unexplained chest pain (13).

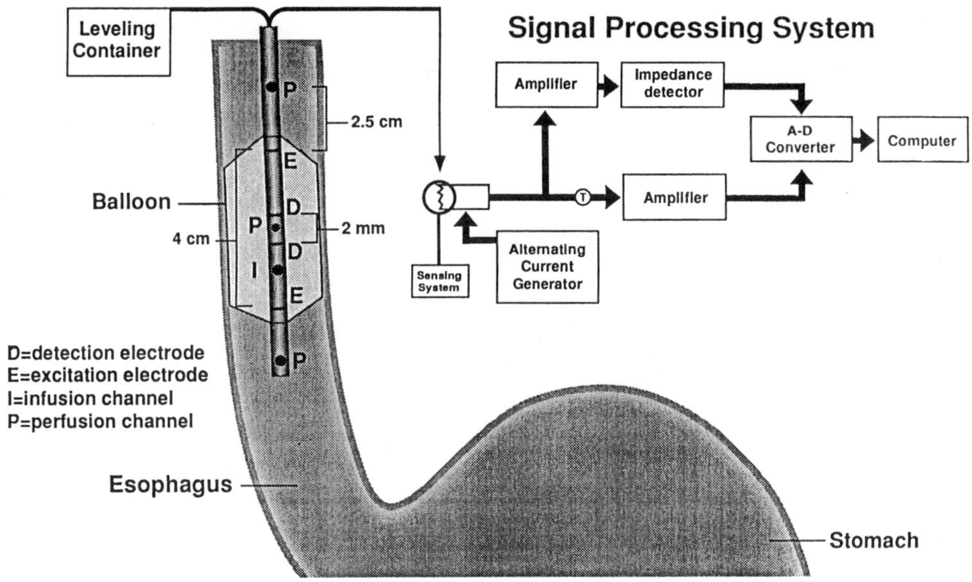

Figure 1. A schematic diagram of the impedance planimetry system.

Because it provides comprehensive evaluation of esophageal function, impedance planimetry may prove to be a useful technique for evaluating the therapeutic potential of drugs that may modify visceral sensation. However, this technique is not suitable for assessing biomechanics in large, globular organs such as the stomach. Also, the ambient temperature, the conductivity of the electrolyte solution and the distance between electrodes may each influence the measurements (12,14) and should be optimally set, prior to assessing a particular organ.

Barostat

In this system (Figure 3), a highly compliant PVC balloon is placed inside a hollow organ such as the stomach or the rectum and stepwise or ramp balloon distentions are performed using a barostat. The barostat (Figure 4) is a computer-controlled balloon-distend-

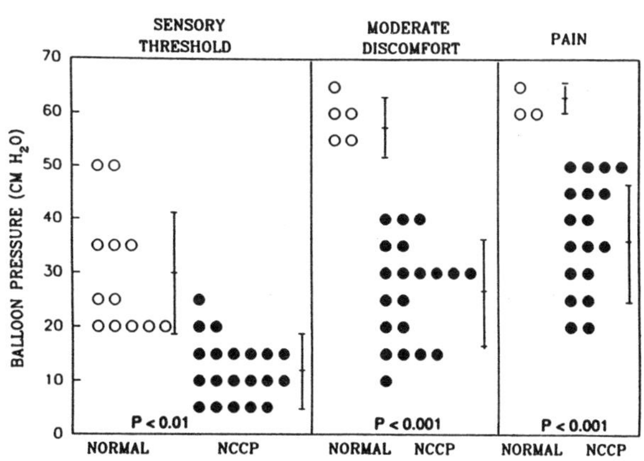

Figure 2. Patients with unexplained chest pain have lower sensory thresholds for first perception, moderate discomfort and pain when compared to normal subjects. i.e., they have a hypersensitive esophagus. (With permission from Rao, et al. Reference 13).

Figure 3. A barostat distending system along with an infinitely compliant polyvinyl chloride balloon mounted on a probe.

ing device which consists of a pump, often a piston located inside a hollow cylinder, a motor, and pressure transducers (17–19). At preset pressures, the piston drives air out of the cylinder into a highly compliant balloon which then distends the lumen. Thereafter, intraballoon pressures and intraballoon volumes are continuously recorded and displayed on a monitor. The primary function of the barostat is to maintain a constant intraballoon pressure or volume. When the hollow organ contracts or its tone increases, the balloon is compressed. This increases the intraballoon pressure. In order to maintain the preset pressure, the pump withdraws air from the balloon into the cylinder. This changes the intraballoon volume which is displayed on the recorder as a decrease in balloon volume. In contrast, when the muscle wall relaxes, the intraballoon pressure falls. In order to maintain the preset pressure the pump pushes air to expand the balloon. Consequently, the intraballoon volume increases. Thus, by plotting the changes in balloon volume, indirectly, the barostat method can assess the tonic changes of the luminal wall.

The barostat has been used to examine the tonic changes of the esophagus (20), stomach (17), colon (21), and rectum (22, 23) and to examine the tonic changes of the muscle wall after pharmacomodulation (19). Using the barostat, it has been shown that IBS patients have lower sensory thresholds (22, 23). Also, repetitive stimulation has been shown to induce rectosigmoid hyperalgesia and allodynia (22). A few investigators have modified the barostat probe to include perfusion side holes that are located proximal and distal to the balloon (21). This arrangement can provide additional information regarding luminal pressure changes.

A modification of the barostat, the tenso-stat has also been developed for examining the biomechanics of the luminal wall (24). Here, the lumen can be distended up to a desired level of wall tension which is calculated from the intraluminal pressure and the volume infused (Laplace's Law) (14). Using this device, it has been shown that perception of gastric distension is chiefly mediated by tension receptors (24).

Today, the barostat is widely used in clinical trials and in research studies. The computerized and semi-automated distention technique allows for repeated balloon distentions

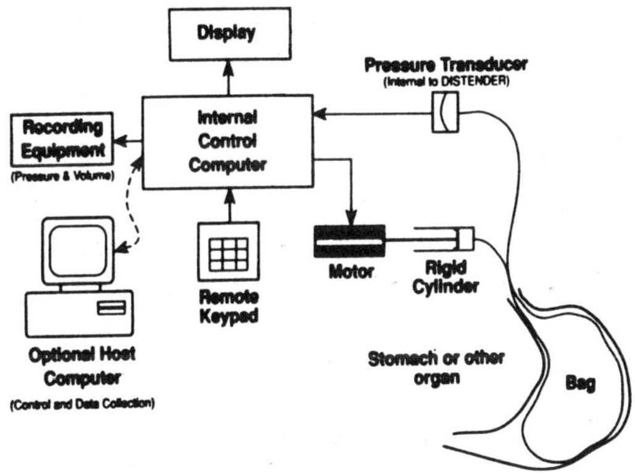

Figure 4. A schematic diagram of the barostat distending devise.

and is clearly a less demanding and more user-friendly technique. Also, newer software permits easy analysis of data (19). However, there are some limitations. The barostat may be suitable for examining the sensory responses and the tonic changes in hollow globular organs such as the stomach or the urinary bladder but is unsuitable for examining cylindrical organs such as the duodenum and colon (14). It also cannot assess segmental changes within an organ but provides an aggregate tone of the luminal wall.

Pitfalls during Balloon Distention Tests. There is no uniform method or technique of performing balloon distention tests. Hence, several aspects must be considered when interpreting the results of these tests. Whether balloon distention is performed by using increments of pressure or volume is critical. Stepwise graded balloon distention, using increments of pressure may be more reproducible and more accurate for evoking visceral pain than increments of volume, particularly in organs with a small lumen such as the esophagus (9). But in organs with a large lumen such as the rectum, it may be less important (25). Also, the characteristics of the visceral sensory response may vary depending on which segment of the gut is examined. For example, in the rectum, luminal distention is commonly perceived as an urge to defecate and pain is an uncommon sensation (25). Similarly, distention of the gastric fundus may produce nausea or retching (26) rather than pain. Hence, symptom descriptors for nociception can be organ specific and must be considered when evaluating visceral pain.

Another pitfall that should be considered is the method of performing balloon distention, i.e., phasic vs. ramp distention (10) or graded increments in balloon distention vs. tracking the sensory responses (19) Previously, it has been shown that the threshold for visceral perception is much lower when stepwise or intermittent phasic distention is used as opposed to ramp distention (10). This may be due to accommodation of the mucosal tension receptors. Although graded increments in balloon distention often provides a distinct threshold for sensory responses, several investigators have suggested that the tracking technique for evaluating visceral pain may be better (18). This is based on the premise that subjective responses of pain are variable and the surest way to ascertain that a particular threshold causes pain is to randomly decrease or increase subsequent balloon distentions and track the sensory responses. However, in unpublished observations, we have seen that the tracking technique, particularly with the barostat, may alter the resting wall tension (Figure 5). Also, by repeated distentions, one can enhance the afferent neuronal

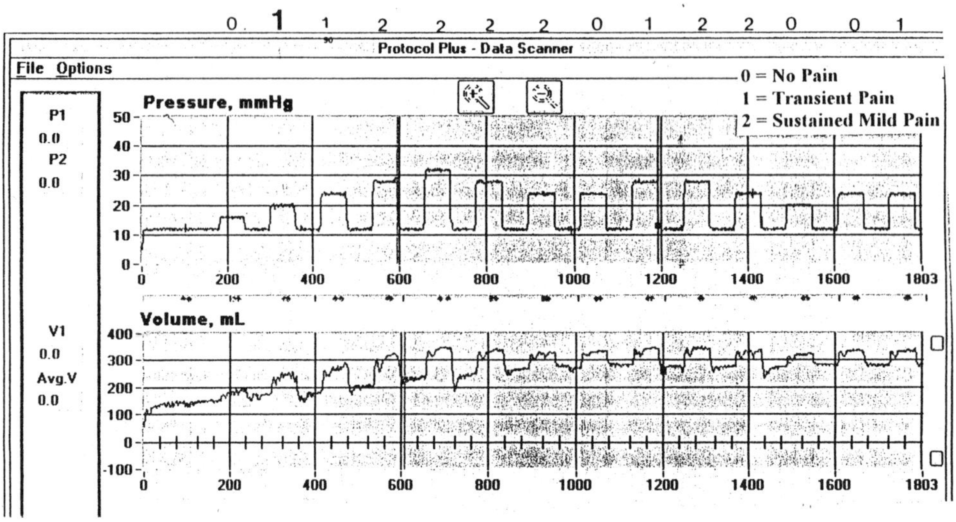

Figure 5. A barostat recording from the rectum, using the tracking method of evoking visceral sensation. During balloon distentions, it can be seen that the pressure threshold for the same sensation (pain) varied between 24–32 mmHg. Repeated distentions lowered the threshold i.e sensitized the rectum. Regardless, the intraballoon volume increased suggesting that the baseline tone had decreased. This illustrates the discrepancy between pressure, volume, tone and visceral sensation.

firing rate and thereby sensitize the mucosal tension receptors (1,26,27). It is therefore conceivable that tracking may induce physiological changes that could potentially lower the thresholds for sensory perception. Finally, the type of balloon used, i.e., latex versus silicon rubber vs. PVC balloons may also influence the sensory and biomechanical responses (18,28). Hence, the assessment of visceral sensation can be influenced by many factors and each of these must be considered when interpreting or comparing data.

Electrical Stimulation

Electrical stimulation of the rectum or the anal canal has been used for testing visceral sensation (29) as well as for eliciting somatosensory evoked potentials (30). This test provides an assessment of the neuronal integrity between the peripheral sensory receptors and the central nervous system. Electrical stimulation of the esophagus and the small intestine (31) has also been used to evoke visceral sensory responses, but these techniques are at present research tools. In the USA, there are regulatory constraints for using intraluminal electrical stimulation tests. Also, transmucosal electrical nerve stimulation may depolarize all the afferents within the gut wall, whereas balloon distentions may selectively stimulate the mechano receptors (26). This and other drawbacks such as safety and electrical isolation of the patient could limit the use of this technique for clinical research.

Chemical Stimulation

It has been suggested that the gut mucosa may contain specific chemoreceptors or osmoreceptors (32). When sensitized or stimulated excessively, these receptors may mediate visceral pain. For example, esophageal chest pain that is secondary to acid reflux may be mediated by chemoreceptors in the distal esophagus. Two popular tests that have been used to assess esophageal visceral pain may be triggered by chemoreceptors. The Bernstein test (33) which consists of infusing randomly either O.1N HC1 or 0.9% NaCl into the esophagus in a single blinded fashion has been used for many years to examine the association between acid reflux and chest pain (34). Although simple, it is a more useful test for examining the sensitivity of the esophageal mucosa to acid perfusion than for testing a patient with chest pain and acid reflux (34, 35, 36). In contrast, ambulatory 24 hour pH test is a superior test for identifying acid reflux and for evaluating the correlation between symptoms and the occurrence of acid reflux (34).

Another situation where chemical stimulation may be useful is in the assessment of unexplained right upper quadrant pain. In these patients, duodenal infusion of lipid has been shown to reproduce right upper quadrant pain (37). This presumably is a consequence of stimulating the sensitized chemoreceptors in the duodenum (32, 38, 39). However, further studies are required to characterize the relationship between chemical stimulation and the chemoreceptors that mediate right upper quadrant pain before this form of testing can be advocated as a diagnostic tool.

TECHNIQUES FOR ASSESSING THE LOCUS OF VISCERAL PAIN

Although there are several elegant methods for evoking visceral pain, the precise locus of this sensation remains in dispute. Visceral pain may arise from mechanoreceptors in the myenteric or submucosal plexus or from abnormal afferent neuronal function (sensitization of the afferent neuron), or from abnormal processing or relay within the spinal cord or the thalamus or from heightened perception within the sensory cortex (Figure 6). Visceral pain may also be due to altered efferent regulation of the sensory input to the brain. Recently, several techniques have been used for evaluating the locus of visceral pain (Table 2).

Electroencephalography (EEG)

This is a time-honored test of examining the cerebral electrical activity at rest and after provocative maneuvers such as hyperventilation. In neurology, this technique has been primarily used for evaluating epilepsy. But the same technique can also be used for

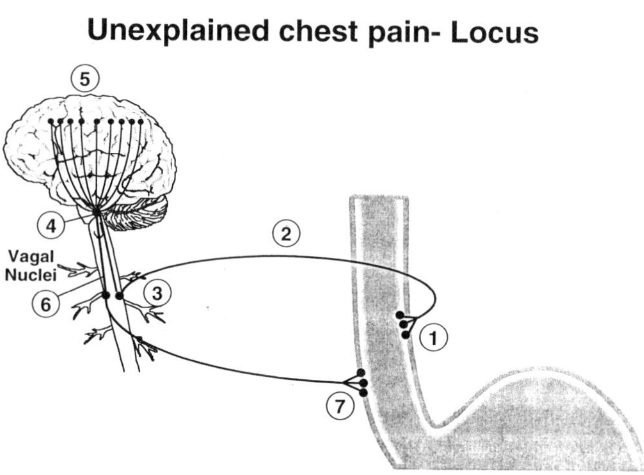

Figure 6. Schematic diagram depicting the possible potential loci for esophageal visceral pain; 1) Mechanoreceptors in the luminal wall 2) Sensory afferent nerves 3) Spinal and spinothalamic processing of neuronal signals 4) Thalamic gating 5) Sensory cortex 6) Descending neural inhibitory pathways 7) Efferent nerves ± modulation by the autonomic nervous system.

Table 2. Techniques of evaluating the locus of visceral pain

• Electroencephalography (EEG)	• Positron Emission Tomography (PET) scanning
• Magnetoencephalography (MEG)	• Functional magnetic resonance imaging (FMRI)
• Somatosensory evoked potentials	• Pharmacomodulation

examining the changes in cerebral electrical activity after mechanical or electrical stimulation of the gastrointestinal tract. Once the gut-brain axis has been carefully mapped out by other sophisticated techniques, it may be possible to apply simpler and more widely available techniques such as the EEG for evoking and diagnosing visceral pain. Currently, its utility is being evaluated in research studies (40).

Magnetoencephalography (MEG)

This technique is an adaptation of the standard electroencephalography (EEG). Here, instead of recording electrical activity from the central nervous system, the MEG detects small magnetic fluxes that are generated by neural activity (41). Unlike electrical activity, the advantage of this technique is that there is less likelihood of interference by intervening tissues. Hence, it can provide a clear recording and a better localization of neuronal dysfunction. Using this technique, it has been suggested that magnetic fluxes emanating from areas that are 2–3 mm in diameter can be accurately detected (40,41). However, it can only detect activity close to the cerebral surface and does not provide information regarding deeper structures.

Somatosensory Evoked Potentials

This method consists of detecting the spinal or cerebral responses that occur after stimulating the gastrointestinal tract. The time of onset, the magnitude, and the characteristics of the neural response can be examined. In pilot studies, cerebral evoked potentials have been obtained after electrical stimulation of the rectum (30). But as a clinical tool, the most promising technique appears to be the recording of cerebral evoked potentials after repeated (at least 25–50) balloon distentions of the rectum, at pressures ranging from 10–25 mm Hg. Usually, these distentions are performed at thresholds that are either not perceived or are not painful (42, 43). For the purposes of this test, a computer controlled balloon inflating device is used (42, 43). The normal response consists of three positive peaks (p1-p3) and two negative peaks (n1-n2) with a variable latency (onset time). In children with encopresis, it has been shown that the cerebral evoked potentials are delayed (43). This suggests that these patients have impaired conduction through the afferent neuronal tracts. These studies are painstaking and are largely confined to research laboratories. Unless one can develop a reliable computer program that averages the cerebral evoked responses and generates a meaningful profile, its use for clinical purposes may be limited.

Positron Emission Tomography (PET) Scanning

This technique involves scanning the cerebral cortex for positron emitting isotopes. Because neural activation causes changes in cerebral blood flow—a process which is closely coupled with cerebral glucose consumption and neuronal Na/K ATPase activity—by assessing the regional cerebral blood flow, PET scanning indirectly measures the neuronal rate of firing (40, 44).

In order to examine the cerebral cortical responses that occur in association with chest pain and myocardial ischemia, PET scanning was performed in patients with ischemic heart disease after infusing dobutamine (45). Although focal cerebral activation was seen in patients with silent ischemia, a more extensive pattern of cerebral activation which included the cingulate gyrus and the basal frontal area was seen in patients who reported chest pain (45). Recently, PET scanning has also been performed in patients with IBS after rectal balloon distention (46) and in normal subjects after painful esophageal distention (44). These studies have confirmed that the cingulate gyrus is always activated during painful distention (40, 44, 46). Additionally, the right anterior insular cortex may also be involved (40). These novel studies suggest that perturbations of selected areas of the cerebral cortex may regulate the emotional responses associated with visceral pain. It is therefore possible that the central processing of visceral sensory information may be disturbed in patients with visceral hyperalgesia. However, it is unclear whether the cortical activation is a nonspecific response to nociceptive events such as pain or is a consequence of heightened visceral perception. Although, PET scanning may prove to be a useful research tool for examining the locus of pain, it is unlikely to become a clinical diagnostic tool because of the enormous cost of performing this study and the use of radiation emission.

Functional Magnetic Resonance Imaging (FMRI)

This is an adaptation of the standard technique of magnetic resonance imaging that has been successfully used for diagnostic neuroradiology. It identifies foci of altered blood flow, i.e., it reveals functional changes, unlike the standard MRI scan which displays structural changes. A recent study has showed that FMRI may be a useful method of studying the activation of somatosensory cortex (47). Unlike topographic magnetoencephalography, which is limited to detecting superficial cerebral electrical activity, FMRI can detect changes from deeper structures i.e., those that are located more than 1 cm from the cortical surface (47). Thus, this novel technique promises to be a very useful method of examining the brain-gut axis. The advantages of this technique are: a) the lack of radiation, b) the quantification of cerebral activity after electrical or mechanoreceptor stimulation of the gut and c) the ability to perform repeated evaluations within the same subject. However, like other radiological techniques, its use may be limited by the availability of equipment and the high costs of the procedure.

Pharmacomodulation

The rationale here is to administer receptor-sensitive and region-specific drugs that may modulate visceral sensory responses. Theoretically, this is a very attractive clinical research tool. However, our knowledge regarding the origin of visceral sensation and the mediators of visceral pain, particularly in man, is still primitive. But there are many studies in progress that are testing the role of specific 5 HT_3 and 5 HT_4 receptor antagonists (48), Kappa opioid receptor agonist (49) and other drugs for modulating visceral pain (48). At present, pharmacomodulation remains a tool of the enthusiastic bench or clinical investigator. Hopefully, one day, with the availability of receptor-selective drugs, pharmacomodulation may prove to be a useful technique for defining the locus of visceral pain.

CONCLUSIONS

In summary, visceral hyperalgesia has come of age and is increasingly being accepted as an important mechanism for unexplained symptoms arising from the gut. Several exciting studies are in progress both to develop better and more reliable techniques of evoking pain as well as to pinpoint the origin and locus of visceral pain. Over the next few years, it is likely that noninvasive, user friendly and inexpensive methods of evoking visceral pain will become available. These techniques will undoubtedly revolutionize our knowledge of functional gastrointestinal disorders. Such advances could then pave the way for developing novel therapies for treating visceral pain.

REFERENCES

1. Rao SSC. Visceral Hyperalgesia: The key for unraveling functional gastrointestinal disorders. Dig Dis 1996;14:271–275.
2. Mayer, EA, Gebhart GF. Functional bowel disorders and the visceral hyperalgesia hypothesis. In: EA Mayer and HE Raybould (Eds.), Basic and Clinical Aspects of Chronic Abdominal Pain, Elsevier, Amsterdam, 1993, pp. 3–28.
3. Ritchie JA. Pain from distension of the pelvic colon by inflating a balloon in the irritable colon syndrome. Gut 1973;16:125–132.
4. Sun WM, Edwards CA, Prior A, Rao SSC, Read NW. The effect of nicardipine on anorectal motility in normal human volunteers and in patients with irritable bowel syndrome. Dig Dis Sci 1990;35(7):885–890.
5. Swarbrick CT, Bat T, Hegarty JE, Williams CB, Dawson AM. Site of pain from the irritable bowel. Lancet 1980;11:443–446.
6. Richter JE, Barish CF, Castell DO. Abnormal sensory perception in patients with esophageal chest pain. Gastroenterology 1986;91:845–852.
7. Cannon RO, Cattau EL, Yakhe PN, Maher KM, Schenke WH, Benjamin SB, Epstein SE. Coronary flow reserve esophageal motility and chest pain in patients with angiographically normal coronary arteries. Am J Med 1990;88:217–222.
8. Nevens F, Janssens J, Piessens J, Ghillebert G, DeGeest H, Vantrappen G. Prospective study on prevalence of esophageal chest pain in patients referred on an elective basis to a cardiac unit for suspected myocardial ischemia. Dig Dis Sci 1991; 36:229–235.
9. Patel, R., Rao SSC: Esophageal biomechanical and sensory properties vary along its length. Am J Physiol 1998;275: G187–G191.
10. Sun WM, Read NW, Prior A, Daly J, Cheah K, Grundy D. Sensory and motor response to rectal distention vary according to rate and pattern of balloon inflation. Gastroenterology 1990; 99:1008–1015.
11. Rao SSC, Hayek B, Summers RW. Impedance planimetry: an integrated approach for assessing sensory, active and passive biomechanical properties of the human esophagus. Am J Gastroenterol 1995;90:431–438.
12. Gregersen H, Andersen MB. Impedance measuring system for quantification of cross-sectional area in the gastrointestinal trait. Med Biol Eng Comput 1991;29:108–110.
13. Rao SSC, Gregersen H, Hayek B, Summers RW, Christensen J. Unexplained chest pain: The hypersensitive, hyperreactive, and poorly compliant esophagus. Ann Intern Med 1996; 124:950–958.
14. Gregersen H, Kassab G. Biomechanics of the gastrointestinal tract. Neurogastroenterol Mot 1996;8:1–20.
15. Gregerson H, Orvar K, Christensen J. Biomechanical wall properties and tone during phase I and phase II of the MMC. Am J Physiol 1992; 263:G795–801.
16. Rasmussen OO, Sorensen M, Tetzschner T, Christiansen J. Dynamic anal manometry: Physiological variations and pathophysiological findings in fecal incontinence. Gastroenterology 1992;103:103–113.
17. Azpiroz F, Malagelada JR. Gastric tone measured by an electronic barostat in health and past surgical gastroparesis. Gastroenterology 1987;92:934–943.
18. Whitehead WE, Delvaux M. Standardization of procedures for testing smooth muscle tone and sensory thresholds in the gastrointestinal trait. Dig Dis Sci 1997; 42: 223–241.
19. Delvaux M. The barostat: a tool to evaluate visceral sensitivity. In: Neurogastroenterology: Ed Corazziari E. Walter deGruyter, Berlin, N. York. 1996, pp. 115–133.
20. Mayrand S, Diamant NE. Measurement of human esophageal tone in vivo. Gastroenterology 1993; 105:1411–20.
21. O'Brien MD, Camilleri M, vonderOhe MR, Phillips SF, Pemberton JH, Prather CM, Wiste JA, Hanson RB. Motility and tone of the left colon in constipation: a role in clinical practice. 1996. 91:2532–2538.

22. Munakata J, Naliboff B, Harraf F, Kodner A, Lembo T, Chang L, Silverman DH, Mayer EA. Repetitive sigmoid stimulation induces rectal hyperalgesia in patients with irritable bowel syndrome. Gastroenterology 1997;112:55–63.
23. Whitehead WE, Holtkotter B, Enck P, Hoelzl R, Holmes KD, Anthony J, Shabsin HS, Schuster MM. Tolerance for rectosigmoid distention in irritable bowel syndrome. Gastroenterology 1990;98:1187–1192.
24. Distrutti E, Soldevilla A, Azpiroz F, Malagelada JR. Gastric mechanoreceptors response characterized by a computerized tenso stat in humans. Gastroenterology 1996;110:A657.
25. Hammer HF, Phillips SF, Camilleri M, Hanson RB. Rectal tone, distensibility, and perception: reproducibility and response to different distensions. Am J Physiol 1998;274:G584–90.
26. Azpiroz F. Functional dyspepsia. In: Neurogastroenterology: Ed Corazziari E. Walter deGruyter, Berlin, N. York, 1996, pp. 57–64.
27. Rao SSC, Welcher KD, Leistikow JS. Obstructive Defecation: A failure of rectoanal coordination. Am J Gastroenterol 1998;329:1905–11.
28. Lasch H, DeVault KR, Castell DO. Intraesophageal balloon distention in the evaluation of sensory thresholds: studies on reproducibility and comparison of balloon composition. Am J Gastroenterol 1994; 89:1185–1190.
29. Rogers J. Testing for and the role of anal and rectal sensation. In: Bailliere's Clinical Gastroenterology: Ed Henry M. Bailliere Tindall (WB Saunders) London, UK, 1992; 6:179–191.
30. Loening V, Read NW, Yamada T. Cerebral evoked potentials after rectal stimulation. Electroenceph Clin Neurophysiol 1991;80–490–495.
31. Accarino AM, Azpiroz F, Malagelada J-R: Selective dysfunction of mechanosensitive intestinal afferents in the irritable bowel syndrome. Gastroenterology 1995;108:636–643.
32. Rao SSC, and Schulze-Delrieu K. Normal gastrointestinal motility of stomach and duodenum. In: An illustrated guide to gastrointestinal motility . (2nd Edition) Kumar D and Wingate DL (Eds), Churchill Livingstone, London UK 1993, pp. 373–392.
33. Bernstein L. and Baker L. A clinical test for esophagitis, Gastroenterology 1958; 34:760–781.
34. Richter JE, Johns DN, Wu WC and Castell DO. Are esophageal motility abnormalities produced during the intraesophageal acid perfusion test? JAMA 1985; 253:1914–1917.
35. Hewson EG, Sinclair JW, Dalton CG, Wo WC, Castell DO and Richter JE. Acid Perfusion Test. Does it have a role in the assessment of noncardiac chest pain. Gut 1989;30:305–310.
36. Rao SSC. Esophageal (NonCardiac) Chest Pain: Visceral Pain: Visceral hyperalgesia, reflux disorder or motor disorder. In: Visceral Pain, GF Gebhart (Ed), IASP Press, Seattle 1997, pp. 351–371.
37. Barbera R, Feinle C, Read NW. Abnormal sensitivity to duodenal lipid infusion in patients with functional dyspepsia. Eur Gastroenterol & Hep 1995; 7:1051–7.
38. Barbera R, Feinle C. Read NW. Nutrient-specific modulation of gastric mechanosensitivity in patients with functional dyspepsia. Dig Dis & Sci 1995, 40.1636–41.
39. Feinle C, Grundy D, Schwizer W, Fried M. Involvement of duodenal receptors in the induction of gastrointestinal (GI) sensations by gastric distension and duodenal lipid. Neurogastroenterol & Motil, 1998;10:A72.
40. Aziz Q, Thompson DG. Brain-gut axis in health and diseas. Gastroenterology 1998;114:559–578.
41. Furlong PL, Aziz Q, Singh K, Holliday I, Barnes I, Harding GFA, Thompson DG. Localisation of cortical centres for human esophageal sensation using magnetoencephalography (abstr). Gastroenterology 1995;108:A726.
42. Loening-Baucke V, Anderson RH, Yamada T, Zhu YX. Study of the afferent pathways from the rectum with a new distention control device. Neurology 1995;45:1510–1516.
43. Loening-Baucke V, Yamada T. Is the afferent pathway from the rectum impaired in children with chronic constipation and encopresis? Gastroenterology 1995; 109–397–403.
44. Aziz Q, Andersson JLR, Valind S, Sundin A, Hamdy S, Jones AKP, Foster ER, Langstrom B, and Thompson DG. Identification of human brain loci processing esophageal sensation using positron emission tomograph. J Gastroenterology 1997;113:50–59.
45. Rosen SD, Paulesu E, Nihoyannopoulos P, Tousoulis D, Frackowiak RS, Frith CD, Jones T, Camici PG. Silent ischemia as a central problem: Regional brain activation compared in silent and painful myocardial ischemia. Ann Intern Med 1996;124:939–949.
46. Munakata J, Silverman DHS, Naliboff B, Matin K, Mayer EA. Anterior cingulate cortical activity levels during expected pain coorelate with sensitized rectal perception thresholds in IBS and healthy subjects. Gastroenterology 1997;112:A794.
47. Enck P, Binkofski F, Schnitzler A, Aziz Q, Freiling T. Functional magnetic resonance brain imaging of the esophagus following distention. Neurogastroenterol & Motil 1998;10;69.

48. Sanger GJ. 5-Hydroxytryptamine and functional bowel disorder. Neurogastroenterol Motil 1996; 8:319–331.
49. Junien JL, Riviere P. The hypersensitive gut-peripheral kappa agonists as a new pharmalogical approach. Alim Pharmacol & Therap 1995;9:117–126.

CLINICAL APPROACH TO NON-CARDIAC CHEST PAIN

Donald O. Castell*

Department of Medicine
Graduate Hospital
Philadelphia, Pennsylvania

INTRODUCTION

What is the likelihood that recurring substernal chest pain might be related to an esophageal abnormality? There are a number of potential mechanisms for what I prefer to call unexplained chest pain because I respect the concept of an entity such as microvascular angina that may be a source of unrecognized heart disease, although not fixed coronary artery disease, in some of these patients. Esophageal etiologies such as GERD or spastic type abnormal contractions must be considered. Finally, we must not forget the interaction of the nervous system, particularly with syndromes such as visceral hyperalgesia.

CASE STUDY

A 52-year-old salesman, was awakened at 2 AM with severe, squeezing chest pain. Previous night at 10 PM he had a hamburger, french fries, and a chocolate shake. He had similar episodes of pain during the past two years. But on this occasion, antacids did not help. Hence, he presented to the emergency room. His father died of a M.I., aged 55 and his brother has angina pectoris. Once during a family argument, after a heavy meal, patient had similar pain which was relieved by a nitroglycerin tablet given to him by his brother. He jogs regularly and occasionally experiences similar pain on exertion O/E , he was anxious but healthy. P.R. = 95/min., chest and abdomen exam was normal. Serial EKG's, cardiac enzymes and coronary arteriogram was normal. Now he is referred for further evaluation of his chest pain.

* Phone: 215-893-7128, Fax: 215-893-2475

Gastrointestinal Motility, edited by Rao *et al.*
Kluwer Academic / Plenum Publishers, New York, 1999.

1. What is your next diagnostic step?

   A. Barium esophagram and UGI.
   B. Endoscopy.
   C. Esophageal motility.
   D. Ambulatory pH study.
   E. Trial of anti-reflux therapy.
   F. Provocative test (Tensilon; Bernstein; or balloon).

2. Upper endoscopy showed mild, patchy erythema around the cardia but no ulceration. He was advised life-style measures and placed on H_2 blockers. Two months later his symptoms are marginally improved. How would you manage this patient?

   A. Double the dose of H_2 blocker.
   B. Repeat endoscopy.
   C. Start proton pump inhibitor.
   D. Order a 24-hour ambulatory pH study.
   E. Order esophageal manometry.

3. Which of the following is the most effective therapeutic option for Noncardiac chest pain:

   A. Acid suppression.
   B. Promotility agents.
   C. Imipramine.
   D. Calcium channel blockers.
   E. Esophageal myotomy.

4. In noncardiac chest pain, which statement is not true?

   A. The estimated annual incidence is 180,000 cases.
   B. The most common esophageal motility disorder is DES.
   C. The most useful diagnostic test is 24-hour pH test.
   D. Patients may have both dysmotility and GERD.
   E. Esophageal motility may be normal.

PREVALENCE

At Graduate Hospital we recently reviewed patients seen in the emergency room complaining of chest pain over a 10-month period to ascertain what percentage of them ruled out for heart disease. The data showed that 76% were ruled out and sent home, either in the emergency department (61%) or after admission (15%) and only 24% actually were shown to have coronary artery disease as a cause of their pain (Figure 1). Thus, the majority of patients seen in an urban emergency room with chest pain are likely to rule out for cardiac disease. An important aspect of this entity was identified in the study by Ockene and colleagues from the University of Massachusetts (1). They followed a group of patients with unexplained chest pain for 16 months after they had a negative angiogram and were told that the pain was not due to their heart. Unfortunately, 51% of them were still unable to work, 47% of them still had limited activity because of their pain, and 44% of them still thought that they had heart disease. So the message to the patient that they do not have heart disease does not necessarily change the impact on their life.

Figure 1. Final diagnosis for 635 patients presenting to the emergency room at Graduate Hospital with chest pain. Note that 389 patients (61 %) had cardiac disease ruled out (r/o) in the emergency room (ER). Of the 246 patients that were initially admitted to the hospital because of their pain, 97 (39% of the total) subsequently were ruled out for cardiac disease, bringing the total patients with unexplained chest pain (UCP) to 486 (76.5%) of the initial 635 presenting to the emergency room. Only 149 patients (23.5%) were considered to have coronary artery disease (CAD) as their final diagnosis; 63 of these patients were diagnosed as angina without myocardial infarction (MI) and 86 were diagnosed as having an acute MI.

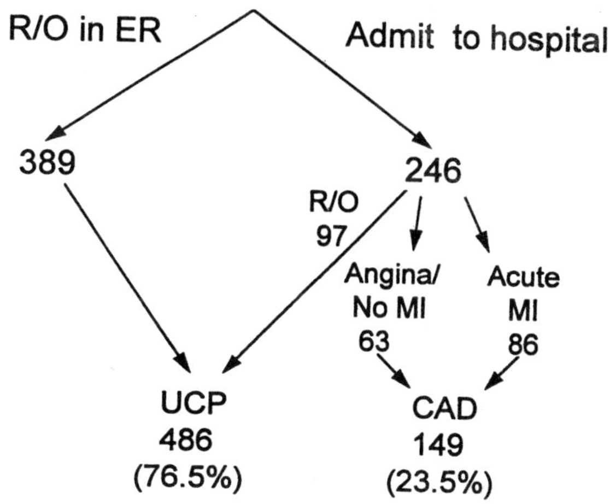

ETIOLOGY

Esophageal Dysmotility

For many years there has been a great deal of interest regarding esophageal dysmotility as a cause of chest pain. Over a century ago, William Osler wrote about spasm of the esophagus in the first edition of his textbook, describing it as a "spasmodic stricture of the gullet seen in hysterical patients and hypochondriacs which may come on abruptly and be associated with emotional disturbances and substernal pain" (2). Thus, Osler suggested 100 years ago that perhaps a spasm in the esophagus might cause pain beneath the sternum and that there may be a psychological component to it. The first motility studies in these patients were reported from the Mayo Clinic by Schmidt and colleagues describing 45 patients with undiagnosed chest pain (3). Most of the patients 35/45 (78%) had some abnormality of their motility study. Esophageal spasm was not very common 6/45 (13%). A hiatal hernia pattern, either a low LES pressure or a double high pressure zone was more common 12/45 (27%) and the hypertensive lower esophageal sphincter was the most common manometric diagnosis reported 14/45 (31%). This manometric abnormality was described from the Mayo Clinic by Code and colleagues in 1960 (4), but it may resemble what we now call a nutcracker esophagus; that is, hypertensive peristaltic waves. Schmidt et al also made the point that they could not clinically distinguish the pain pattern of an ischemic heart event from that of an esophageal motility disorder. Even today, we can not separate pain from the esophagus or the heart with absolute certainty because they share the same innervation.

In 1979, Benjamin, et al first recognized the particular motility pattern of very large peristaltic waves, not the simultaneous contractions of spasms, and used the term nutcracker esophagus (5). From our laboratory, Katz et al reported the largest series of patients with unexplained chest pain (6). After coronary artery disease was excluded, esophageal manometry in these 910 patients revealed that 28% had some abnormality in their motility tracing, defined in Figure 2. The high amplitude peristaltic waves of the nutcracker esophagus were the most common. The category of nonspecific motility disorder, which we do now classify as ineffective esophageal motility, was the second most common. Typical manometric findings of diffuse esophageal spasm were quite unusual; 10% of the 28%, or 2.8% of the entire group of patients. There were a few patients with

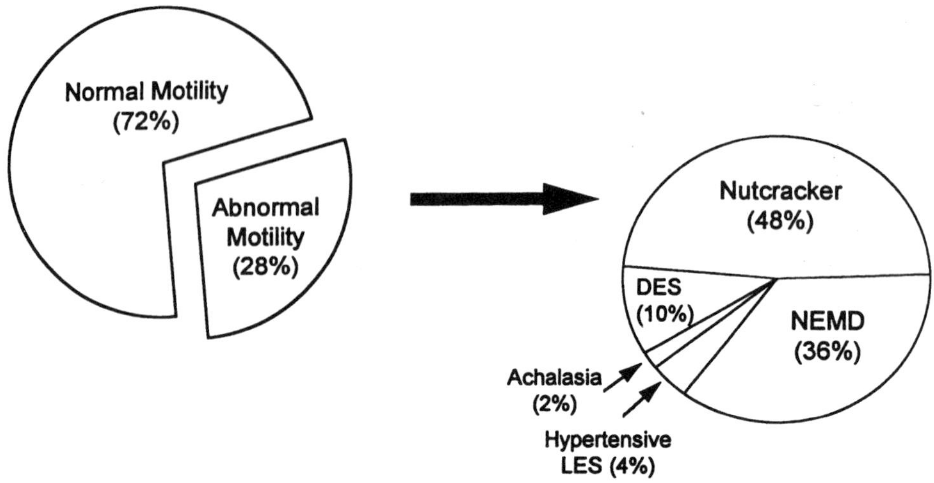

Figure 2. Results of esophageal motility testing in 910 patients with chest pain who were evaluated after coronary artery disease had been excluded. Twenty-eight percent of patients had an abnormal motility test as shown in the pie on the left. The pie diagram on the right illustrates the distribution of motility abnormalities found among these patients. High amplitude peristalsis (nutcracker esophagus) was the most common finding, closely followed by non-specific esophageal motility disorder (NEMD). Note that only 10% of the 28% of patients with abnormal motility (i.e. 2.8% of the entire group of patients with non-cardiac chest pain) had a motility pattern that was considered typical for diffuse esophageal spasm (DES). Achalasia and the hypertensive lower esophageal sphincter (LES) provided the remainder of the manometric abnormalities seen in this study.

hypertensive LES only, but about half of the nutcracker patients also had a hypertensive LES. The major problem with these findings is that most of these patients do not have pain when the manometric study is being done. It is difficult to be sure that the motility pattern is an explanation for the pain when it is found in a pain-free patient; it is only a *probable* cause. This is why investigators over the years have tried to reproduce the pain by provoking it with a Bernstein test, Tensilon, or balloon distention; or have recorded esophageal events over extended periods of time to identify an esophageal abnormality at the time of spontaneous pain. Adding the Bernstein and Tensilon test to motility studies in the 910 patients revealed that only 7% of patients were Bernstein test positive and 24% were positive to Tensilon stimulation, leaving an additional 21% that still had baseline abnormal manometry and negative provocative testing (Figure 3). Because only approximately half the patients had an abnormality, the search continued to find an ambulatory technique that might be applied to identify esophageal events during spontaneous pain episodes.

Excessive Acid Reflux

Intraesophageal pH testing has become a standard diagnostic approach for these patients, using analysis of both the total amount of acid exposure (i.e., pH<4) and the symptom association. Intra-esophageal ambulatory pressure monitoring is only occasionally useful in these patients. DeMeester, et al in 1982 first reported on ambulatory pH monitoring in 50 patients discharged from the coronary care unit with a diagnosis of "noncardiac pain" (7). Forty-six percent had an abnormal pH. Ten years later Hewson and Richter reported on 100 patients with noncardiac chest pain, 87 of whom had pain during the 24-

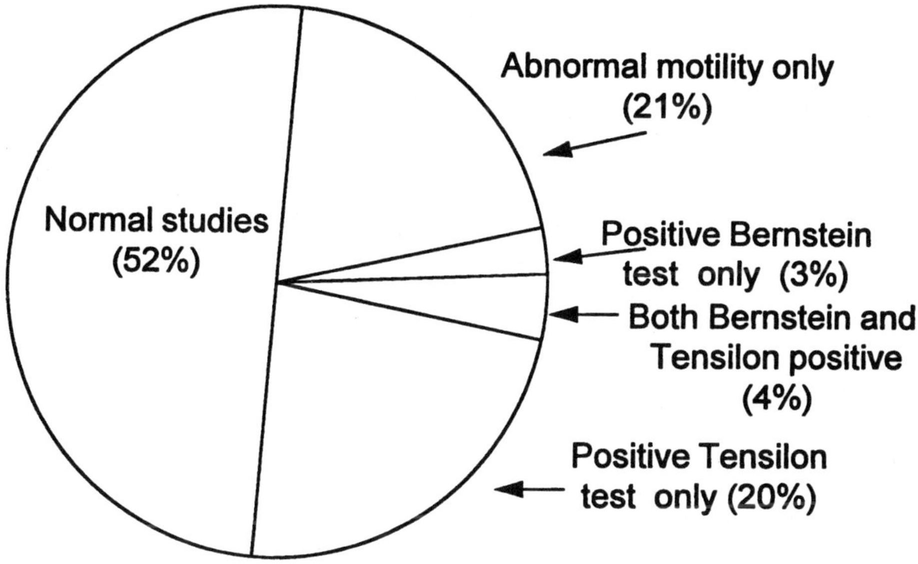

Figure 3. For the same group of 910 patients with non-cardiac chest pain shown in Figure 2, this pie diagram illustrates the frequency of findings when esophageal motility and provocative testing were combined. A positive pain response following edrophonium (Tensilon) injection was noted in 24% of patients, 4% also having a positive acid perfusion (Bernstein) test. Only 7% of patients had a positive Bernstein test with 21% remaining with abnormal motility without a positive pain response to either acid or edrophonium stimulation. Slightly over half of the patients had negative results to esophageal motility and provocative tests.

hour pH study (8). Abnormal total acid exposure was detected in 48%, almost the identical frequency that DeMeester had reported. In addition, a positive symptom index was found in 2 additional patients for a 50% total. Much of the literature from around the world would suggest that approximately 50% of patients with unexplained chest pain may have reflux. In our experience at Graduate Hospital, it is 46%.

Visceral Hyperalgesia

The original study using intra-esophageal balloon distention to identify the pain threshold in 50 patients with unexplained chest pain compared to 30 healthy controls revealed that the patient's cumulative pain curve showed a shift to the left indicating visceral hyperalgesia (9). Using impedance planimetry Rao has shown more recently that patients with unexplained chest pain had pain at lower intra-esophageal pressures than healthy controls (10). In addition, studies using evoked potentials from the human esophagus during esophageal distention showed that patients with unexplained chest pain had their pain at a less clear level of evoked potential than controls (11). It was concluded that the patient had abnormal central CNS processing such that even though they were getting less strong afferent signals to the brain, the patients were perceiving pain when the controls did not. To complete the story and return to Osler and the psychological components, Beitman and colleagues have implicated panic attacks in these patients (12). Evaluating 94 patients with normal coronary angiograms, a structured psychological interview within 24 hours of the angiogram revealed that 1/3 of them met the diagnostic criteria for panic attack.

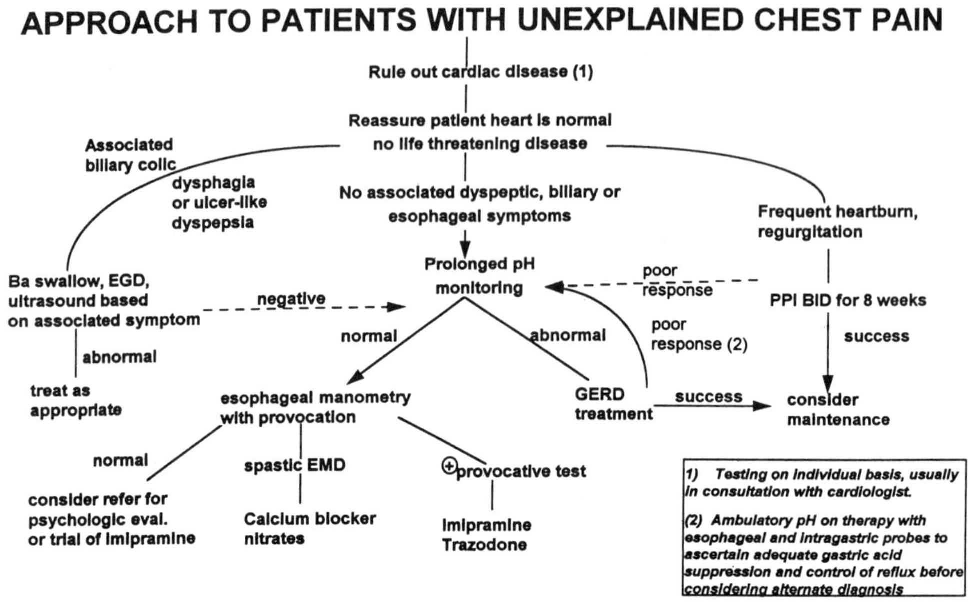

Figure 4. Algorithm for a suggested approach to patients with unexplained chest pain in which various potential therapies (including empiric treatment of reflux) and diagnostic testing are included.

SUMMARY AND MANAGEMENT

The algorithm shown in Figure 4 provides a suggested approach to the patient with chest pain. The first step, of course, is to ensure that the patient does not have coronary artery disease. It is extremely important to reassure the patient that they do not have heart disease and that they do not have a life-threatening problem. If they have definite reflux symptoms, it is appropriate to give a trial of anti-reflux therapy with a proton pump inhibitor twice a day for 8 weeks. If that is successful, a diagnosis of GE reflux is probably correct. If the patient fails to respond, or has no symptoms that suggest reflux with their chest pain, an ambulatory pH study should be performed. If it is abnormal, treat for reflux. If the treatment response is not good, consider repeating the pH study *on medication* to identify those patients who may not have adequate control of acid and who may need a larger dose of PPI. If the pH study is normal, reflux is excluded and manometry and provocative tests are indicated. If a manometry is abnormal indicating DES or a hypercontractile abnormality, a calcium blocker or a nitrate may be effective. If the provocative tests are positive, the patient may have visceral hyperalgesia syndrome and imipramine or trazodone may be effective. If all of these tests are normal one should consider a psychological etiology

ANSWER KEY AND EXPLANATIONS

1. In the absence of dysphagia, barium esophagram is not useful in the evaluation of a patient with chest pain and is much less sensitive for detecting esophagitis compared to endoscopy. Esophageal motility is an option largely with a view to identify an underlying motility disorder such as diffuse spasm or nutcracker

esophagus. But this should be the next step, i.e., one must exclude GERD before proceeding to motility studies. Ambulatory pH is again useful but not when there is endoscopic evidence for reflux disease. Hence this could be considered if endoscopy is negative. Trial of antireflux therapy is appropriate, but in this man with persistent symptoms, one should exclude Barret's esophagus and severe esophagitis as this would alter your treatment plan. Provocative test is an option after excluding GERD. Hence the best option is **B**.

2. Doubling the dose of H_2 blocker may provide more relief but is more costly and less effective than PPI. Repeating endoscopy is unlikely to be beneficial and certainly an expensive option. A 24 hour pH study could help to document the severity of reflux or the adequacy of treatment and could be an option, particularly if one is contemplating surgery. Esophageal manometry may be useful to identify achalasia, ineffective esophageal peristalsis or poor motor function. The best option is to start the patient on PPI therapy as it has been shown to be the most useful drug in the management of GERD. The best option is **C**.
3. In the presence of reflux disease the best options are a PPI or combination of PPI with cisapride. Calcium channel blockers can reduce the amplitude of peristaltic contractions in patients with spasm or nutcracker esophagus but do not influence the symptom of pain in most patients. Imipramine has been shown to be helpful in the treatment of patients with visceral hyperalgesia and is the best option. Esophageal myotomy has been tried in the past for patients with spastic esophageal disorders but has not been shown to be effective in the long term. The best option is **C**.
4. All of the statements are true except option B. Diffuse esophageal spasm may account for 10% or less number of patients presenting to an esophageal motility laboratory with chest pain. Normal or nonspecific dysmotility is more common. The best option is **B**.

REFERENCES

1. Ockene IS, Shay MJ, Alpert JS, et al: Unexplained chest pain in patients with normal coronary arteriograms: A follow-up study of functional status. N Engl J Med 1980; 303:1249–1252.
2. Osler W: Spasm of the oesophagus. In: The Principles and Practice of Medicine. W Osler (ed). Appleton and Co. New York. 1892. pp. 340–341.
3. Schmidt CD, Jones HD, Hunt JC, et al: The value of the esophageal motility test in evaluation of thoracic pain problems. Dis Chest 1962; 41:303–314.
4. Code CF, Schlegel JF, Kelly ML, et al: Hypertensive gastroesophageal sphincter. Proc Mayo Clinic 1960; 35:391–399.
5. Benjamin SB, Gerhardt DC, Castell DO: High amplitude, peristaltic esophageal contractions associated with chest pain and/or dysphagia. Gastroenterology 1979; 77:478–483.
6. Katz PO, Dalton CB, Richter JE, Wu WC, Castell DO: Esophageal testing in patients with non-cardiac chest pain and/or dysphagia. Ann Intern Med 1987; 106:593–597.
7. DeMeester TR, O'Sullivan GC, Bermudez G, et al: Esophageal function in patients with angina-type chest pain and normal coronary angiograms. Ann Surg 1982; 196:488–498.
8. Hewson EG, Sinclair JW, Dalton CB, Richter JE: Twenty-four hour esophageal pH monitoring: The most useful test for evaluating non-cardiac chest pain. Am J Med 1991; 90:576–583.
9. Richter JE, Barish CF, Castell DO: Abnormal sensory perception in patients with esophageal chest pain. Gastroenterology 1986; 91:845–852.
10. Rao S, Gregersen H, Hayek B, Summers R, Christensen J: Unexplained chest pain: The hypersensitive, hyperreactive and poorly compliant esophagus. Ann Intern Med 1996; 124:950–958.

11. Smout A, DeVore M, Dalton C, Castell D: Cerebral potentials evoked by esophageal distention in patients with noncardiac chest pain. Gut 1992; 33:298–302.
12. Beitman BD, Basha I, Flaker G, et al: Atypical or nonanginal chest pain. Panic disorder or coronary artery disease? Arch Intern Med 1987; 147:1548–1552.

28

CLINICAL APPROACH TO NON-ULCER DYSPEPSIA

Mark Pimentel and Henry C. Lin

GI Motility Program
Department of Medicine
Cedars-Sinai Medical Center and
UCLA School of Medicine
Los Angeles, California, 90048

INTRODUCTION

The first modern description of dyspepsia (1), described it as an "episodic or persistent" abdominal pain in the absence of bleeding or jaundice. Since then, various descriptions have been used (2, 3). These definitions include symptoms of peptic ulcer disease in a subject without an ulcer (2) and upper abdominal pain with a lack of identifiable cause (3). In 1989, Barbara, et al. once again tried to define NUD (dyspepsia) as episodic or persistent abdominal symptoms often related to feeding, which patients or physicians believe to be due to disorders of the proximal portion of the digestive tract (4). More recently, groups have tried to sub-categorize NUD into categories such as reflux-like, dysmotility-like, ulcer-like (5,6) and even irritable bowel-like (7).

These attempts at categorization and sub-categorization of dyspeptic patients illustrate the frustration with studying and treating patients with NUD. As there is tremendous variation in the definition of NUD, studies become very difficult to interpret due to heterogeneity of study populations. Between 1995 and 1998, when "non-ulcer dyspepsia" is searched for on Medline, 14 out of 49 references (29%), are review articles (6, 8–20). The large number of reviews highlights the existence of varying opinions and inconsistent interpretations of a complicated literature. It also reflects the frustration faced by practicing gastroenterologists who are caring for patients with symptoms that cannot easily be categorized.

Notwithstanding all the ongoing efforts to establish a definition and treatment strategy, cost-effective medicine has been the focus of the 1990's. This is important in diseases that lack clear characterization such as NUD. This chapter will attempt to cover the relevant literature in support of the etiology, appropriate work up, management and outcomes of patients suffering with NUD.

Gastrointestinal Motility, edited by Rao *et al.*
Kluwer Academic / Plenum Publishers, New York, 1999.

CASE STUDY

A 28-year-old female presents with 11-year history of nausea/vomiting and epigastric distress. Previously she was evaluated in the East Coast with UGI X-ray and gastric emptying study which were normal.

Subsequently, she underwent a negative exploratory laparotomy. She has continued to complain of daily nausea and vomiting accompanied by epigastric pain (even on an empty stomach). There is no history of heartburn. Recently, she was treated for volume depletion. She has lost 12 lbs over the past few weeks. Her current medications include cisapride 10 mg TID and lansoprazole 30 mg QD. PE: She is thin. Abdomen exam was unremarkable.

1. What is your next step in the management of this patient?

   A. Abdominal ultrasound.
   B. Upper gastrointestinal endoscopy.
   C. CT scan of abdomen.
   D. Increase current dose of medications.

2. EGD was normal. Her symptoms persisted even though she has stopped all PO intake. At this stage, which of the following options would you chose?

   A. Gastric emptying study.
   B. 24-hour pH study.
   C. Antroduodenal manometry.
   D. Electrogastrography.

3. A 24 hour pH study showed that a pH < 4.0 was present in the lower esophagus for 42.1% of the total period of study (normal $< 5\%$). At this stage, which of the following treatment options would you recommend?

   A. Increase lansoprazole.
   B. Switch to omeprazole.
   C. Switch from proton pump inhibitor to H2 blocker.
   D. Increase cisapride.

EPIDEMIOLOGY AND NATURAL HISTORY OF NUD

Functional disorders of the gastrointestinal tract are extremely common in the general population. Research indicates that up to 40% of visits to gastroenterologists can be categorized as functional disorders (21–23). In a 1980 study, Thompson, et al., administered questionnaires to 301 normal subjects to establish the prevalence of functional disorders in the normal population. Approximately 30% of patients deemed normal were found to have symptoms compatible with functional gastrointestinal disease (24). Therefore, functional symptoms are extremely common. Based on other studies, from 30 to 70% of the normal population have symptoms consistent with functional bowel disorders at some point in time (7, 24, 25). These data also suggest that the symptoms and their reporting are not consistent (25). Accordingly, it is very difficult to accurately determine the natural history of NUD in a population.

CLASSIFICATION SYSTEMS FOR NUD

Efforts have recently been made to sub-classify NUD. Two of these endeavors include the Colins-Jones Criteria (6) and the Rome Working Group (26). These classifications systems were undertaken to narrow the criteria and purify the grouping of NUD from a more heterogeneous array of upper GI symptoms. From the Rome Working Group, the classification included a reflux-like, motility-like and a general category (26). In another study of 213 patients with dyspepsia, 64% had ulcer-like, 31% had dysmotility-like and 38% had reflux-like dyspepsia by preset definitions (7). In this study, there was a tremendous overlap in symptoms which made the categories difficult to use. In addition, 29–41% of patients in these subgroups also met the criteria for IBS.

The potential importance of such assignments is to improve the success of therapy by customizing treatment for patients with different pathophysiology for their symptoms and to reduce the cost of workup. The ability to better identify patients (by clinical assessment alone) with peptic ulcer disease, reflux or other motility disorder and target them for an appropriate diagnostic test could be a significant cost saving. Yet as the studies imply, these subclassifications of NUD are not able to achieve these goals since the complaints retain significant overlap.

PATHOPHYSIOLOGY

As emphasized in the previous pages, NUD is a heterogeneous group of conditions with poorly understood etiologies. The most commonly attributed causes of NUD include, a motility disorder, visceral hypersensitivity, gastroesophageal reflux disease and an association between NUD and *Helicobacter pylori*. To date, none of these associations can be used to explain all patients with NUD. However, motility disorders and GERD appear to be well represented. The following is a description of the literature to support these hypotheses.

Disordered Upper Gastrointestinal Motility

In the last 20 years, multiple studies have tried to define motility disturbances in patients with NUD. In 1985, Malagelada, et al. performed antroduodenal motility studies on 104 patients referred to the Mayo Clinic with NUD (27). Of these patients, 21 were found to have normal motility studies despite persistence of symptoms. A second group of 43 patients had abnormal gastric motility with normal intestinal motility. Some of these gastric motility disturbances included normal frequency but high amplitude contractions, normal amplitude but high frequency contractions and both frequency and amplitude disturbances. A third group was defined as having abnormalities in both gastric and intestinal motility measurements. Table 1 outlines these reported disturbances. In total, 80% of patients with NUD had poorly understood, but clearly detectable motility abnormalities. Other authors have shown similar motility abnormalities associated with NUD (28,29). The same study also showed that 62% of patients without identifiable motility abnormalities reported a psychiatric history. This is compared to 23% and 3% in patients with only gastric abnormalities and both intestinal and gastric changes. Therefore, in NUD without an identifiable organic cause, a psychiatric illness may be another contributor to dyspeptic symptoms.

In addition to the manometric alterations seen in patients with NUD, disordered gastric emptying is also found in some subgroups. In patients with symptoms consistent with NUD 30–65% have abnormally delayed gastric emptying (30–32). As with all studies in NUD the definition of NUD in each of these three studies differed making comparison impossible. However, the results were all consistent.

Electrogastrographers have also tried to identify subgroups of patients with NUD and characterize abnormalities in slow wave rhythm disturbances. In one study of adults

Table 1. Manometric abnormalities in subjects with NUD

Manometric group (number of subjects)	Abnormalities detected (number of subjects)
1. No abnormalities (21).	No abnormalities detected.
2. "No abnormalities" (8).	Simultaneous contractions in antrum and duodenum associated with symptoms.
3. Gastric motility normal/Intestinal motility normal (43).	a. Normal frequency but decreased amplitude of antral contractions (14).
	b. Normal amplitude but decreased frequency of antral contractions (8).
	c. Abnormal amplitude and frequency of antral contractions (21).
4. Abnormal gastric and intestinal motility (32).	a. Bursts of tonic and clonic contractions (13).
	b. Abnormal or absent MMC (13).
	c. Constant duodenal phasic contractions (3).

Adapted from data in Malagelada JR, Stanghellini V. (Reference 27)

with NUD, 68.75% had episodes of dysrhythmias as compared to 20% in controls (33). Others have confirmed the high rate of gastric arrhythmias in these patients (34–37). In patients with true peptic ulcer disease, however, tachygastria and a decrease in dominant frequency have also been reported (38). These changes were restricted to those patients presenting with nausea and vomiting as their primary complaint. Unfortunately, until EGG is better established as a diagnostic tool its availability will be limited to specialized centers and research protocols.

Lastly, colonic motility has also been assessed in an Indian study, (39). Twenty-five subjects with NUD had a transit time of 9 hours as compared to that of 15.8 hours in matched controls (39). The authors speculate that NUD may be a generalized motility disorder rather than limited to the upper gastrointestinal tract.

Gastroesophageal Reflux

The literature looking at the association with gastroesophageal reflux disease (GERD) and NUD is more consistent and easier to interpret. Approximately 20–60% of patients with NUD have GERD (40,41). The Rome Working Group concluded that GERD should be sought in patients with NUD (26). Studies have also suggested that H_2-receptor antagonists improve certain subsets of patients with NUD (42–45). In these studies GERD was not determined to be present by objective means in order to exclude these patients. It is therefore suggested that at least the portion of patients with NUD who respond to acid suppressive therapy may be patients with GERD. Since a 24-hour pH study has a sensitivity of 96% and specificity of 98% for the diagnosis of GERD patients (46), it is an important test when NUD is suspected. Additionally, as patients with GERD have little or no luminal findings during endoscopy (47–49) and histology may not be revealing (50), a negative endoscopy would not rule out GERD in these patients.

The majority of all previous research in NUD has been conducted without the documentation of GERD by 24-hour pH testing. Although there is data on the empiric use of H_2-receptor antagonists in NUD (42–45), there is little data on the use of more effective acid eradication with proton pump inhibitors, which has become the standard treatment for GERD. In the non-cardiac chest pain literature, GERD is also seen in as high as 62% of patients. (51–55). In this group of patients, omeprazole is found to be very effective treatment for the 62% of patients with confirmed GERD (56).

The role of GERD as a subgroup of NUD appears clear. A reasonable approach to NUD patients would therefore include a 24-hour pH test. However, a trial of acid-suppressive therapy may also be attempted such as the omeprazole test in non-cardiac chest pain (54). Perhaps an appropriate work up for GERD will eliminate many patients considered to have functional dyspepsia.

Visceral Perception in NUD

In recent years, visceral afferent pathways and their aberrations have been addressed in many of the functional disorders. The majority of work to date has come from the irritable bowel syndrome literature. The limited work in patients with functional dyspepsia have shown that these patients have decreased sensory thresholds for gastric balloon distension than controls (56–59). Up to a 4-fold increase in abdominal discomfort was found in relationship to gastric balloon distension in patients (56).

Recently, investigators have assessed the relationship between the response to luminal nutrients and gastric sensitivity in subjects with NUD. In a seminal study, intraduodenal infusion of lipids during gastric balloon distension lowered sensory thresholds in patients with NUD compared to controls and to NUD patients infused with intraduodenal saline infusion (60). The significance of this observation is that patients with NUD frequently have post-prandial symptoms.

Visceral sensitivity as a possible mechanism for NUD brings the promise of new pharmacologic intervention. In the last two years, new peripheral acting opioid agonists have offered hope to researchers. Specifically, fedotozine, a κ-agonist has been used successfully in alleviating symptoms in these patients (61–63). However, it is not effective for all patients.

Helicobacter and NUD

The discovery of *Helicobacter pylori* (HP) as a cause of peptic ulcer disease in the 1980's sparked enthusiasm about a possible role of this organism in NUD. In a summary of this literature by Talley (64), results of published trials of *H. pylori* eradication in NUD are mixed. Half of the trials reviewed imply a benefit of eradication on NUD symptoms, while the remaining suggest no effect. Many of the responses to *H. pylori* eradication may be related to the beneficial effect of concomitantly administered acid suppressive drugs in patients where GERD was not definitively excluded. Any literature on HP eradication must also take into account the high placebo effect in functional disorders which will be discussed later in this chapter.

DIAGNOSIS OF NUD

Figure 1 summarizes the overlap of conditions with NUD. Given this heterogeneity, patients need various investigations to rule out obvious causes. This poses a substantial cost issue. A sensible strategy for NUD work up is shown in Figure 2. In a recent decision analysis (65), endoscopy was compared to empiric therapy for NUD. They found that the additional costs associated with an initial endoscopy would be a mere $39.90, indicating that endoscopy could be used as the first step. Additionally, as GERD could represent as much as 60% of NUD, 24-hour pH testing would be an important part of the workup. Once GERD has been ruled out, the minority of NUD could then be considered for more intense investigations such as motility and visceral testing.

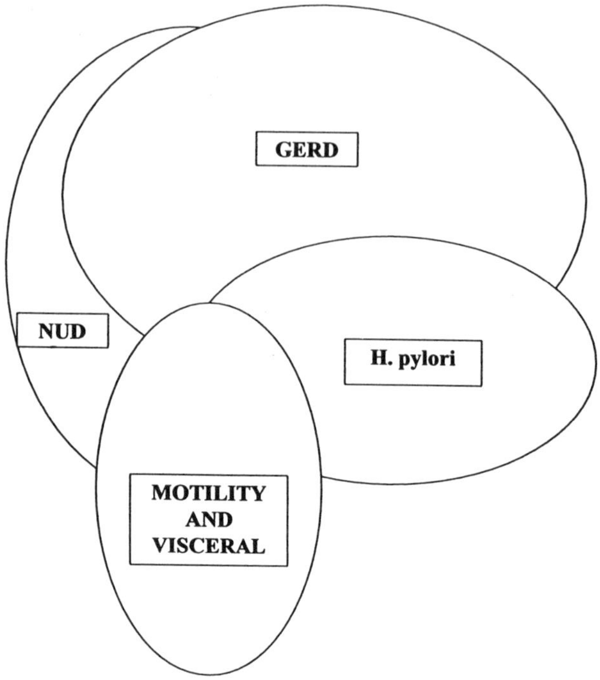

Figure 1. The interrelationships of NUD.

TREATMENT OF NUD

Since NUD is a complex entity with multiple potential causes, any single therapy would not be effective in the entire cohort. One of the key problems in studying any treatment of NUD is the placebo effect. In studies of *H. pylori* treatment in NUD, the placebo effect is as high as 70% (66–69). Without a better understanding of the underlying mechanism of NUD that allows us to select a customized treatment for patients, it is not surprising that no treatment to date is considered highly effective for NUD.

In addition, since 40–60 % of patients with NUD have GERD on 24-hour pH testing it is likely that the same response rate should be seen using therapies directed at acid suppression. This response to empiric acid-suppressive therapy in patients with NUD has been shown to be associated with GERD (70).

In consideration of NUD as a motility problem, attempts have been made to use prokinetic agents in the management of this condition. Once again, however, the exclusion of GERD in these patients was not undertaken since patients with GERD can respond to prokinetic drugs. The possibility of a beneficial effect independent of GERD must still be pursued since prokinetic drugs such as domperidone or cisapride substantially improve symptoms that are not GERD-like (71, 72). This effect also occurred in certain groups but not others (72), implying the heterogeneity of the population that are responsive to prokinetic therapy.

For the presumed visceral afferent dysfunction component of NUD, investigational treatment has turned to peripheral opioid agonists for symptom control. This therapy appears to be very successful based on preliminary studies (61–63). The concern with this therapy remains the identification of patients who are likely to respond. Do these opioid

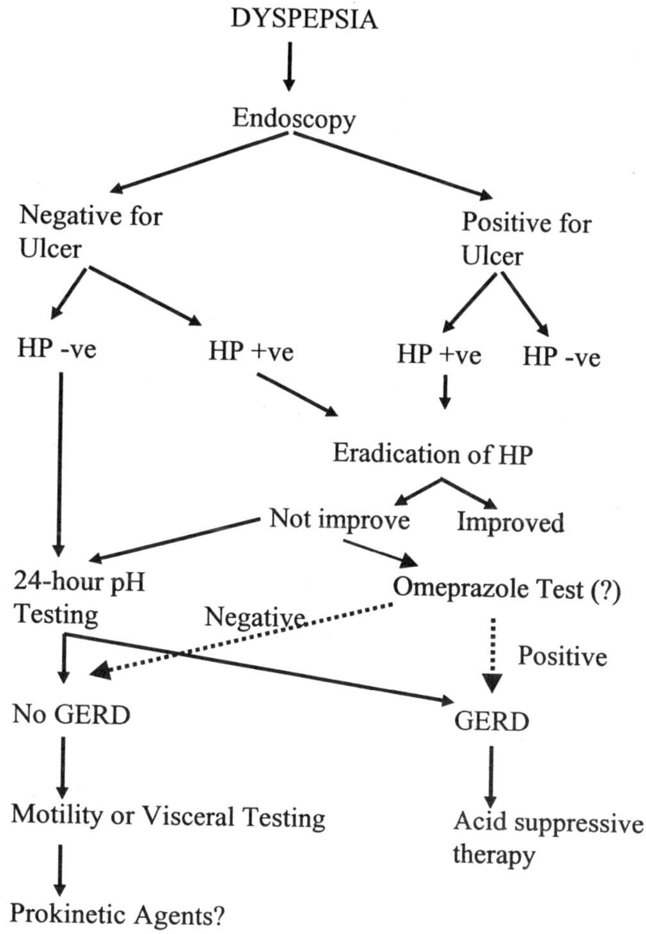

Figure 2. Possible future strategy in the workup and management of non-ulcer dyspepsia.

agonists simply relieve symptoms or do they treat the underlying condition? This has yet to be considered. A proposed algorithm for the evaluation and treatment is shown in Figure 2.

CONCLUSIONS

It is clear from the literature that NUD remains a complex and controversial topic. As is typical of any chronic condition without adequate therapy, it remains frustrating for gastroenterologists trying to manage these complaints. What seems clear is that patients with NUD are a heterogeneous group and certainly include a group of GERD, visceral hypersensitivity, peptic related/ Helicobacter related and patients with poorly characterized (primary or secondary) motility abnormalities. The future of NUD requires better clinical criteria for separation of these subgroups of NUD and to establish validated cost-effective algorithms for the diagnosis and customized management for them.

ANSWER KEY AND EXPLANATIONS

1. Because the patient is already receiving maximum possible therapy for acid peptic or GERD and yet has persistent symptoms, you are obliged to reevaluate this

patient. She has already had a negative laparotomy and hence a serious organic disease such as gall stones or pancreatic cancer would be most unlikely. Therefore an ultrasound or CAT scan would be superfluous. An endoscopy would be reasonable although it is unlikely that one would find serious disease. Antroduodenal manometry would be an appropriate test for this patient but this option is not given. Similarly excessive reflux is also a possibility. The best option is **B.**

2. A gastric emptying study was performed previously and was reported to be normal. Hence it would be unlikely to be abnormal once again. Antroduodenal manometry would be an appropriate test but is not widely available. Another possibility here is refractory reflux disease. Therefore a 24 hr pH test which is more widely available would be a reasonable option. EGG is not available widely and at present still remains a research tool. The best option is **B**.
3. Increasing lansoprazole would be an appropriate option since she appears to have significant reflux. Occasionally switching to another PPI may be useful although there are no studies that show benefits of one over the other. It is useful to remind patients to take PPI, at least one-half hour before meals as it is most effective then. Option C and D are unlikely to help and could potentially worsen her symptoms. Because she is a young person with refractory symptoms, surgical options such as Nissen –fundoplication may also be appropriate. The best option is **A**.

REFERENCES

1. Crean GP, Land WI, Beattie AD, et al. Ulcer-like dyspepsia. Scand J Gastroenterol 1982;17(Suppl 79):9–15.
2. Thompson WE. Non-ulcer dyspepsia. Can Med Assoc J 1984;13:565–9.
3. Lagarde SP, Spiro HM. Non-ulcer dyspepsia. Clin Gastroenterol 1984;13:437–46.
4. Barbara L, Camilleri M, Corinaldesi R, et al. Definition and investigation of dyspepsia: Consensus of an ad hoc working party. Dig Dis Sci 1989;34:1272–76.
5. Colin-Jones DG. Management of dyspepsia: report of a working party. Lancet 1988;1:576–9.
6. Colin-Jones DG. Dyspepsia Update. Scand J Gastroenterol 1995;30 (Suppl 210);32–5.
7. Talley NJ, Zinsmeister AR, Schleck CD, Melton LJ. Dyspepsia and dyspepsia subgroups: A population-based study. Gastro 1992:102;1259–68.
8. Sahay P, Axon AT. Non-ulcer dyspepsia: Does Helicobacter matter. Postgrad Med J 1995;71:262–4.
9. Buckley M, O'Morain C. Prevalence of Helicobacter pylori in non-ulcer dyspepsia. Aliment Pharm Ther 1995;9:53–8.
10. Porro CG, Parente F. Nature of non-ulcer dyspepsia and related conditions. Baillieres Clin Gastro 1995;9:549–62.
11. Talley NJ. Modern management of dyspepsia. Aust Fam Phys 1996;25:47–52.
12. O'Morain C, Buckley M. Helicobacter pylori and dyspepsia. Scand J Gastroenterol 1996;Suppl 214:28–30.
13. Armstrong D. Helicobacter pylori infection and dyspepsia. Scand J Gastroenterol 1996;Suppl 215:38–47.
14. Talley NJ. Helicobacter pylori and non-ulcer dyspepsia. Scand J Gastroenterol 1996;Suppl 220:19–22.
15. Laheji RJ, Jansen JB, van de Lisdonk E, et al. Review article: Symptom improvement through eradication of Helicobacter pylori in patients with non-ulcer dyspepsia. Aliment Pharm Ther 1996;10:843–50.
16. Veldhuyzen van Zanten SJ. The role of Helicobacter pylori infection in non-ulcer dyspepsia. Aliment Pharm Ther 1997;11(Suppl 1):63–9.
17. Bobrzynski A. Hormonal, secretory and morphological alterations in gastric mucosa in the course of Helicobacter pylori eradication in patients with duodenal ulcer and non-ulcer dypepsia. J Physiol Pharm 1997;48(Suppl 3):1–56.
18. Froelich F, Gonvers JJ, Beglinger C. Has the time now arrived to eradicate Helicobacter pylori routinely in non-ulcer dyspepsia. Eur J Gastro Hep 1997;9:917.
19. de Boer WA. Helicobacter pylori and non-ulcer dyspepsia. Scand J Gastroenterol 1997;32:535–40.
20. Agreus L, Talley NJ. Dyspepsia: Current understanding and management. Ann Rev Med 1998;49:475–93.

21. Thompson WG, Dotevall G, Drossman DA, et al. Irritable bowel syndrome: guidelines for the diagnosis. Gastroenterol Int 1989;2:92–95.
22. Drossman DA, Thompson WG, Talley NJ, et al. Identification of sub-groups of functional gastrointestinal disorders. Gastroenterol Int 1990;3:159–72.
23. Sandler RS. Epidemiology of irritable bowel syndrome in the US. Gastro 1990:99:409–15.
24. Thompson WG, Heaton KW. Functional bowel disorders in apparently healthy people. Gastro 1980:79:283–8.
25. Talley NJ, Weaver AL, Zinsmeister AR, et al. Onset and disappearance of gastrointestinal symptoms and functional gastrointestinal disorders. Am J Epidemiol 1992;136:165–77.
26. Talley NJ, Colin-Jones D, Koch KJ, et al. Functional dyspepsia: a classification with guidelines for diagnoses and management. Gastroenterol Int 1991;4:145–60.
27. Malagelada JR, Stanghellini V. Manometric evaluation of functional upper gut symptoms. Gastro 1985;88:1223–31.
28. Rees WDW, Miller LJ, Malagelada JR. Dyspepsia, antral motor dysfunction, and gastric stasis of solids. Gastro 1980;78:360–65.
29. Kerlin P. Postprandial antral hypomotility in patients with idiopathic nausea and vomiting. Gut 1989;30:54–9.
30. Scott AM, Kellow JE, Shuter B, et al. Intragastric distribution and gastric emptying of solids and liquids in functional dyspepsia. Dig Dis Sci 1993;12:2247–54
31. Chang CS, Chen GH, Kao CH, et al. Gastric clearance of radiopaque markers in non-ulcer dyspepsia patients. Scand J Gastroenterol 1996;31:136–9.
32. Mittal BR, Dhiman RK, Maini A, et al. Gastric emptying in patients with non ulcer dyspepsia (dysmotility type). Trop Gastroenterol 1997;18:67–9.
33. Lopez GA, Lopez DGA, Samiento A, Andrusch A. Abnormalities in gastric mechanic sensitivity, gastric emptying and electrogastrography in non organic dyspepsia (NOD). Acta Gastro Latinoamer 1997;27:7.
34. Cucchiara S, Riezzo G, Minella R, et al. Electrogastrography in non-ulcer dyspepsia. Arch Dis Child 1992;67:613.
35. Pfaffenbach B, Adamek RJ, Bartholomaus C, Wegener M. Gastric dysrhythmias and delayed gastric emptying in patients with functional dyspepsia. Dig Dis Sciences 1997;42:2094.
36. Parkman HP, Miller MA, Trate D, et al. Electrogastrography and gastric emptying scintigraphy are complementary for assessment of dyspepsia. J Clin Gastro 1997;24:214.
37. Thor P, Lorens K, Tabor S, et al. Dysfunction in gastric myoelectrical and motor activity in Helicobacter pylori positive gastritis patients with non-ulcer dyspepsia. J Physiol Pharm 1996;47:469.
38. Geldof H, van der Schee EJ, Smout AJPM, et al. Myoelectrical activity of the stomach in gastric ulcer patients: An electrogastrographic study. J Gastro Motility 1989;1:122.
39. Nabar AA, Bhatia SJ, Abraham P, et al. Total and segmental colonic transit time in non ulcer dypepsia. Ind J Gastro 1995;12:131–3.
40. Pfeiffer A, Aronbayev J, Schmidt T, et al. Gastric emptying, esophageal 24-hour pH and gastric potential difference measurements in non-ulcer dyspepsia. Gastroenterol Clin Biol 1992;16:395–400.
41. Klauser AG, Voderholzer WA, Knesewitsch PA, et al. What is behind dyspepsia? Dig Dis Sci 1993;38:147–54.
42. Gotthard R, Bodemar G. Brodin U, et al. Treatment with cimetidine, antacid, or placebo in patients with dyspepsia of unknown origin. Scand J Gastroenterol 1988;23:7–18.
43. Johannessen T, Fjosne U, Kleveland PM, et al. Cimetidine responders in non-ulcer dyspepsia. Scand J Gastroenterol 1988;23:327–36.
44. Farup PG, Larsen S, Ulshagen K, et al. Ranitidine for non-ulcer dyspepsia. A clinical study of the symptomatic effect of ranitidine and a classification and characterization of the responders to treatment. Scand J Gastroenterol 1991;26:1209–16.
45. Johannessen T, Petersen H, Kristensen P, et al. Cimetidine on-demand in dyspepsia. Experience with randomized controlled single-subject trials. Scand J Gastroenterol 1992;27:189–95.
46. Jamieson JR, Stein HJ, DeMeester TR, et al. Ambulatory 24-H Esophageal pH monitoring: Normal values, optimal threshold, specificity, sensitivity, and reproducibility. Am J Gastro 1992;87:1102–11.
47. Heading RC. Epidemiology of oesophageal reflux disease. Scand J Gastroenterol 1989;24(suppl 168):33–7.
48. Schindlbeck NE, Klauser AG, Berghammer G, et al. Three year follow up of patients with gastrooesophageal reflux disease. Gut 1992;33:1016–9.
49. Klauser AG, Voderholzer WA, Knesewitsch PA. Et al. What is behind dyspepsia? Dig Dis Sci 1993;38:147–54.

50. Schindlbeck NE Wiebecke B, Klauser AG, et al. Diagnostic value of histology in non-erosive reflux disease. Gut 1996;39:151–4.
51. Hewson EG, Sinclair JW, Dalton CB, et al. Twenty-four hour esophageal pH monitoring. The most useful test for evaluating noncardiac chest pain. Am J med 1991;90:576–83.
52. Janssen J, Vantrappen G, Ghillibert G. 24-hour recording of esophageal pressure and pH in patients with noncardiac chest pain. Gastro 1986;90:1978–84.
53. Decaestecker JS. Blackwell JN, Brown J et al. The oesophagus as a cause of recurrent chest pain: which patients should be investigated and which tests should be used? Lancet 1985;2:1143–46.
54. Fass R, Fennerty B, Ofman JJ, et al. The clinical and economic value of a short course of omeprazole in patients with noncardiac chest pain. Gastro 1998;115:42–9.
55. Achem SR, Kolts BE, MacMath T, et al. Effects of omeprazole versus placebo in treatment of noncardiac chest pain and gastroesophageal reflux. Dig Dis Sci 1997;42:2138–45.
56. Mearin F, Cucala M, Azpiroz F, et al. The origin of symptoms on the brain-gut axis in functional dyspepsia. Gastro 1991;101:999–1006.
57. Lemann M, Dederding JP, Flourie B, et al. Abnormal perception of visceral pain in response to gastric distension in chronic idiopathic dyspepsia. Dig Dis Sci 1991;36:1249–54.
58. Bradette M, Pare P, Douville P, et al. Visceral perception in health and functional dyspepsia. Dig Dis Sci 1991;36:52–58.
59. Coffin B, Azpiroz F, Guarner Et al. Selective gastric hypersensitivity and reflex hyporeactivity in functional dyspepsia. Gastro 1994;107:1345–51.
60. Barbera R, Feinle C, Read NW. Nutrient-specific modulation of gastric mechanosensitivity in patients with functional dyspepsia. Dig Dis Sci 1995;40:1636–41.
61. Fraitag B, Homerin M, Hecketsweiler P. Double-blind dose-response multicenter comparison of fedotozine and placebo in treatment of nonulcer dyspepsia. Dig Dis Sci 1994;39:1072–77.
62. Dapoigny M, Abitbol JL, Fraitag B. Efficacy of peripheral kappa agonist fedotozine versus placebo in treatment of irritable bowel syndrome: A multicenter dose response study. Dig Dis Sci 1995;40:2244–48.
63. Read NW, Abitbol JL, Bardhan KD, et al. Efficacy and safety of the peripheral kappa agonist fedotozine versus placebo in the treatment of functional dyspepsia. Gut 1997;41:664–8.
64. Talley NJ. A critique of therapeutic trials in Helicobacter pylori-positive functional dyspepsia. Gastro 1994;106:1174–83.
65. Silverstein MD, Petterson T, Talley NJ. Initial endoscopy or empirical therapy with or without testing for Helicobacter pylori for dyspepsia: a decision analysis. Gastro 1996;110:72–83.
66. McNulty CAM, Gearty JC, Crump B, et al. *Campylobacter pyloridis* and associated gastritis: Investigator blind placebo-controlled trial of bismuth salicylate and erythromycin ethyl-succinate. BMJ 1986;293:645–9.
67. Lambert JR, Dunn K, Borromeo M, Korman MG, et al. *Campylobacter pylori*-a role in non-ulcer dyspepsia? Scand J Gastroenterol 1989;24(Suppl 160):7–13.
68. Loffeld RJLF, Potters HVJP, Toberingh E, et al. Campylobacter associated gastritis in patients with non-ulcer dyspepsia: A double blind placebo controlled trial with colloidal bismuth subcitrate. Gut 1989;30:1206–12.
69. Frazzoni M, Lonardo A, Grisendi A, et al. Are routine duodenal and antral biopsies useful in the management of "functional" dyspepsia? A diagnostic and therapeutic study. J Clin Gastroenterol 1993;17:101–8.
70. Farup PG, Hovde O, Breder O. Are frequent short gastro-oesophageal reflux episodes the cause of symptoms in patients with non-ulcer dyspepsia responding to treatment with ranitidine. Scand J Gastro 1995;30:829–32.
71. Urbain JC, Siegel JA, Debie NC, Pauwels SP. Effect of cisapride on gastric emptying in dyspeptic patients. Dig Dis Sci 1988;33:779–83.
72. Cutts TF, Abel TL, Karas JG, Kuns J. Symptom improvement from prokinetic therapy corresponds to improved quality of life in patients with severe dyspepsia. Dig Dis Sci 1996;41:1369–78.

29

CLINICAL APPROACH TO COLORECTAL PAIN

William J. Snape, Jr.*

University of California, Irvine
Director, Bowel Disease and Motility Center
Long Beach Memorial Medical Center

INTRODUCTION

Chronic abdominal pain (>3 months) is a common symptom of functional colonic disease occurring in more than 20% of the population (1). Most often the pain presents in the left lower quadrant, but it also occurs in the right lower quadrant. However, the abdominal pain can be over any part of the anatomic location of the colon. The prevalence of abdominal pain associated with functional disorders of the gastrointestinal tract and visceral hypersensitivity is between 20% and 30% of otherwise healthy people (1).

CASE STUDY

A 28-year-old woman presents with a chief complaint of moderately severe {7 out of 10} left lower quadrant abdominal pain associated with constipation for the past year. The pain is worsened by eating and is improved following a bowel movement. She has had right shoulder pain and left hip pain for the past six months.

1. The next step in the work-up is?

   A. Flexible sigmoidoscopy.
   B. Colonoscopy.
   C. Barium enema.
   D. CBC, sedimentation rate, thyroid function test, stool hemoccult.

2. The CBC {hematocrit = 30%, white count 5200 per cubic mm}, sedimentation rate {15 mm}, and stool hemoccult were normal excluding systemic arthritis

* Phone: 562-595-5421, Fax: 562-426-2862

Gastrointestinal Motility, edited by Rao *et al.*
Kluwer Academic / Plenum Publishers, New York, 1999.

and severe gastrointestinal disease. Thyroid function was normal. What is the next step in the evaluation?

A. Flexible sigmoidoscopy.
B. Barium enema.
C. Initiate therapy with fiber supplement (methyl cellulose)
D. Cisapride.

3. After beginning a fiber supplement, the patient had increasing difficulty with abdominal bloating. What is the next therapeutic step?

A. Anal manometry.
B. Amitriptyline.
C. Pelvic ultrasound.
D. Colonoscopy.

PATHOPHYSIOLOGY AND TREATMENT OF COLORECTAL PAIN

Diagnosis of Exclusion

When a patient presents with colorectal pain, the first duty of the physician is to exclude a serious underlying organic disease. This requires the physician to elicit alarm symptoms including a decreased appetite, malaise or weight loss. It is also important to look for signs of gastrointestinal bleeding or recent onset in altered bowel habit.

Once alarm symptoms have been excluded, dietary components that may exacerbate the symptoms should be identified. Often this includes patient ingestion of excessive amounts of fat and/or milk. Sorbitol, used as sweetener in many dietary foods, may exacerbate abdominal bloating and discomfort.

The work-up of each patient presenting with chronic abdominal pain should include a search for an occult source of gastrointestinal bleeding. Depending on the severity of the patient's symptoms as well as the age of the patient, abdominal x-rays including abdominal series or a CAT scan may be necessary. Often an upper GI or upper endoscopy may be required to further define upper gut disease. Barium enema or colonoscopy can be performed to further define lower gut disease. The pelvic structures may be the cause of apparent colorectal pain. Gynecologic disease and/or anorectal abscesses may present as colorectal pain that will be identified only by either ultrasound of the pelvis and rectum or extension of a CAT scan into the pelvis.

Visceral Hypersensitivity Syndromes

Once a structural cause of the patient's symptoms has been excluded, tests of visceral hypersensitivity can be performed. The most readily accessible test includes distention of the rectum. Normal sensitivity to rectal distention has been well defined in many studies, which allows an estimate of the patient's level of hypersensitivity (2–8).

If the patient is believed to have visceral hypersensitivity, they are also more likely to have one of the functional bowel syndromes, including non-cardiac chest pain, non-ulcer dyspepsia, irritable bowel syndrome, or proctalgia fugax (see Table 1).

These symptom complexes will be associated with altered gut function depending on the anatomic location of the disorder. Patients with functional GI disease have a decreased threshold to sensory stimulation in multiple sites of the GI tract (9). These patients

Table 1. List of common visceral hypersensitivity syndromes

Functional gastrointestinal disorders	Other
Non-cardiac chest pain	Fibromyalgia
Non-ulcer dyspepsia	Migraine syndrome
Irritable bowel syndrome	Gynecologic pain
Proctalgia fugax	Dyspareunia
Coccygodynia	Dysmenorrhea
	Interstitial cystitis

may also have fibromyalgia and increased incidence of migraine headaches. Gynecologic pain is common in many of the female patients with visceral hypersensitivity (10, 11). These symptoms may include painful menses or dyspareunia. There also appears to be an increased incidence of interstitial cystitis in patients with visceral hypersensitivity and functional bowel complaints (12).

Anorectal pain can range from a dull throbbing discomfort that feels as though the patient were sitting on a ball to a severe sharp pain. Both extremes can be incapacitating to the patient. Proctalgia fugax consists of a lancinating pain of short duration. The anorectal pain can be associated with anxiety alone, or a specific motility disorder. A lax anal sphincter and spasm of the levator ani have been identified in proctalgia fugax. High amplitude sigmoid contractions also occur in proctalgia fugax (13, 14). A generalized visceral hypersensitivity throughout the colon associated with the irritable bowel syndrome may contribute to anorectal pain (15). Coccygodynia is pain over the coccyx, which may be misinterpreted as an anorectal pain rather than from a musculoskeletal source. Anal abscess and fissure are anatomic disorders that may lead to severe anorectal pain. An anal abscess identified by physical examination or anal ultrasound can be treated rapidly by drainage. Anal fissure may be associated with a hypertensive anal sphincter (16). If the fissure cannot be healed by stool softeners, laxatives and local anesthetic/steroid preparations an anal sphincterotomy may be needed. Rectal prolapse, especially if associated with a solitary rectal ulcer, also may be a cause of anorectal pain.

Lower abdominal pain is caused by a different group of disease states. Irritable bowel syndrome is common to both lower abdominal as well as anorectal discomfort. Lower abdominal discomfort may be associated with painful diverticular disease. Gynecologic disorders, pelvic inflammatory disease or endometriosis must still be considered in lower abdominal pain. For lower abdominal pain, both carcinomas of the colon and diverticulitis must be considered. Patients with symptomatic diverticular disease are younger and they have higher pressures than patients with asymptomatic diverticular disease (17–19).

Figure 1 shows the underlying pathophysiology for colorectal pain. The pain may be secondary to either a psychosocial dysfunction, a motility disorder or visceral hypersensitivity. There is interactivity between each of these categories. Due to the complexity of the pathogenesis of colorectal pain the physician and the patient should identify realistic goals for the success in therapy. Otherwise both the physician and the patient will become frustrated.

Psychological Dysfunction

Emotional stress associated with psychosocial dysfunction may alter the motility within the distal colon (20, 21). It may also alter the recognition of the pain after balloon

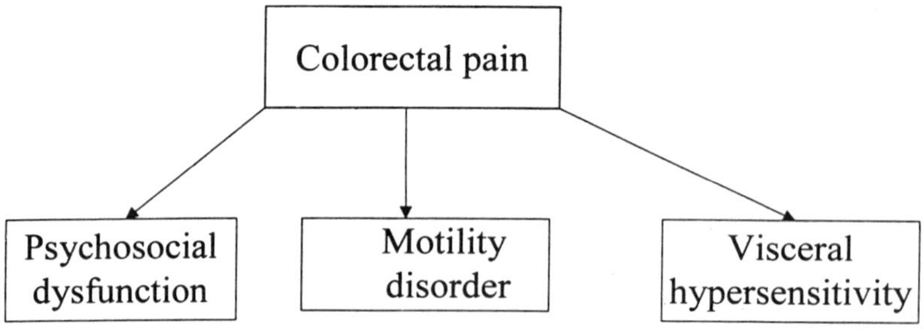

Figure 1. Colorectal pain is caused by varying contributions from psychosocial dysfunction, a motility disorder of the intestinal tract, or visceral hypersensitivity due to an increased in afferent nerve sensitivity.

distention of the rectum (22). The questions in Figure 2 will help to characterize the patient's psychosocial status. If the patient's answers suggest a psychosocial dysfunction, further evaluation and treatment of the patient's psychiatric disease is important. In those patients who have psychosocial dysfunction, psychoactive medications including the antidepressants may be useful. In some patients formal psychotherapy may control symptoms (23, 24). Hypnotherapy has also been useful in some patients (25).

Colorectal Dysmotility

In many patients visceral hypersensitivity and increased motility are intertwined in the underlying pathophysiology (Figure 3). Recent studies have suggested that altered afferent sensation might occur in 60% of patients who have functional abdominal pain. The type of sensory stimulus can determine the degree of hypersensitivity (26, 27). Patients with altered afferent sensation may either have a normal motility, or they may have a concomitant motility disturbance consisting of either decreased relaxation or increased contractions of the colonic smooth muscle. In patients with normal motility, visceral hypersensitivity appears to be the primary cause of their abdominal pain. Central neuroplasticity and increased sensitivity maintains visceral hypersensitivity even after the initiating pain stimuli have stopped (28, 29)

- What is the patient's life history of illness?
- Why is the patient coming now?
- Is there a history of traumatic life events?
- What is the patient's understanding of the illness?
- What is the impact of the pain on quality of life?
- Is there an associated psychiatric diagnosis?
- What is the role of family's culture?
- What are the patient's psychosocial resources?

Figure 2. A concise list of questions that help determine the psychosocial status of the patient presenting with colorectal pain.

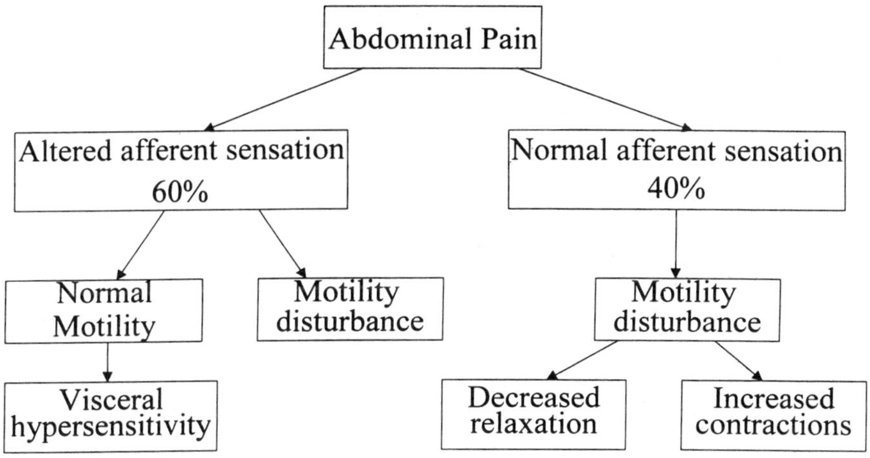

Figure 3. A flow diagram showing the potential causes of colorectal pain which provides a background for diagnostic evaluation and eventual treatment.

If the patient has a colorectal motility dysfunction, dietary manipulation may be useful. Fiber supplement and decreased fat ingestion are my initial therapeutic trial, since this regimen may decrease colonic contractions and intraluminal pressure generation. (30). Fiber supplementation increases the diameter and decreases intraluminal pressure through LaPlace's Law. Fiber improves symptoms best in patients with normal transit constipation (31). Fiber also will bind bile salts that have been implicated in some patients with diarrhea predominant irritable bowel syndrome. In patients unresponsive to changes in diet smooth muscle relaxants including anticholinergics, calcium channel blockers or nitrates may be useful for decreasing smooth muscle contraction and decreasing symptomatic abdominal pain (32; 33).

Colonic Visceral Hyperalgesia

In patients with possible visceral hypersensitivity, objective measurements should be obtained. Balloon distention of the rectum with a computer-controlled barostat is a useful way to measure visceral sensitivity. This technique allows the investigator to measure responses to both ramp distention as well as phasic distention of the colon. Repetitive sigmoid distention has been demonstrated to cause pain in patients with irritable bowel syndrome, but not in normal subjects (34, 35).

Table 2 shows peripheral pain mediators that may be involved in causing visceral hypersensitivity (36). Multiple agents are involved in causing visceral pain. These compounds will be targets for future therapeutic investigation. Table 3 shows the drugs potentially useful in the treatment of visceral hypersensitivity.

Tricyclic antidepressants may decrease visceral hypersensitivity by decreasing activation of the serotonin receptors. The serotonin type III antagonists have also been shown to decrease visceral hypersensitivity. Granisetron and possibly ondansetron increase the visceral afferent threshold (37–39). Calcium channel blockers may be effective, by decreasing the effect of exciting agents on neuromembrane excitability. Decreasing calcium movement across neural membranes may decrease pain perception. Octreotide increases the threshold for visceral perception in the irritable bowel syndrome without modifying

Table 2. Peripheral mediators of visceral pain

Agents acting directly on afferent nerves	Agents acting indirectly on afferent nerves
1. Prostaglandin E2, I2	1. Noradrenaline
2. diHETE	2. Interleukin-1, -8, -6
3. Adenosine, ATP	3. TNF-alpha
4. Bradykinin	4. Bradykinin
5. Serotonin	5. Leukotriene B4
6. CGRP, Substance P	6. Complement 5A
	7. Substance P, VIP

compliance of the bowel wall (3, 40, 41). The peripheral kappa opioid receptor agonist, fedotozine, improves pain and bloating in the irritable bowel syndrome (42). Oxytocin increases the threshold for visceral perception at doses greater than 20 mU per minute. However, the drug must be given intravenously (43). There is some evidence that alpha-2 agonists such as clonidine and CCK A-B antagonists may have some benefit on visceral pain.

CONCLUSIONS

Visceral hypersensitivity is a common cause of distress in patients with functional bowel disease. Recently there have been many advances in our understanding of abdominal pain that has allowed further work on new drug development to treat these patients. We require new drugs that adequately inhibit visceral pain through their effects on specific pain mediators and at specific sites.

ANSWER KEY AND EXPLANATIONS

1. In a young person the risk of serious colorectal disorder is very low. Also, her symptoms have been present for over a year without weight loss. Hence, the first step would be to exclude any systemic problems, emotional issues and dietary factors rather than expensive and invasive investigations. Reassurance together with excluding systemic problems would be most appropriate. The best option is **D.**
2. Because the patient seems to have constipation predominate IBS rather than diarrhea, flex sigmoidoscopy or Barium enema is unnecessary. Hence options A and B would be inappropriate although some may argue that a flexible sigmoidoscopy may be helpful reassurance. Cisapride is not the first line or the most effective therapy for constipation. Hence, laxatives together with fiber supplements would be most helpful. The best option is **C.**

Table 3. Pharmacologic treatment of visceral hypersensitivity

Tricyclic antidepressant.	Kappa opioid receptor agonist
5-HT3 antagonist.	Oxytocin.
Calcium channel blockers.	Alpha-2-agonists.
Octreotide.	

3. Amitriptyline was begun at a dose of 10 mg tid. Although this is a useful drug for visceral hyperalgesia it has anticholinergic effects and could make her constipation worse. The patient is young and most likely has the irritable bowel syndrome with constipation. Further, objective evaluation of her colonic and anorectal function will be required to identify an anatomic or physiologic dysfunction. Also gynecologic evaluation may be required. The best option is **B**.

REFERENCES

1. Thompson WG, Heaton KW. Functional bowel disorders in apparently healthy people. Gastroenterology. 1980;79:283–8.
2. Bradette M, Delvaux M, Staumont G, Fioramonti J, Bueno L, Frexinos J. Evaluation of colonic sensory thresholds in IBS patients using a barostat. Definition of optimal conditions and comparison with healthy subjects. Dig.Dis.Sci. 1994;39:449–57.
3. Bradette M, Delvaux M, Staumont G, Fioramonti J, Bueno L, Frexinos J. Octreotide increases thresholds of colonic visceral perception in IBS patients without modifying muscle tone. Dig.Dis.Sci. 1994;39:1171–8.
4. De Medici A, Badiali D, Corazziari E, Bausano G, Anzini F. Rectal sensitivity in chronic constipation. Dig.Dis.Sci. 1989;34:747–53.
5. Prior A, Read NW. Reduction of rectal sensitivity and post-prandial motility by granisetron, a 5 HT3-receptor antagonist, in patients with irritable bowel syndrome. Aliment Pharmacol Thera. 1993;7:175–80.
6. Whitehead WE, Crowell MD, Davidoff AL, Palsson OS, Schuster MM. Pain from rectal distension in women with irritable bowel syndrome: relationship to sexual abuse. Dig.Dis.Sci. 1997;42:796–804.
7. Whitehead WE, Holtkotter B, Enck P, et al. Tolerance for rectosigmoid distention in irritable bowel syndrome. Gastroenterology. 1990;98:1187–92.
8. Zighelboim J, Talley NJ, Phillips SF, Harmsen WS, Zinsmeister AR. Visceral perception in irritable bowel syndrome. Rectal and gastric responses to distension and serotonin type 3 antagonism. Dig.Dis.Sci. 1995;40:819–27.
9. Trimble KC, Farouk R, Pryde A, Douglas S, Heading Rc. Heightened visceral sensation in functional gastrointestinal disease is not site-specific. Evidence for a generalized disorder of gut sensitivity. Dig.Dis.Sci. 1995;40:1607–13.
10. Guthrie E, Creed FH, Whorwell PJ. Severe sexual dysfunction in women with the irritable bowel syndrome: comparison with inflammatory bowel disease and duodenal ulceration. B..M.J. 1987;295:577–8.
11. Prior A, Whorwell PJ. Gynaecological consultation in patients with the irritable bowel syndrome. Gut. 1989;30:996–8.
12. Clauw DJ, Schmidt M, Radulovic D, Singer A , Katz P, Bresette J. The relationship between fibromyalgia and interstitial cystitis . J.Psychiatr.Res. 1997;31:125–31.
13. Harvey RF. Colonic motility in proctalgia fugax. Lancet. 1979;2:713–4.
14. Thompson WG. Proctalgia fugax. Dig.Dis.Sci. 1981;26:1121–3.
15. Ritchie J. Pain from distension of the pelvic colon by inflating a balloon in the irritable colon syndrome. Gut. 1973;14:125–32.
16. Nothman BJ, Schuster MM. Internal anal sphincter derangement with anal fissures. Gastroenterology. 1974;67:216–20.
17. Cortesini C, Pantalone D. Usefulness of colonic motility study in identifying patients at risk for complicated diverticular disease. Dis.Colon Rectum. 1991;34:339–42.
18. Schauer PR, Ramos R, Ghiatas AA, Sirinek KR. Virulent diverticular disease in young obese men. Am.J.Surg. 1992;164:443–6; discuss.
19. Trotman IF, Misiewicz JJ. Sigmoid motility in diverticular disease and the irritable bowel syndrome. Gut. 1988;29:218–22.
20. Camilleri M, Neri M. Motility disorders and stress. [Review]. Dig. Dis. Sci. 1989;34:1777–86.
21. Narducci F, Snape WJ, Jr., Battle WM, London RL, Cohen S. Increased colonic motility during exposure to a stressful situation. Dig.Dis.Sci. 1985;30:40–4.
22. Ford MJ, Camilleri M, Zinsmeister AR, Hanson RB. Psychosensory modulation of colonic sensation in the human transverse and sigmoid colon. Gastroenterology. 1995;109:1772–80.
23. Creed F. Irritable bowel or irritable mind? Psychological treatment is essential for some. BMJ. 1994;309:1647–8.

24. Guthrie E, Creed F, Dawson D, Tomenson B. A controlled trial of psychological treatment for the irritable bowel syndrome. Gastroenterology. 1991;100:450–7.
25. Whorwell PJ, Prior A, Faragher EB. Controlled trial of hypnotherapy in the treatment of severe refractory irritable-bowel syndrome. Lancet. 1984;2:1232–4.
26. Lembo T, Munakata J, Naliboff B, Fullerton S, Mayer EA. Sigmoid afferent mechanisms in patients with irritable bowel syndrome. Dig.Dis.Sci. 1997;42:1112–20.
27. Mertz H, Naliboff B, Munakata J, Niazi N, Mayer EA. Altered rectal perception is a biological marker of patients with irritable bowel syndrome. Gastroenterology. 1995;109:40–52.
28. Drossman DA. Chronic functional abdominal pain. Am.J.Gastroenterol. 1996;91:2270–81.
29. Mayer EA, Gebhart GF. Basic and clinical aspects of visceral hyperalgesia [see comments]. Gastroenterology. 1994;107:271–93.
30. Wright SH, Snape WJ, Jr., Battle WM, Cohen S , London RL. Effect of dietary components on gastrocolonic response. Am.J.Physiol. 1980;238:G228-G232
31. Voderholzer WA, Schatke W, Muhldorfer BE, Klauser AG, Birkner B, Muller-Lissner SA. Clinical response to dietary fiber treatment of chronic constipation. Am.J.Gastroenterol. 1997;92:95–8.
32. Ritchie JA, Truelove SC. Comparison of various treatments of irritable bowel syndrome. Br.Med.J. 1980;281:1317–9.
33. Ritchie JA, Truelove SC. Treatment of irritable bowel syndrome with lorazepam, hyoscine butylbromide, and ispaghula husk. B.M.J. 1979;376–8.
34. Lembo T, Munakata J, Mertz H, et al. Evidence for the hypersensitivity of lumbar splanchnic afferents in irritable bowel syndrome [see comments]. Gastroenterology. 1994;107:1686–96.
35. Munakata J, Naliboff B, Harraf F, et al. Repetitive sigmoid stimulation induces rectal hyperalgesia in patients with irritable bowel syndrome. Gastroenterology. 1997;112:55–63.
36. Bueno L, Fioramonti J, Delvaux M, Frexinos J . Mediators and pharmacology of visceral sensitivity: from basic to clinical investigations. Gastroenterology. 1997;112:1714–43.
37. Banner SE, Carter M, Sanger GJ. 5-Hydroxytryptamine3 receptor antagonism modulates a noxious visceral pseudoaffective reflex. Neuropharmacology. 1995;34:263–7.
38. Goldberg PA, Kamm MA, Setti-Carraro P, van der Sijp JR, Roth C. Modification of visceral sensitivity and pain in irritable bowel syndrome by 5-HT3 antagonism (ondansetron). Digestion. 1996;57:478–83.
39. Hammer J, Phillips SF, Talley NJ, Camilleri M. Effect of a 5HT3-antagonist (ondansetron) on rectal sensitivity and compliance in health and the irritable bowel syndrome. Aliment.Pharmacol.Ther. 1993;7:543–51.
40. Chey WD, Beydoun A, Roberts DJ, Hasler WL, Owyang C. Octreotide reduces perception of rectal electrical stimulation by spinal afferent pathway inhibition. Am.J.Physiol. 1995;269:G821–6.
41. Hasler WL, Soudah HC, Owyang C. A somatostatin analogue inhibits afferent pathways mediating perception of rectal distention. Gastroenterology. 1993;104:1390–7.
42. DaPoigny M, Abitbol JL, Fraitag B. Efficacy of peripheral kappa agonist fedotozine versus placebo in treatment of irritable bowel syndrome. A multicenter dose- response study. Dig.Dis.Sci. 1995;40:2244–9.
43. Louvel D, Delvaux M, Felez A, et al. Oxytocin increases thresholds of colonic visceral perception in patients with irritable bowel syndrome [see comments]. Gut. 1996;39:741–7.

CONTRIBUTORS

Dr. Satish S. C. Rao
Associate Professor
Department of Internal Medicine
University of Iowa Hospitals and Clinics
200 Hawkins Drive
Iowa City, IA 52242
Phone: 319/353-6602
FAX: 319/353-6399

Dr. Bruce P. Brown
Associate Professor
Department of Radiology
University of Iowa Hospitals and Clinics
200 Hawkins Drive
Iowa City, IA 52242
Phone: 319/356-4374
FAX: 319/356-2220

Dr. Donald O. Castell
Professor and Chairman
Department of Medicine, Pepper Pavilion
The Graduate Hospital, Suite 501
1800 Lombard Street
Philadelphia, PA 19146-1497
Phone: 215/893-7128
FAX: 215/893-2472

Dr. Jeffrey Conklin
Associate Professor
Department of Internal Medicine
University of Iowa Hospitals and Clinics
200 Hawkins Drive
Iowa City, IA 52242
Phone: 319/356-2131
FAX: 319/353-6399

Dr. Carlo Di Lorenzo
Associate Professor
Division of Pediatrics-Gastroenterology
Children's Hospital of Pittsburgh
3705 Fifth Avenue at DeSoto Street
Pittsburgh, PA 15213-2583
Phone: 412/692-5180
FAX: 412/692-7355

Dr. G.F. Gebhart
Professor and Head
Department of Pharmacology
University of Iowa
Iowa City, IA 52242
Phone: 319/335-7946
FAX: 319/335-8930

Dr. Frederick Johlin
Associate Professor
Department of Internal Medicine
University of Iowa Hospitals and Clinics
200 Hawkins Drive
Iowa City, IA 52242
Phone: 319/356-4030
FAX: 319/353-6399

Dr. Glen A. Lehman
Professor
Department of Internal Medicine
Indiana University Medical Center
550 N. University Blvd.
Suite 2300
Indianapolis, IN 46202-5244
Phone: 317/274-4821
FAX: 317/278-9164

Dr. Henry C. Lin
Associate Professor
Department of Internal Medicine
Cedars-Sinai Medical Center
8700 Beverly Blvd.
Suite 7511
Los Angeles, CA 90048
Phone: 310/855-6141
FAX: 310/967-0125

Dr. Joseph Murray
Associate Professor
Gastroenterology and Hepatology
Mayo Clinic
200 First Street, SW
Rochester, MN 55905
Phone: 507/284-2467
FAX: 507/284-0538

Dr. Retta Pelsang
Assistant Professor
Department of Radiology
University of Iowa Hospitals and Clinics
200 Hawkins Drive
Iowa City, IA 52242
Phone: 319/356-3767
FAX: 319/356-2220

Dr. John H. Pemberton
Professor of Surgery
Mayo Clinic
200 First Street, S.W.
Rochester, MN 55905
Phone: 507/284-2359
FAX: 507/284-1794

Dr. Mark Pimental
Fellow — Gastroenterology
Department of Internal Medicine
Cedars-Sinai Medical Center
8700 Beverly Blvd.
Suite 7511
Los Angeles, CA 90048
Phone: 310-855-6141
FAX: 310-967-0125

Dr. Konrad Schulze-Delrieu
Professor
Department of Internal Medicine
University of Iowa Hospitals and Clinics
200 Hawkins Drive
Iowa City, Iowa 52242
Phone: 319/356-2793

Dr. William Snape, Jr.
Clinical Professor of Medicine
University of California-Irvine
2650 Elm Avenue, # 201
Long Beach, CA 90806-1600
Phone: 562/595-5421
FAX: 562/426-2862

Dr. Robert Summers
Professor
Department of Internal Medicine
University of Iowa Hospitals and Clinics
200 Hawkins Drive, Iowa City, IA 52242
Phone: 319/356-2130
FAX: 319/353-6399

INDEX